International Perspectives in Geography

AJG Library

Volume 12

Aim and Scope

The AJG Library is published by Springer under the auspices of the Association of Japanese Geographers. This is a scholarly series of international standing. Given the multidisciplinary nature of geography, the objective of the series is to provide an invaluable source of information not only for geographers, but also for students, researchers, teachers, administrators, and professionals outside the discipline. Strong emphasis is placed on the theoretical and empirical understanding of the changing relationships between nature and human activities. The overall aim of the series is to provide readers throughout the world with stimulating and up-to-date scientific outcomes mainly by Japanese and other Asian geographers. Thus, an "Asian" flavor different from the Western way of thinking may be reflected in this series. The AJG Library will be available both in print and online via SpringerLink.

About the AJG

The Association of Japanese Geographers (AJG), founded in 1925, is one of the largest and leading organizations on geographical research in Asia and the Pacific Rim today, with around 3000 members. AJG is devoted to promoting research on various aspects of human and physical geography and contributing to academic development through exchanges of information and knowledge with relevant internal and external academic communities. Members are tackling contemporary issues such as global warming, air/water pollution, natural disasters, rapid urbanization, irregular land-use changes, and regional disparities through comprehensive investigation into the earth and its people. In addition, to make the next generation aware of these academic achievements, the members are engaged in teaching and outreach activities of spreading geographical awareness. With the recent developments and much improved international linkages, AJG launches the publication of the AJG Library series in 2012.

More information about this series at http://www.springer.com/series/10223

Hitoshi Miyazawa · Teruo Hatakeyama
Editors

Community-Based Integrated Care and the Inclusive Society

Recent Social Security Reform in Japan

Springer

Editors
Hitoshi Miyazawa
Ochanomizu University
Tokyo, Japan

Teruo Hatakeyama
Naruto University of Education
Naruto, Tokushima, Japan

ISSN 2197-7798 ISSN 2197-7801 (electronic)
International Perspectives in Geography
ISBN 978-981-33-4472-3 ISBN 978-981-33-4473-0 (eBook)
https://doi.org/10.1007/978-981-33-4473-0

This Springer imprint is published by the registered company Springer Nature Singapore Pte Ltd.
The registered company address is: 152 Beach Road, #21-01/04 Gateway East, Singapore 189721, Singapore

Regions and prefectures in Japan

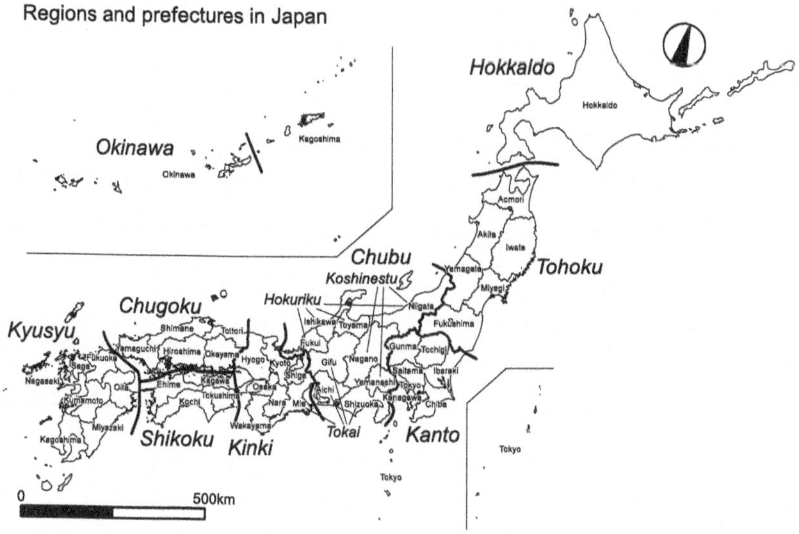

Municipalities and major cities in Japan

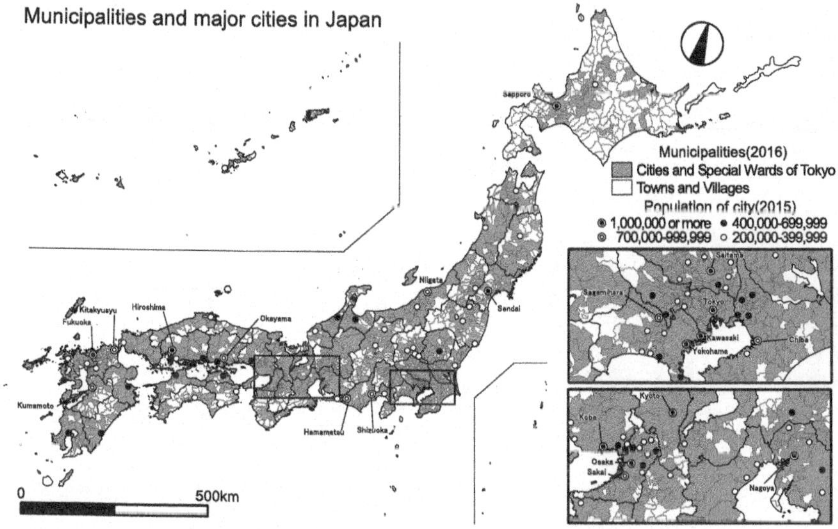

Preface

This book attempts to discuss the building of "comprehensive community support systems," which constitutes a key issue in social security reforms in Japan. All papers in the book were written by members of the Association of Japanese Geographers Study Group "Regional Issues Related to the Birthrate Decline and Population Aging" based on research related to the building of comprehensive community support systems in Japan. The authors undertook this research for the following reasons.

In social security reforms in Japan, a "community-based integrated care system" is being built for 2025, when all members of the baby-boom generation will be aged 75 years or older, as a comprehensive community support system for elderly people. The system provides specialized support such as nursing care, medical care, in addition to health, appropriate housing, and various means of livelihood support in an integrated manner to enable elderly people to continue to live in the area in which they have lived for a long time. The community-based integrated care system must be built by municipal governments taking the initiative and in accordance with local circumstances. Furthermore, the building of "all-generation, all-people comprehensive support" is underway as the foundation for a "community-based inclusive society" that is regarded as the image of a desirable society in Japan in the era following 2025. "All-generation, all-people comprehensive support" makes it possible to provide "whole" support by generalizing the idea of a community-based integrated care system and by extending the range of people included in the system. In addition, various actors in the community, including local residents, must be involved in the building of comprehensive support systems as their "own business." Regional variations will occur in these efforts. An approach from geography is expected for the vision of comprehensive community support systems.

Geographers in Japan researching social security have mainly analyzed regional differences in service demand for nursing care and child care since the latter half of the 1990s (Miyazawa ed. 2005; Sugiura 2005, 2017; Kukimoto 2016). Furthermore, social security and welfare are explained as important matters in a recently published encyclopedia of geography and books presenting surveys of research trends in geography (Human Geographical Society of Japan ed. 2013; Japan Association of Economic Geographers ed. 2018). In Japan, research related to social security has been established in geography. However, a difficulty in the field is that cooperation among researchers who individually undertake research in the subfields of nursing care, child care, or medical care is weak (Miyazawa 2013). Such cooperation must be strengthened to advance future research. If that proves to be impossible, then geography might fail to play a role in approaching regional issues under the circumstances of an aging population with a declining birth rate and population decline because the system of research will diverge from the direction of social security reforms in Japan to be studied.

Fortunately, the authors had an opportunity to address the challenge presented above with the support of Grants-in-Aid for Scientific Research in Japan that is "Research to Build Community-based Care Systems through 'Geography of Social Security'." It is a 4-year research project with two purposes. The first is to systematize research on social security in geography in Japan. The second is to undertake research related to key policies for social security in Japan. Based on these purposes, many geographers undertaking research related to social security participated in the research project and conducted succeeding research, mutually cooperating. In the first 2 years, the researchers analyzed regional differences in the development levels of various services and social resources that would condition the realization of comprehensive community support systems through geo-visualization using GIS. The results have been published as an atlas with the themes of welfare, medical care, and health levels in Japan (Miyazawa ed. 2017). In the latter 2 years, choosing local governments with different regional characteristics, we conducted empirical research to uncover characteristics of comprehensive community support systems, building processes, and challenges in the respective local governments.

Papers in the book were composed based on research into "community-based integrated care systems" and "community-based inclusive society" that was conducted mainly during the past 2 years. Non-Japanese readers throughout the world must know the characteristics of social security in Japan and the trends of the reforms to understand the contents of the book clearly. To support its use as a reference, this book includes some contents of the atlas described above in the form of reorganizing them. With that intention, the book comprises three parts. The composition of the book will be introduced briefly below.

First Part, "Mapping Social Security in Japan," comprises five chapters that clarify the characteristics of social security in Japan.

Chapter "Welfare Regime in Japan and Recent Social Security Reform", written by Hitoshi Miyazawa, is the introductory part of the book, describing the characteristics of the welfare regime in Japan as well as recent social security reforms. This chapter will help readers understand the importance of the building of community-based integrated care systems and the realization of a community-based inclusive society in social security reforms in Japan, tracing the processes by which they became key policies.

Chapter "Temporal Changes and Regional Differences in Demographic Structure and Health in Japan", written by Masakazu Yamauchi and Tomoya Hanibuchi, elucidates the characteristics of population and households in Japan, as well as Japanese people's life expectancy and health-related activities, based on regional statistics.

The demographic characteristics and health levels are important to understand the needs for social security.

In the building of community-based integrated care systems, cooperation between nursing care and medical care is strongly and particularly required. Consequently, chapter "Nursing Care Provision Systems for Elderly People and Geographical Distribution of Services in Japan", written by Teruo Hatakeyama and others, explains long-term care insurance, which is central to nursing care provision systems in Japan and which clarifies regional differences in care services by the type of service. The succeeding chapter "Medical Care Provision System and Geographical Distribution of Medical Resources in Japan", written by Tsutomu Nakamura and others, describes medical care provision systems in Japan and clarifies the distribution of medical care resources.

Chapter "Securing of Health, Medical, and Welfare Personnel and the Geographical Distribution in Japan", written by Hiroyasu Kamo and Akihito Nakajo, explains welfare and medical care providers in Japan and their geographical distribution. Additionally, it presents a discussion that the emphasis is placed on foreign workers and resident mutual aid in the local community to make up for the recent shortage of specialists.

By reading these chapters, readers will understand that comprehensive support systems in Japan must be built on the premise that the service infrastructure and human resources are regionally distributed unevenly to a great extent and are insufficient in many regions.

Second Part, "Community-Based Integrated Care Systems in Japan," includes six chapters related to the community-based integrated care system: a comprehensive community support system for elderly people.

Chapter "Regional Variation in the Community-Based Integrated Care Systems in Japan", presented by Teruo Hatakeyama and Hitoshi Miyazawa, typifies community-based integrated care systems in local governments based on the spatial structure using results from a questionnaire. Additionally, it explains such variations in community-based integrated care systems from the perspective of regional characteristics of the respective local governments. The findings help readers understand the positioning of case studies in the next and succeeding chapters. That is true because, in the next and the succeeding two chapters, the authors analyze local governance related to the building of community-based integrated care systems by type shown in this chapter.

Chapters "Community-Based Integrated Care Systems in Municipalities Having One Comprehensive Community Support Center" through Chapter "Community-Based Integrated Care Systems in Municipalities Having Multiple Comprehensive Community Support Centers" present analyses of local governance over the building of community-based integrated care systems. Chapter "Community-Based Integrated Care Systems in Municipalities Having One Comprehensive Community Support Center", written by Tsutomu Nakamura and Teruo Hatakeyama, examines local governments that have a single comprehensive community support center. Chapter "Community-Based Integrated Care Systems in Municipalities Having Sub-branches of Comprehensive Community Support Centers", written by Teruo

Hatakeyama and others, describes local governments that have a comprehensive community support center with an attached subcenter or branch. Chapter "Community-Based Integrated Care Systems in Municipalities Having Multiple Comprehensive Community Support Centers", written by Teruo Hatakeyama, presents an examination of local governments that have more than one comprehensive community support center. By reading these three chapters, readers learn the need for decentralized local governance to build community-based integrated care systems that better reflect the regional characteristics of local governments. Additionally, they will understand that challenges exist for the realization, including the security of financial resources and personnel and the building of meta-governance by local governments to unify decentralized local governance.

Chapters "Regeneration of Housing Estates by the Community-Based Integrated Care Systems" and "Integrated Care Systems Established to Strengthen Community Disaster Resilience" present discussions of the building of community-based integrated care systems in cooperation with community development. Chapter "Regeneration of Housing Estates by the Community-Based Integrated Care Systems", written by Yoshimichi Yui and others, elucidates the importance and challenges of the building of community-based integrated care systems that cooperate with the restoration of housing complexes in the suburbs of metropolitan areas, where aging of residents and deterioration of housing have come to pose severe difficulties in recent years. Chapter "Integrated Care Systems Established to Strengthen Community Disaster Resilience", written by Hitoshi Miyazawa and Haruko Kikuchi, explains the importance of introducing community-based integrated care systems to form resilient communities against disasters. By reading these chapters, readers will again understand that the community-based integrated care system is a system supporting the welfare of residents in a broad sense to realize "continuing to live in the area where they have lived for a long time."

Third Part, "A Prospect of Community-Based Inclusive Society in Japan," explains the prospects of building comprehensive support systems by expanding and deepening community-based integrated care to realize a community-based inclusive society. The part comprises the following five chapters.

Chapter "Establishing Community-Based Integrated Support Systems for Pregnancy, Childbirth, and Childcare in Japan: Focusing on Regional differences", written by Hitoshi Miyazawa and Kanoko Tada, takes up local governments in urban and rural areas in terms of comprehensive support in the fields of maternal and child health and parenting and presents a consideration of regional backgrounds that engender differences between their support systems. Results reveal that differences in regional characteristics between urban and rural areas will particularly influence differences in interprofessional cooperative systems and cooperative local residents' organizations.

Chapter "Creation of Social Ties for Prevention of Isolation of Elderly Public Assistance Recipients: The Case of a Project for the Provision of "a Place of One's Own" in Nishinari Ward, Osaka City", written by Nanami Inada, describes analyses of support programs that center on providing places for elderly public assistance recipients in inner cities of metropolitan areas as comprehensive support for people who need it. The author concludes that the support programs are expected to be able to form networks in the community and to prevent elderly public assistance recipients from social isolation.

Chapter "Current Situation and Challenges of "Inclusive Care": An Investigation of the "Community-Based Inclusive Station Program" in Saga Prefecture", written by Mikoto Kukimoto, takes up "inclusive care," which has drawn attention as a service extending across user categories, such as elderly people, children, and children (persons) with disabilities. The inclusive style of care is expected to be expanded to realize a community-based inclusive society. However, the author's investigation clarifies challenges, including that relief of the service providers' burden is necessary for the practice.

Chapter "Formation of Comprehensive Community Welfare Bases in Urban Areas", written by Ryo Koizumi and others, discusses the importance of bases for comprehensive community welfare in urban areas based on practices of advanced local governments. In all the example local governments, bases for comprehensive community welfare are shown to be important facilities that provide comprehensive service and support for diverse residents' diverse problems and deal with consultation in a one-stop manner.

The final Chapter "Development of Community Welfare Activities with Resident Participation and Their Importance in Hilly and Mountainous Areas", written by Akihito Nakajo, discusses resident participative welfare activities in hilly and mountainous areas as support undertaken with the idea of "own business." In hilly and mountainous areas, the expectations for "community-based inclusive society" are likely to be raised because social relationships based on traditional social norms have been maintained. However, with population decline and aging, residents' mutual relationships are weakening. The author points out that one challenge to be considered is to secure supporters from outside the community.

Readers will understand the importance of community-based inclusive society by reading the chapters in Third Part. At the same time, they will learn that the idea is apt to go-ahead for the realization and there are many practical challenges to be solved.

By explaining how this book was produced and introducing the outlines of respective papers, the editors have tried to convey the importance of this book. We hope that many people will read the book.

Finally, we thank the many people who cooperated in our study. We express our appreciation for Grants-in-Aid for Scientific Research (No.15H01783) received from the Japan Society for the Promotion of Science for research that became the basis of this book. We also express our gratitude for Akashi Shoten, the Association of Japanese Geographers, and Japan Association of Surveyors that allowed us to include already published papers in the book. Our thanks also go to professionals at Fastek, Ltd. who undertook proofreading of our first English draft and translation of papers from Japanese to English.

Tokyo, Japan
Naruto, Japan
September 2019

Hitoshi Miyazawa
Teruo Hatakeyama

References

Human Geographical Society of Japan (ed) (2013) The dictionary of human geography. Maruzen Publishing, Tokyo (in Japanese)

Japan Association of Economic Geographers (ed) (2018) Key Words in Economic Geography. Hara Shobo, Tokyo (in Japanese)

Kukimoto M (2016) Geography of child care and parenting support: focusing on "regional differences" in welfare service demand. Akashi Shoten, Tokyo (in Japanese)

Miyazawa H (ed) (2005) Applied cartography and GIS for welfare and well-being. Kokon Shoin, Tokyo (in Japanese)

Miyazawa H (ed) (2017) Mapping health, medical care, and welfare in Japan. Akashi Shoten, Tokyo (in Japanese)

Sugiura S (2005) Geographical analysis of social services for the elderly in Japan. Kokon Shoin, Tokyo (in Japanese)

Sugiura S (2017) Geography of long-term care administration and finance: a possibility of municipal cooperation in a post-growth society. Akashi Shoten, Tokyo (in Japanese)

Contents

Mapping Social Security in Japan

Welfare Regime in Japan and Recent Social Security Reform

Hitoshi Miyazawa

Abstract This chapter presents the characteristics of Japan's welfare regime and outlines trends of recent social security reform. Japan, as a welfare state, has been characterized by the low level of public welfare and emphasis placed on informal care undertaken mainly by families. However, although the number of elderly people in need of care has increased rapidly, the diversification of families, individualization, and aging populations with declining birthrate have progressed, thereby causing weakening of the welfare function of the family. As adaptation measures, a long-term care insurance system to promote the socialization of care and the marketization of service provisions have been introduced. Even so, it has swung back again to a trend in which the roles of municipalities, residents' mutual aid, and voluntary activities are valued. The movement is called "(re-) regionalization": a characteristic of the recent welfare regime in Japan. This movement is apparent from the fact that the country formulated a key policy, realization of "a community-based inclusive society," in the 2010s, along with the establishment of community-based integrated care systems. Unique integrated support systems have begun to be constructed throughout Japan according to the features of the respective regions. Simultaneously, many new difficulties have emerged. Grasping the trends and actual situations represents an extreme geographical challenge.

Keywords Community-based inclusive society · Community-based integrated care system · Long-term care insurance · Social security reform · Welfare regime in Japan

---

H. Miyazawa (✉)
Ochanomizu University, Tokyo, Japan
e-mail: miyazawa.hitoshi@ocha.ac.jp

H. Miyazawa and T. Hatakeyama (eds.), *Community-Based Integrated Care and the Inclusive Society*, International Perspectives in Geography 12,
https://doi.org/10.1007/978-981-33-4473-0_1

1 Post-war Welfare States

International comparative studies of welfare states have clarified that social security systems and mechanisms differ from one country to another. One study is Esping-Andersen's Welfare Regime Typology. His welfare regime typology using three indicators of de-commodification, social stratification, and de-familization clarified the nature of welfare states in developed countries based on relationships among the state, markets, and family. He concluded that three types of welfare regimes exist in post-war developed countries (Esping-Andersen 1990, 1999). They are the social democratic regime (typified mainly by Scandinavian countries), the conservative corporatist regime (mainly Continental European countries), and the liberal regime (an example is the United States).

However, influenced by the end of economic growth in the first half of the 1970s and afterward and the globalization that has developed rapidly since the 1980s, the foundations of those welfare states are being shaken. For this reason, developed countries have entered a phase in which they are groping for a new welfare state model (Takegawa 2007; Tanaka 2017). Under the circumstances, the liberal welfare states' way of thinking, with emphasis on individual self-responsibility, welfare supply through the market, and public welfare supply based on residualism, has come to push forward each country's social policies toward marketization in response to the political trends of neoliberalism. Nevertheless, it has become evident that the liberal regime tends to increase social disparities and causes a loss of social integration. Social policies of European countries integrated in the EU came to set a goal of social inclusion that integrates all people into society through workfare and respect for diversity, not to assign too much importance to conventional ex-post welfare (Tanaka 2017).

As described, after World War II, developed countries have improved and enhanced systems as welfare states. However, in the fourth quarter of the twentieth century, developed countries faced a turning point and came to pursue a new welfare state model. International comparative studies of welfare states have cast Japan as a hybrid regime of the liberal regime and the conservative corporatist regime (Esping-Andersen 1997; Uzuhashi 1997), or a regime peculiar to eastern Asia.[1] The peculiarity for Western developed countries has been pointed out as a characteristic of welfare regime in Japan. When the twentieth century was approaching its end, the existing foundations of social security systems in Japan began to shake; various institutional reforms came to be undertaken. The purpose of this chapter is to describe the characteristics of the welfare regime in Japan as the introduction of this book, as well as to outline the recent trend of social security reform toward the arrival of a full-fledged aging society.

[1] However, many discussions have addressed whether a welfare regime common among eastern Asian countries can be found or not (Goodman et al. 1998; Holliday and Wilding 2003).

2 Formation of Welfare State in Japan and the Characteristics

The embryonic period for social security policies in Japan was from the 1920s through the 1940s, when workmen's accident compensation insurance, medical insurance, and pension insurance were introduced. However, not all people joined social insurance at the time. In 1961, national health insurance and the national pension were operationalized, which made medical care and pension into universal coverage systems at last. Then, in the 1970s, social security benefits were raised (Fig. 1). The take-off period for Japan as a welfare state is said to have been the first half of the 1970s (Takegawa 2007).

However, the level of social security benefits in Japan was not at all high by international standards; it continued to be lower than those of the United States, the representative liberal regime (Fig. 2). Some point out the reasons as follows (Takegawa 2007). The 1970s, when Japan entered the take-off period as a welfare state, was also a time when high economic growth came to an end and financial restrictions became tighter. Under the circumstances, the "Japanese-style welfare society" advocated in the national economic policy guidelines in 1979 was an ideology aimed at "low welfare, low burden" by emphasizing individuals' self-help efforts and the solidarity of family and local community. In the 1980s, social security was reviewed in light of

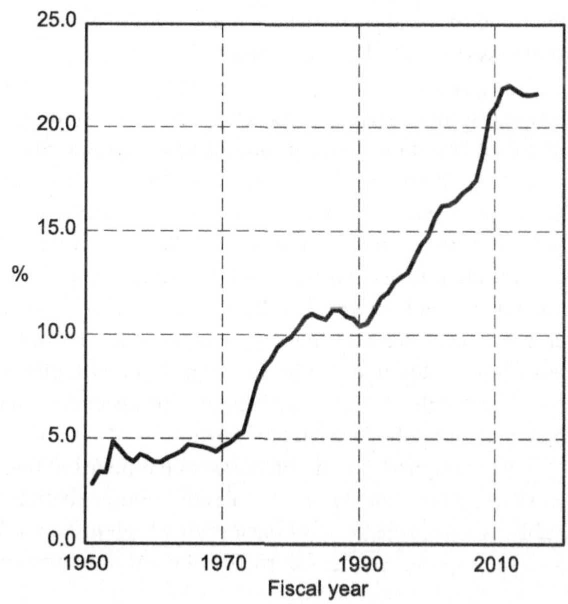

Fig. 1 Changes in social security benefits (as a percentage of GDP) in Japan (Created based on National Institute of Population and Social Security Research 2018)

Fig. 2 International comparison of social expenditure as a percentage of GDP (Created based on National Institute of Population and Social Security Research 2018)

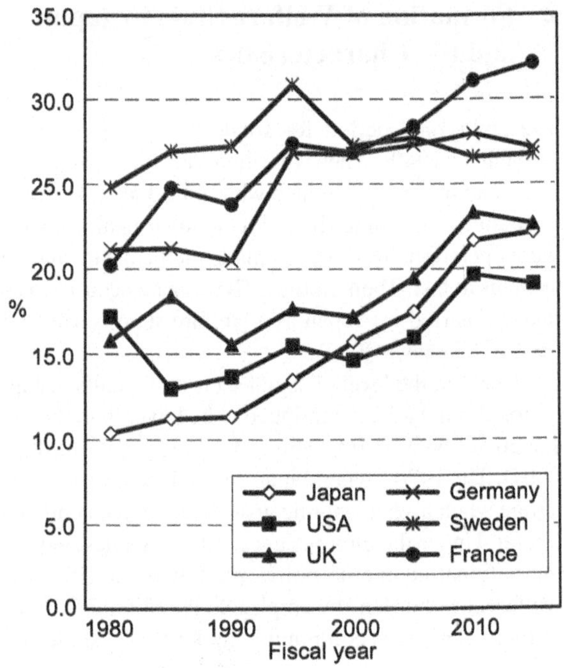

the trend of fiscal reconstruction and administrative reform, particularly addressing suppression of public expenditures.[2]

Takegawa (2007) explains the characteristics of the welfare regime in Japan formed in those days as described below, devoting particular attention to the three points of "politics" "redistribution" and "regulations."

Politics in Japan were characterized by the weak social democrats. Although labor movements and social democratic parties have played a major role in forming welfare states in Europe, it was bureaucrats who played the role in Japan. The conservative Liberal Democratic Party, which was a ruling party for many years, has approved the practice and the role. Furthermore, social security benefits in Japan have hovered at a low level because the suppression of the burden on the people was supported politically. Although the level of universal benefits, such as pension and medical care, was high, the level of benefits for special needs such as disabilities, family, housing, and public assistance was low.

Japan assigned priority on horizontal redistribution among regions through public works projects rather than vertical redistribution between classes. The expenditure for public works projects was considerably high in Japan. Public works projects created jobs and suppressed the occurrence of unemployment and poverty. By securing men's

[2]To be more specific, the following were implemented: Charging for home-help services (1982); Charging for medical expenses for elderly people (1983); Introduction of a copayment system for persons insured by medical insurance (1984); and Reducing the portion of state subsidy for safeguarding expenses for social welfare facilities to 50% (1986).

employment and stabilizing livelihood, women stayed at home as full-time housewives and provided in-home care (child care and nursing care) free of charge. This led to saving of social security benefits.

At least until the 1980s in Japan, although economic regulations were tight, social regulations were loose. The country protected and nurtured high-growth sectors and provided considerable protection and support for low-growth sectors lacking in competitiveness (agriculture, distribution industry, etc.), which also guaranteed employment. Nevertheless, employment discrimination on the grounds of sex or age remained as it was in Japan for a long time. Additionally, the prohibition of discrimination against persons with disabilities has not been common. Moreover, care for people who need welfare response has been left almost entirely to their families.

As described above, Japan as a welfare state has been characterized by its emphasis on a low level of public welfare and informal care undertaken mainly by families. For this reason, Japan has been regarded as a hybrid regime of liberalism and conservatism.

3 Trend of Social Security Reform in Japan

The welfare regime in Japan has faced a turning point since the late twentieth century because many conditions underpinning and justifying the Japanese welfare regime are vanishing (Takegawa 2007).

First, Japan began to be required to recover fiscal discipline to avoid capital flight during globalization. This constraint forced the country to address the downsizing of public works projects in addition to the suppression of social expenditures. Furthermore, the government implementation of policies to protect industry came to be criticized internationally from the perspective of free competition and free trade. After the government eased labor regulations, labor became increasingly unstable; corporate welfare declined. Moreover, the burdens of taxes and social insurance on corporations were reduced. These changes seriously shook job creation and employment protection that had been brought about by economic policies and the welfare regime in Japan that have thereby suppressed and substituted social expenditure. Secondly, family functions have been changing rapidly in recent years. The conventional welfare regime in Japan has been supported by the family that is genderized based on sex-role specialization. However, the rapidly progressing diversification of family, individualization, and an aging population with declining birthrate (see Chapter "Temporal Changes and Regional Differences in Demographic Structure and Health in Japan") weaken the welfare function of the family.

For the new risks associated with the economic and social changes, needs for social security have diversified and expanded. In response to this, institutional reforms started at the end of the 1980s. Hereinafter, we outline the trends of social security reform in Japan responding to an aging society related to the theme of this book.[3]

The mode of thought underlying social security reform has already been proposed in discussions related to reviewing welfare in the 1980s. The Provisional Council for the Promotion of Administrative Reform, as established as an advisory body to the prime minister, listed the following as high-priority policies for future social security in the final report released in 1990: suppression of the national burden rate; provision of services by the private sector to enable free choice; promotion of people's self-help efforts; emphasis on functions of local community; and establishment of a comprehensive system through organic cooperation between health care, medical care, and welfare as well as between employment policies and housing policies (Provisional Council for the Promotion of Administrative Reform and Committee for the Promotion of Administrative and Financial Reforms 1990). In 1989, a council under the Ministry of Health and Welfare reported emphasis on the roles of municipalities, the improvement of in-home welfare, and the development of private welfare services as characteristics of social welfare in the future (Joint Planning Working Group of Three Welfare-related Councils under the Ministry of Health and Welfare, Ministry of Health, Labour and Welfare 1989). Furthermore, in 1986, the Japan National Council of Social Welfare proposed the following as a basic plan for social welfare reform: revision of the roles of the central and local governments; universalization and generalization of social welfare; expansion of selective use of services; promotion of in-home welfare; diversification of welfare supply systems; and promotion of cooperation between social welfare and neighboring fields (Advisory Panel for Basic Plan for Social Welfare, Japan National Council of Social Welfare 1986).

All suggestions were made by anticipating a super-aging society in the twenty-first century. The "Gold Plan" was developed as a policy and was implemented to enhance measures for elderly people for 10 years from 1989.[4] It specifically examined the urgent implementation of measures for in-home welfare in municipalities and urgent improvement of care facilities. For the realization, eight social welfare-related laws were revised in 1990. The revision particularly stipulated emphasis on the roles of municipalities and the promotion of in-home welfare.[5] In the 1990s, with countermeasures against a declining birthrate, social expenditures resulting from those programs increased greatly (Fig. 1).

In this phase, however, the introduction of contract use as a substitute for the safeguards system and the entry of for-profit corporations into the service business

---

[3]The author used Kojima (2005) and materials in Japan Social Security Documents IV (1980–2000) published by the National Institute of Population and Social Security Research (2005) as a reference to write thereafter in this section.

[4]The official name of the Gold Plan is "A 10-Year Strategy of Health and Welfare for the Aged." Because population aging had become more advanced in 1994 than initially expected, the "Gold Plan" was fully revised as the "New Gold Plan." The numerical targets were raised.

[5]To be precise, it stipulated the legislation of in-home welfare services, the transfer of authority over admission measures for elderly people and persons with physical disabilities to villages and towns, and requirement of the formulation of elderly health care and welfare plan for municipalities and prefectures.

were not implemented except for some services. The safeguard system was the main system at that time to provide welfare services in which the administration determines the needs of a person in need of welfare services and selects services for the user. A discussion of whether or not welfare should be supplied through the market mechanism continued throughout 1990 under the influence of policy debates on deregulation aimed at boosting the economy after the collapse of the bubble economy. It was long-term care insurance started in 2000 that realized it.

The Ministry of Health and Welfare started a discussion of long-term care insurance in the first half of the 1990s to respond to an increase in care needs with the progress of population aging and the weakening of the family that has provided care to elderly people. The Council on Health and Welfare for the Elderly under the ministry began debate in 1995 and released a draft of the system in the following year. In 1997, the Long-Term Care Insurance Act was enacted after an agreement was reached between ruling parties. The characteristics of long-term care insurance in Japan are the following (see also Chapter "Nursing Care Provision Systems for Elderly People and Geographical Distribution of Services in Japan"). (1) It adopts a social insurance system and provides a system by which society as a whole supports care for elderly people. (2) It applies the principle that users pay according to the benefits they receive. (3) As a user-centered system to support elderly people's independence, it allows users by their own choice to receive welfare services and further health and medical care services in a comprehensive manner based on contracts with different providers, including private enterprises.

In 1997, the Child Welfare Act was also revised. Users became able to choose a nursery center, which was impossible under the safeguards system. Furthermore, in 2000, NPOs and commercial companies were permitted to establish nursery centers in addition to municipalities and social welfare corporations. Such reforms in the fields of elderly and child welfare took the initiative in the "basic structural reform of social welfare" in Japan and led to reforms of social welfare in other fields. For example, in the field of welfare for persons with disabilities, a support funding system was started in 2003, in which users use services chosen by themselves under contract with providers and the services are subsidized by public funds.[6]

4 Realization of Community-Based Inclusive Society from Community-Based Integrated Care System

4.1 Community-Based Integrated Care System

The long-term care insurance system has been revised five times to date. The building of the "community-based integrated care system" has remained as a key policy after

---

[6]Services were provided based on the Services and Supports for Persons with Disabilities Act from 2006 and based on the General Support for Persons with Disabilities Act from 2013. Further reforms were made.

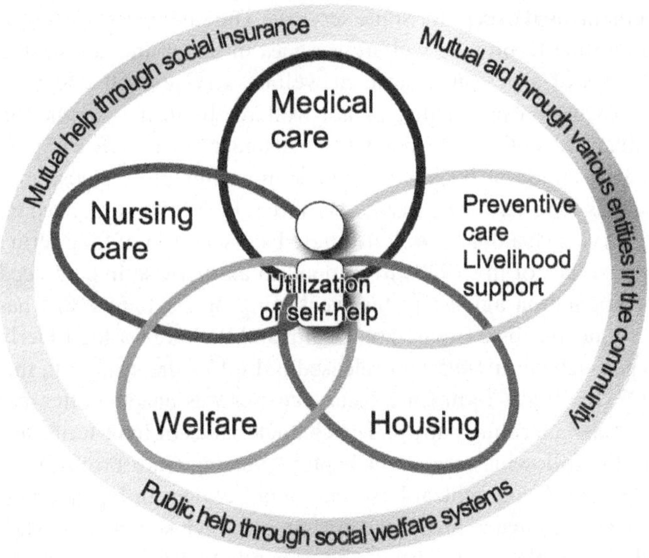

Fig. 3 Community-based integrated care systems in Japan

its 2011 revision. The community-based integrated care system provides nursing, medical, and health care services as well as livelihood support services in an integrated manner for elderly people to live in the community where they have lived for a long time based on stabilization of residence (Fig. 3). To build the system, an appropriate supply of housing and livelihood support of various kinds (safety monitoring, meal delivery, shopping support, etc.) are required in addition to specialized support, including the development of bases for care, the enhancement of cooperation between nursing and medical care, and the promotion of preventive care. Community-based integrated care systems must be built under the initiative of municipal governments according to the circumstances of each community, using small areas in municipalities called a sphere of daily life (junior high school districts or areas within about 30 min of travel time) as one unit (see details in Chapter "Regional Variation in the Community-Based Integrated Care Systems in Japan" through Chapter "Integrated Care Systems Established to Strengthen Community Disaster Resilience").

The community-based integrated care system combines the two concepts of integrated care and community-based care (Morikawa 2014). Integrated care provides an array of services in an integrated manner. Community-based care means building of a mechanism of care for elderly people for each community while using informal resources such as residents' mutual aid and voluntary activities in the local community. The characteristics of community-based integrated care systems in Japan are that they build integrated care on the foundation of community under the leadership of municipal governments (Tsutsui 2014).

As described earlier, the idea of such a community-based integrated care system was already considered in the 1980s. The practice also started in advanced municipalities in the 1970s (Niki 2015).[7] The national government translating this into policies as a "community-based integrated care system" is linked with changes in the goals of the long-term care insurance system; that is, conversion from the marketization of services to planning and coordination by administration. Furthermore, emphasis is placed on preventive care and the promotion of residents' mutual aid in the local community.[8]

The building of community-based integrated care systems has been brought up throughout the 2000s in committees under the Ministry of Health, Labour and Welfare and reports of a private "Study Group on the Community-based Integrated Care" that works by receiving subsidies from the ministry (Study Group on Elderly Care 2003; Study Group on the Community-based Integrated Care 2009). Although the name of "community-based integrated care system" was not used in the third revision of the Long-Term Care Insurance Act in 2011, the idea was stipulated in the law. In the outline of "comprehensive reform of the social security and tax systems"[9] approved by the Cabinet in the next year, the building of community-based integrated care systems was declared with the target date of 2025.[10] In 2013, the "community-based integrated care system" was legally defined in the "Social Security Reform Program Act"[11] for the first time.

---

[7]The practice includes efforts developed mainly by the public Mitsugi General Hospital in Mitsugi Town (present-day Onomichi City), Hiroshima Prefecture (Yamaguchi 1992), those in Nishiaizu Town, Fukushima Prefecture and Mogami Town, Yamagata Prefecture made under the influence (Miyazawa 2006; Yokoyama 2006), and the promotion of community welfare undertaken by social welfare councils and social welfare corporations around the country (Ohashi and Shirasawa eds 2014).

[8]At an early stage of the beginning of long-term care insurance, long-term care benefits and expenses increased. A system was established to control them. The marketization brought geographic disparities to the supply of services (Miyazawa 2003, 2010). Therefore, the systematic development of a service infrastructure by administration came to be required. It was institutionalized in the first revision of long-term care insurance in 2005 as the creation of community-based services that municipal governments play a central role in developing and the establishment of comprehensive community support centers that become the bases for preventive care services (see details in Chapters "Nursing Care Provision Systems for Elderly People and Geographical Distribution of Services in Japan" and "Regional Variation in the Community-based Integrated Care Systems in Japan"). Preventive care services using mutual aid between residents were enshrined into law in the fourth revision in 2014.

[9]This is a reform to achieve the two goals of social security enhancement and stabilization concurrently with sound finance.

[10]In 2025, the baby-boom generation will become the elderly generation in the latter stage of life (75 years and over): 1 in every 5 people will be aged 75 years or older. Therefore, the year was set as the immediate target year for social security reform.

[11]The official name is the "Act on Promotion of Reforms of Establishment of Sustainable Social Security Systems." The act clearly presents a comprehensive picture of the reform of the social security system in Japan and explains how to carry it out based on a 2013 Cabinet decision on the promotion of social security system reform.

The "Act on Promotion of Comprehensive Securing of Medical Care and Nursing Care"[12] enacted in 2014 aimed at developing tax systems and laws to build community-based integrated care systems. In the revision of the Medical Care Act associated with this, prefectures were required to formulate a "community health care vision" that determines the reorganization of hospital beds based on the estimated future number of necessary beds as part of the medical care plan (see Chapter "Medical Care Provision System and Geographical Distribution of Medical Resources in Japan"). Because medical care in community-based integrated care systems includes not only clinics and in-home care, but also hospital care, the building must be regarded together with the community health care vision (Niki 2017).

4.2 For Realization of Community-Based Inclusive Society

The idea of a community-based integrated care system that "provides necessary support in an integrated manner" has begun to be realized not only in the field of elderly welfare but also in other fields. For example, after 2015, the support system for the self-reliance of needy persons was established (see Chapter "Creation of Social Ties for Prevention of Isolation of Elderly Public Assistance Recipients: The Case of a Project for the Provision of "A Place of One's Own" in Nishinari Ward, Osaka City"). For the field of child welfare, user support programs for children and parenting households and comprehensive support centers for childrearing generation, which became the bases for the programs, began to be developed (see Chapter "Establishing Community-Based Integrated Support Systems for Pregnancy, Childbirth, and Childcare in Japan: Focusing on Regional Differences"). It is the building of an "all-generation, all-people comprehensive community support system" that is envisioned as the universalization of the idea of such a community-based integrated care system and the expansion of the scope of subjects (Fig. 4).

It refers to a system in which all people in the community, irrespective of age and circumstances, can receive appropriate support according to the needs of the person. The purpose is to revise the support that has been provided through "vertical sectioning" in each field and to provide comprehensive support and services for complex challenges that individuals and households face: the purpose is to make "whole" support possible. This point was raised in the "New Welfare Vision"[13] reported by a project team under the Ministry of Health, Labour and Welfare with a challenge of development and security of human resources who have comprehensiveness capability of undertaking it[14] (Project Team for Reviewing New Welfare

[12] The official name is the "Act on Arrangement of Relevant Acts to Promote Comprehensive Securing of Medical Care and Nursing Care in Areas."

[13] The official name is the "Realization of Welfare Services for Building of Community Where Everyone Supports Each Other: A Vision of Welfare Provision Responding to A New Era."

[14] See Chapter "Securing of Health, Medical, and Welfare Personnel and the Geographical Distribution in Japan" for information related to the system of human resources development in the field of medical and welfare in Japan and challenges for the security.

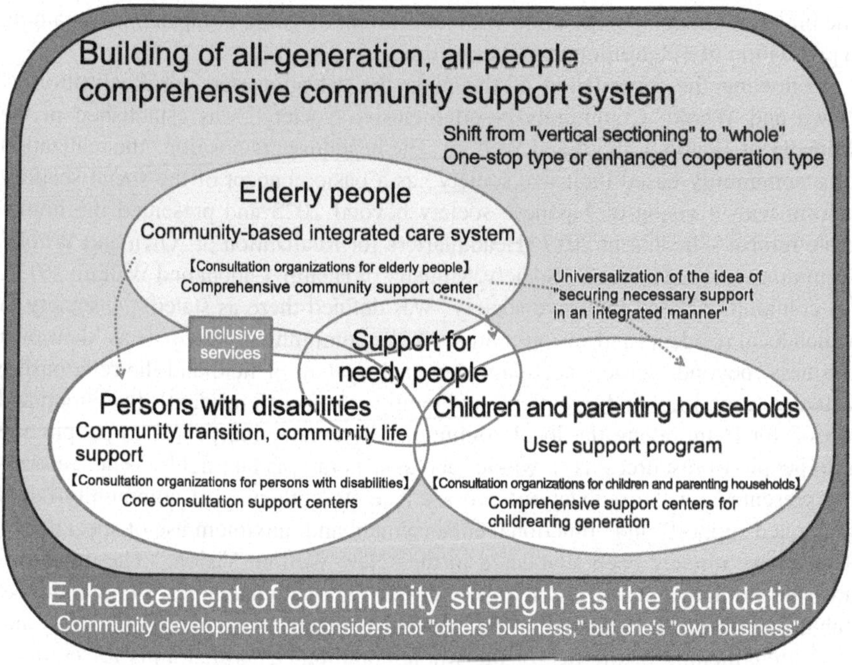

Fig. 4 Enhancement of comprehensive community support systems for the realization of "a community-based inclusive society" (Modified figure published in the Ministry of Health, Labour and Welfare 2017)

Service Systems, Ministry of Health, Labour and Welfare 2015). There are two types of all-generation, all-people comprehensive community support systems: the one-stop type and the enhanced cooperation type. They are selectable according to the actual circumstances of the community. The former set forth the development of intergenerational exchange-type and multifunctional-type welfare facilities that become the bases for the systems (see Chapter "Formation of Comprehensive Community Welfare Bases in Urban Areas"). Furthermore, the "New Welfare Vision" advocates the expansion of systems to "community development" through cooperation with other fields, including employment, agriculture, education, and regional development.

The direction emphasizing on "community development" was further clarified in the form of the enhancement of community strength that becomes the foundation for the "realization of a community-based inclusive society." This idea was first taken up as a national policy in "The Japan's Plan for Dynamic Engagement of All Citizens" approved by the Cabinet in 2016.[15] In fact, the plan is the centerpiece of the Abe administration. It aims at building a new economic and social system to be a society in which anyone can be active at home, workplace, and in the community, applying

[15]However, the idea of "a community-based inclusive society" is deemed to be an idea that researchers in the field of community welfare have advocated since the 1970s (Niki 2017).

the brakes to an aging population with a declining birthrate in Japan and sustaining a population of 100 million in 50 years' time.

Following the formulation of the plan, the "Headquarters for Realization of 'Own and Whole' Community-based Inclusive Society" was established in the Ministry of Health, Labour and Welfare. The headquarters position "the realization of a community-based inclusive society" as a basic concept of the social security reform with a vision of Japanese society beyond 2025 and presented the immediate reform schedule in 2017 (Headquarters for Realization of 'Own and Whole' Community-based Inclusive Society, Ministry of Health, Labour and Welfare 2017). "A community-based inclusive society" was defined there as stated: "A society in which local residents and various entities in the community participate as 'their own business' beyond 'vertical sectioning' in each system or field and the relationship between 'supporter' and 'recipient' and jointly create each resident's livelihood and reason for living along the local community by connecting people to people and also people to resources as 'a whole' across generations and fields." Four reforms are presented for the realization: two are "the strengthening of community-based integrated support" and "functional enhancement and maximum use of specialists." Those have already been advocated in the "New Welfare Vision." The other two are "the strengthening of problem-solving for regional challenges," which aims at enhancing the mutual aid relationship between residents in the local community and "the strengthening of the ties of the whole community," which aims for community development based on cooperation with various fields in addition to the field of welfare, are to enhance community strength for the realization of a community-based inclusive society (see Chapter "Development of Community Welfare Activities with Resident Participation and Their Importance in Hilly and Mountainous Areas").

In 2017, the "Community-based Integrated Care System Enhancement Act"[16] was enacted. It stipulated the promotion of efforts for the realization of a community-based inclusive society as part of the deepening and promotion of community-based integrated care systems. Specifically, it set forth the promotion of community welfare with the philosophy that challenges in life are resolved through cooperation among local residents and people involved in welfare and among organizations concerned. It also obligated municipal governments to undertake efforts to create comprehensive support systems as well as to improve community welfare plans. To make it easier for elderly people and persons with disabilities (children and adults) to receive services in the same service center, inclusive services were newly established both in the long-term care insurance system and the welfare system for persons with disabilities (see Chapter "Current Situation and Challenges of "Inclusive Care": An Investigation of the "Community-based Inclusive Station Program" in Saga Prefecture"). The Act came into force in 2018. Efforts for the realization of a community-based inclusive society have already begun.

[16]The official name is the "Act on Partial Revision of Long-term Care Insurance Act, etc. to Strengthen Community-based Integrated Care Systems."

5 Towards Geographical Research of Comprehensive Community Support

This chapter described characteristics of the conventional welfare regime in Japan and also outlined the trend of social security reforms undertaken to date. The conventional welfare regime in Japan has emphasized a low level of public welfare and informal care. Moreover, it has cast male-earner nuclear family households as hidden assets. With the 1990 revision of eight social welfare-related laws, the roles of municipalities were increasingly emphasized. Furthermore, the function of mutual aid that local communities have has been valued since that time. The introduction of long-term care insurance systems greatly altered the conventional welfare regime in Japan that has assigned importance to such informal resources. The start of long-term care insurance pushed forward the socialization of care, the marketization of service provision, and the centralization of management and operation in systems and standards for fees and services. It influenced the entire field of welfare as "the basic structural reform of social welfare."

However, during the first revision made five years after the long-term care insurance started, a system to control benefits was immediately institutionalized. The planning and coordination function of municipalities was enhanced in a manner contrary to marketization. The roles of informal resources, such as residents' mutual aid and voluntary activities in the local community, came to be re-emphasized. They have been incorporated into public service provision systems. Hiraoka (2018) designates the movement as "(re-) regionalization." Actually, this movement is a recent characteristic of the welfare regime in Japan. It is apparent in the policies of the building of community-based integrated care systems and the realization of a community-based inclusive society that appeared in the 2010s.

As the two policies demonstrate, the social security trend in Japan is to build a comprehensive support system carefully in each small area in a manner of reflecting the features of the area. Diversity will arise there according to differences in geographical conditions such as the distribution of population and social resources and natural characteristics, and from further differences in each local government's planning and coordination skills and local governance (Hatakeyama et al. 2018). For this reason, geography attempts to tackle this theme. This book evaluates attempts at the building of comprehensive community support systems in Japan through analysis of actual circumstances.

References

Advisory Panel for Basic Plan for Social Welfare, Japan National Council of Social Welfare (1986) Suggestions for basic plan of social welfare reform. In: National Institute of Population and Social Security Research (ed) Japan social security documents IV (1980–2000). Report and statistics. National Institute of Population and Social Security Research 13. http://www.ipss.go.jp/publication/j/shiryou/no.13/data/shiryou/syakaifukushi/296.pdf. Accessed 1 May 2019 (in Japanese)

Esping-Andersen G (1990) The three world of welfare capitalism. Basil Blackwell, Oxford
Esping-Andersen G (1997) Hybrid or unique? The Japanese welfare state between Europe and America. J Eur Soc Policy 7:179–189
Esping-Andersen G (1999) Social foundations of postindustrial economies. Oxford University Press, Oxford
Goodman R, White G, Kwon H (eds) (1998) The East Asian welfare model: Welfare orientalism and the state. Routledge, London
Hatakeyama T, Nakamura T, Miyazawa H (2018) Community-based integrated care systems in Japan: Focusing on spatial structures and local governance. E-Journal GEO 13: 486–510 (in Japanese with English abstract)
Headquarters for Realization of 'Own and Whole' Community-based Inclusive Society, Ministry of Health, Labour and Welfare (2017) For realization of Community-based Inclusive Society (immediate reform schedule). https://www.mhlw.go.jp/file/05-Shingikai-12601000-Seisakutoukatsukan-Sanjikanshitsu_Shakaihoshoutantou/0000150615.pdf. Accessed 1 May 2019 (in Japanese)
Hiraoka K (2018) Creation and reform of long-term care insurance system and elderly care regime in Japan. In: Suda Y, Hiraoka K, Morikawa M (eds) Elderly care in East Asia: future of country, community, and family. Toshindo Publishing, Tokyo (in Japanese)
Holliday I, Wilding P (eds) (2003) Welfare capitalism in East Asia: social policy in the tiger economics. Palgrave Macmillan, London
Joint Planning Working Group of Three Welfare-related Councils, Ministry of Health, Labour and Welfare (1989) Way of future social welfare should be (opinion offer): Suggestions to realize a sound longevity and welfare society. In: National Institute of Population and Social Security Research (ed) Japan social security documents IV (1980–2000). Report and statistics. National Institute of Population and Social Security Research 13. http://www.ipss.go.jp/publication/j/shiryou/no.13/data/shiryou/syakaifukushi/376.pdf. Accessed 1 May 2019 (in Japanese)
Kojima S (2005) Social welfare. In: National Institute of Population and Social Security Research (ed) Japan social security documents IV (1980–2000). Report and statistics. National Institute of Population and Social Security Research 13. http://www.ipss.go.jp/publication/j/shiryou/no.13/data/kaidai/12.html. Accessed 1 May 2019 (in Japanese)
Ministry of Health, Labour and Welfare (2017) Toward realization of an inclusive, community-based society. https://www.mhlw.go.jp/stf/newpage_00506.html. Accessed 1 May 2019 (in Japanese)
Miyazawa H (2003) Uneven nursing care service opportunity and the behavior of service providers under the long-term care insurance system: A statistical analysis in the Kanto District. Geogr Rev Jpn 76:59–80 (in Japanese with English abstract)
Miyazawa H (2006) Local revitalization and planning for health and welfare promotion in Nishiaizu Machi, Fukushima Prefecture. Jpn J Hum Geogr 58:235–252 (in Japanese with English abstract)
Miyazawa H (2010) Location and characteristics of pay nursing homes in the Tokyo metropolitan area. E-Journal GEO 4:69–85 (in Japanese with English abstract)
Morikawa M (2014) Towards community-based integrated care: trends and issues in Japan's long-term care policy. Int J Integr Care 14(1):1–10
National Institute of Population and Social Security Research (ed) (2005) Japan social security documents IV(1980–2000) Report and statistics. Nat Inst Popul Soc Secur Res 13. http://www.ipss.go.jp/publication/j/shiryou/no.13/title.html Accessed 1 May 2019 (in Japanese)
National Institute of Population and Social Security Research (2018) The financial statistics of social security in Japan FY 2016. National Institute of Population and Social Security Statistical Report 29. http://www.ipss.go.jp/ss-cost/e/fsss-16/fsss-16.asp Accessed 1 May 2019
Niki R (2015) Community-based integrated care and regional medical cooperation. Keiso Sobo, Tokyo (in Japanese)
Niki R (2017) Community-based integrated care and welfare reform. Keiso Sobo, Tokyo (in Japanese)
Ohashi K, Shirasawa M (eds) (2014) Practice and prospects of community-based integrated care: Learning form efforts in advanced areas. Chuohoki Publishing, Tokyo (in Japanese)

Project Team for Reviewing New Welfare Service Systems, Ministry of Health, Labour and Welfare (2015) Realization of welfare services for building of community where everyone supports each other: A vision of welfare provision responding to a new era. https://www.mhlw.go.jp/file/05-Shingikai-12201000-Shakaiengokyokushougaihokenfukushibu-Kikakuka/bijon.pdf. Accessed 1 May 2019 (in Japanese)

Provisional Council for the Promotion of Administrative Reform and Committee for the Promotion of Administrative and Financial Reforms (1990) Report of committee for the promotion of administrative and financial reforms. In: National institute of population and social security research (ed) Japan social security documents IV (1980–2000). Report and statistics. National institute of population and social security research 13. http://www.ipss.go.jp/publication/j/shiryou/no.13/data/shiryou/syakaifukushi/399.pdf. Accessed 1 May 2019 (in Japanese)

Study Group on Elderly Care (2003) Elderly care in 2015: For establishment of care to support dignity of elderly people. https://www.mhlw.go.jp/topics/kaigo/kentou/15kourei/3.html Accessed 1 May 2019 (in Japanese)

Study Group on the Community-based Integrated Care (2009) Report of study group on the community-based integrated care: summary of main points for future discussion. The 2008 MHLW program for health promotion of elderly. https://www.murc.jp/sp/1509/houkatsu/houkatsu_01_pdf01.pdf. Accessed 1 May 2019 (in Japanese)

Takegawa S (2007) Solidarity and recognition: the welfare state in globalization and individualization. University of Tokyo Press, Tokyo (in Japanese)

Tanaka T (2017) History of welfare politics: Democracy against inequality. Keiso Sobo, Tokyo (in Japanese)

Tsutsui T (2014) Management strategy to achieve community-based integrated care in Japan: Theoretical concept of integrated care and applications. Chuohoki Publishing, Tokyo (in Japanese)

Uzuhashi T (1997) International comparison of modern welfare states: Positioning and prospects of Japanese model. Nippon Hyoron Sha, Tokyo (in Japanese)

Yamaguchi N (1992) Strategy for zero bedridden elderly. Ie-no-hikari Association, Tokyo (in Japanese)

Yokoyama S (2006) Mogami-chou wellness plaza and public care insurance: The arrenngement [sic] of institution and the spread of care insurance. Bulletin of Yamagata University. Social Science 36(2):87–105 (in Japanese)

Temporal Changes and Regional Differences in Demographic Structure and Health in Japan

Masakazu Yamauchi and Tomoya Hanibuchi

Abstract With population decline and aging, household composition is changing in Japan. Both average and healthy life expectancy have increased. Moreover, an increasing number of people are adopting healthy behaviors. Such changes in demographic characteristics and health are extremely important because they specify needs for social security such as medical care, nursing care, and parenting. Importantly, vast regional differences exist in demographic characteristics and health. This chapter first introduces and summarizes demographic characteristics for residents of Japan. Next, it presents various basic data related to future population projections that are expected to influence social security systems in the future. This chapter also presents data related to household family types that are relevant to whether one is able to rely on in-home care. Finally, we draw attention to regional differences in health, specifically addressing life expectancy, health conditions, and health behaviors that are relevant to the demand for medical and nursing care.

Keywords Aging · Health-related behaviors · Life expectancy · Population decline

1 Demographic Characteristics of Japan

Japan, similarly to many eastern and southern European countries, is confronted with a declining population. The number of countries in which the population declined during 2010–2015 was 22, most of which are in eastern and southern Europe. Table 1 presents demographic indexes of 19 countries out of the 22 countries, leaving out the 3 countries with populations of fewer than 100,000 (Andorra, Cook Islands, and Niue). Japan has the following six salient demographic characteristics.

This chapter is a revised version of Hanibuchi (2017a, b, c), and Yamauchi (2017).

M. Yamauchi (✉)
Waseda University, Tokyo, Japan
e-mail: yamauchi-masa@waseda.jp

T. Hanibuchi
Tohoku University, Sendai, Japan

H. Miyazawa and T. Hatakeyama (eds.), *Community-Based Integrated Care and the Inclusive Society*, International Perspectives in Geography 12,
https://doi.org/10.1007/978-981-33-4473-0_2

The first characteristic is that Japan had the largest population and the highest population density among all countries with a declining population during 2010–2015. The population of Japan was 128 million in 2015: the tenth largest population in the world. The population density was 351 people/km^2 in 2015: the 16th highest among countries with a population of 1 million or more.

The second characteristic is that the proportion of urban population is high. The percentage of urban population in Japan was 91.4% in 2015. Several countries had urban populations of more than 90.0% in 2015, including city states such as Singapore and Monaco, and countries situated in arid regions such as Kuwait and Qatar. Among those countries, Japan alone has a population that declined during 2010–2015.

The third characteristic is that the percentage of population aged 65 or older is the highest in the world. Moreover, the aging of the population has progressed rapidly. The percentage of population of 65 or older in Japan was 26.0% in 2015. That figure rose from 7% in 1970 to 14% 24 years later in 1994 and had risen to 21% a further

Table 1 Demographic indexes of depopulation countries

Country	Average annual percentage of population change, 2010–2015	Population, 2015 (1,000)	Population density, 2015 (persons per square km)	Percentage of Population in Urban Areas, 2015
Syria	−2.30	18,735	102	52.2
Georgia	−1.37	3,952	57	57.4
Lithuania	−1.27	2,932	47	67.2
Latvia	−1.23	1,993	32	68.0
Bosnia and Herzegovina	−1.03	3,536	69	47.2
Bulgaria	−0.62	7,177	66	74.0
Romania	−0.56	19,877	86	53.9
Ukraine	−0.50	44,658	77	69.1
Portugal	−0.44	10,418	114	63.5
Croatia	−0.43	4,236	76	56.2
Greece	−0.40	11,218	87	78.0
Serbia	−0.40	8,851	101	55.7
Hungary	−0.29	9,784	108	70.5
Estonia	−0.25	1,315	31	68.4
Spain	−0.17	46,398	93	79.6
Albania	−0.12	2,923	107	57.4
Moldova	−0.09	4,066	124	42.5
Japan	**−0.09**	**127,975**	**351**	**91.4**
Italy	−0.08	59,504	202	69.6

(continued)

Table 1 (continued)

Country	Percentage of total population by 65 and over, both sexes, 2015	Total fertility rate, 2010–2015 (live births per woman)	Life expectancy at birth for both sexes combined, 2010–2015 (years)	Net migration rate, 2010–2015 (%)
Syria	4.0	3.1	69.9	−4.18
Georgia	14.6	2.0	72.8	−1.49
Lithuania	18.7	1.6	74.0	−0.97
Latvia	19.3	1.5	73.9	−0.81
Bosnia and Herzegovina	15.7	1.3	76.3	−0.89
Bulgaria	20.1	1.5	74.3	−0.07
Romania	17.0	1.5	74.8	−0.30
Ukraine	15.9	1.5	71.1	−0.09
Portugal	20.7	1.3	80.5	−0.27
Croatia	18.9	1.5	77.0	−0.15
Greece	19.9	1.3	80.6	−0.29
Serbia	16.3	1.6	74.7	−0.22
Hungary	17.5	1.3	75.4	0.06
Estonia	18.8	1.6	76.8	−0.16
Spain	18.9	1.3	82.5	−0.24
Albania	12.5	1.7	77.7	−0.64
Moldova	9.9	1.3	71.0	−0.05
Japan	**26.0**	**1.4**	**83.3**	**0.06**
Italy	22.4	1.4	82.3	0.09

Created based on United Nations, Department of Economic and Social Affairs, Population Division (2017) and United Nations, Department of Economic and Social Affairs, Population Division (2018)

13 years later in 2007. By contrast, Italy which had the second highest proportion of population aged 65 or older, has experienced gradual population aging. That figure in Italy was 7% in 1927. It rose to 14% 61 years later in 1988 and had risen to 21% a further 24 years later in 2012.

The fourth characteristic is that the total fertility rate is low. The total fertility rate is an index equivalent to the average number of children for a woman. The total fertility rate exceeded 4 in the late 1940s in Japan. However, it decreased rapidly to about 2 in the 1950s and fell gradually after the 1970s. Many of the countries that experienced a population decline during 2010–2015 also had a low total fertility rate.

The fifth characteristic is that the average life expectancy in Japan was the highest in the world during 2010–2015. Among the countries in which the population declined during 2010–2015, the differences between Syria and Japan and between East European countries and Japan are considerably large, whereas the differences between Japan and West European countries are small. The average life expectancy in Japan increased rapidly after World War II. In the process, causes of death shifted from infectious diseases such as gastroenteritis to non-infectious diseases such as cancer. Although the decline of the mortality rate of young people strongly influenced the extension of average life expectancy in the past, that of elderly people has a stronger influence these days.

The sixth characteristic is that the net migration rate is small. The absolute value of the net migration rate in Japan has been consistently small since the 1950s, partly because of a restrictive immigration control policy and economic development after the 1950s. In contrast, in Syria, which is in conflict, and eastern Europe, where economies remain stagnant, population movements to western Europe in recent years are noticeable.

Based on the demographic characteristics in Japan described above, the following section presents organization of geographical characteristics of populations and health conditions in Japan. Section 2 takes up discussion of future population, which is expected to affect social security systems in the future. Section 3 addresses household family types, which are relevant to whether one can rely on in-home care. Section 4 explains life expectancy and health levels that are relevant to demand for medical and nursing care. Section 5 devotes some attention to regional differences in lifestyle habits as causes of non-infectious diseases.

2 Geographical Diversity in Population Changes

2.1 Differences Between Metropolitan and Non-metropolitan Areas

For the analyses described in this section, we divide 47 prefectures of Japan into Metropolitan and Non-metropolitan areas. The Metropolitan Area comprises three major metropolitan areas (the Tokyo metropolitan area, Nagoya metropolitan area, and Osaka metropolitan area) defined by the Statistics Bureau, Ministry of Internal Affairs and Communications Japan (2019), including 11 prefectures: Saitama, Chiba, Tokyo, Kanagawa, Gifu, Aichi, Mie, Kyoto, Osaka, Hyogo, and Nara. The Non-Metropolitan Area includes 36 other prefectures. Figure 1 shows the population and the percentage of population aged 65 or older in the Metropolitan and Non-metropolitan areas. In the Non-metropolitan Area, the population has been declining after reaching a peak in 2000; the rate of population decline is expected to increase. The percentage of population aged 65 or older rose, hitting a record of 28.4% in

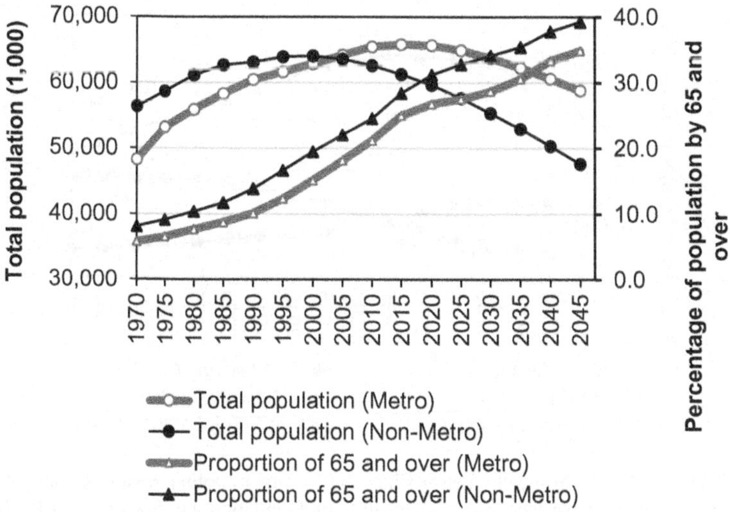

Fig. 1 Population size and percentage of population by 65 and over in metropolitan and Non-metropolitan region (Created based on Population Census of Japan and National Institute of Population and Social Security Research 2018a)

2015. Although the increase is expected to be slow, it is expected to reach 39.3% in 2045.

The Metropolitan Area shows the same change pattern as that of the Non-metropolitan Area, although a time lag is apparent. The population in the Metropolitan Area reached its peak in 2015, but its future decline is expected. The percentage of population aged 65 or older was 25.0% in 2015. It has been projected to become 34.9% in 2045.

The Metropolitan and Non-metropolitan areas show similar change patterns in terms of the population by age (Fig. 2). The population of 0–14-year-olds has decreased in both regions since the 1980s. Although the trend is expected to continue, the decline of the share of the population of 0–14-year-olds is occurring faster in the Non-metropolitan Area. The population of 15–64-year-olds has decreased in both regions since 1995. Although the trend is expected to continue, the rate of decline of the population of 15–64-year-olds is greater in the Non-metropolitan Area. The population of 65 years old and older has increased to date in both areas. It is expected to increase slowly for a while. However, the population of 65 years old and older is expected to decrease in the Non-metropolitan Area after 2040.

2.2 *Differences Among Municipalities*

Municipalities of Japan are local entities including cities (shi), towns (cho), villages (son) and special wards of Tokyo (ku). For the discussion in this section, we use the

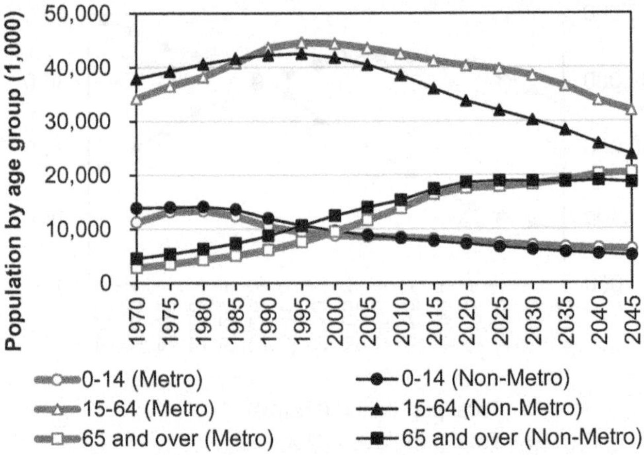

Fig. 2 Populationby age groups in metropolitan and Non-metropolitan region (Created based on Population Census of Japan and National Institute of Population and Social Security Research 2018a)

results of municipal population projections released in 2018 conducted by National Institute of Population and Social Security Research. The geographical units applied to the projection are 1,682 municipalities (778 cities, 713 towns, 168 villages, and 23 special wards of Tokyo) by the boundaries of March 1, 2018, except the municipalities of Fukushima Prefecture. Figure 3 shows the population growth rate during 2015–2045. The map represents a cartogram created to modify municipal areas based on the projected population in 2045. The map colors differentiate municipalities according to their respective population growth rates during 2015–2045. The number of municipalities for which the population growth rate is higher than 0 during 2015–2045 is 94, accounting for 5.6% of all municipalities. Similarly, the number of municipalities for which the population growth rate is −25% –0% is 456, accounting for 27.1% of all municipalities. Municipalities for which the projected total population in 2045 is larger than the total population in 2015 or for which the rate of population decrease is low are located mainly in cities, wards and towns in the Metropolitan Area and, prefectural capital cities and their surrounding cities and towns in the Non-metropolitan Area. They are drawn as large areas on the map because their populations are large.

The population of 0–14-year-old children is expected to decrease in almost all municipalities. Figure 4 shows the growth rate of the population of 0–14-year-olds during 2015–2045. The number of municipalities with a value higher than 0% is 35, accounting for 2.1% of all municipalities, although the number of municipalities with a value below −50% is 777, reaching 46.2% of all municipalities.

Geographical patterns suggest that the growth rate of the population of 0–14-year-olds is high in cities, wards and towns in the Metropolitan Area, and in prefectural capital cities and their surrounding cities and towns in the Non-metropolitan Area. The low rates are particularly remarkable in municipalities in the regions

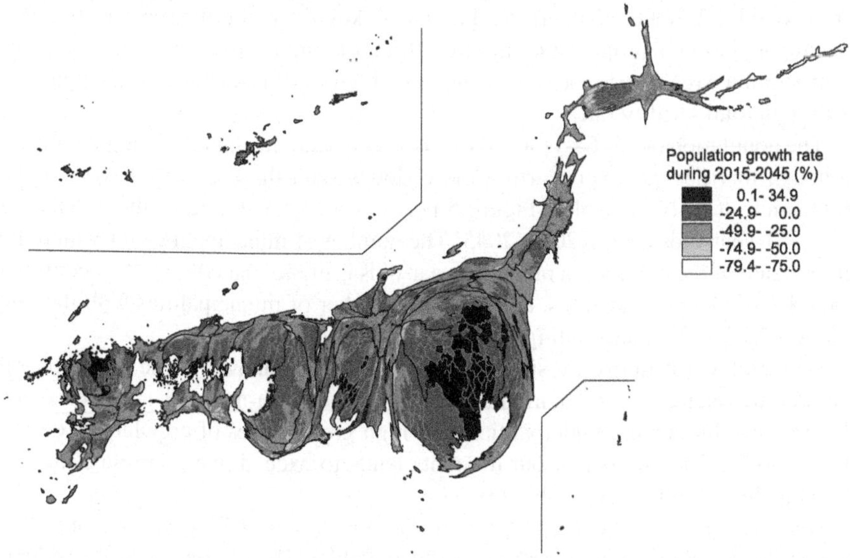

Fig. 3 Population growth rate during 2015–2045 (by municipality). Data for municipalities are those as of March 1, 2018. Municipalities in Fukushima Prefecture are treated as a whole because the prefecture does not provide estimates by municipality. The same applies to Figs. 4, 5, 6 and 7. (Created based on National Institute of Population and Social Security Research 2018a)

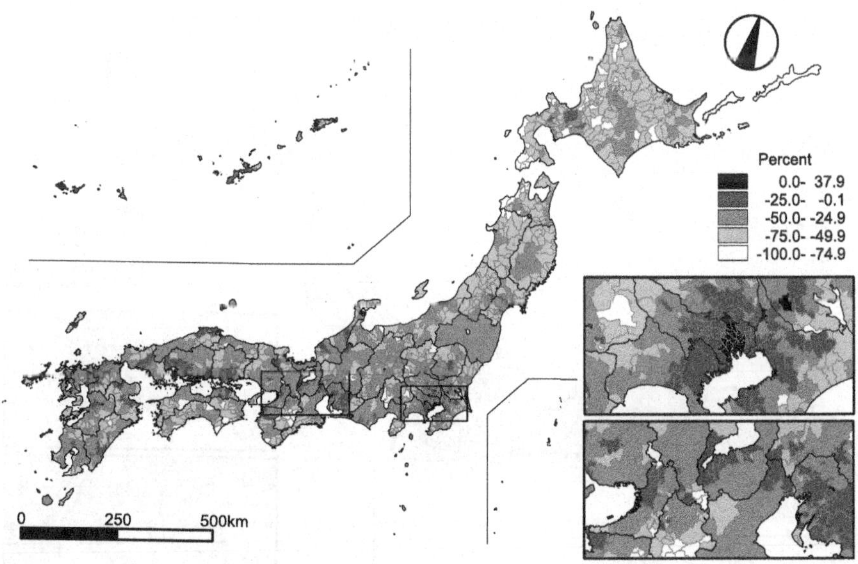

Fig. 4 Growth rate of the population of 0–14 years old during 2015–2045 (by municipality) (Created based on National Institute of Population and Social Security Research 2018a)

of Hokkaido, Tohoku, southern Kinki, and Shikoku because the reproductive age group population is projected to decrease further in there. However, even in a Non-metropolitan Area such as the Kyushu region, a high value is shown for municipalities with high total fertility rates.

The population of 15–64-year-olds is also decreasing in almost all municipalities, although that decrease is occurring more slowly than the decrease shown for the population of 0–14-year-olds. Figure 5 portrays the growth rate of the population of 15–64-year-olds during 2015–2045. The number of municipalities for which the growth rate of the population of 15–64-year-olds is higher than 0% is 24, accounting for 1.4% of all municipalities, although the number of municipalities with the rate below −50% is 729, accounting for 43.3% of all municipalities.

Geographical patterns are similar to those in Fig. 4 above. These geographical patterns are related to internal migration. The number of in-migrants tends to exceed that of out-migrants in municipalities with high growth rates of populations of 15–64-year-olds. The number of out-migrants tends to exceed that of in-migrants for municipalities with low growth rates.

Here we devote particular attention to the population of 75 years old and older among the population who are 65 years old and older. The reason is that the proportion of the population of 75 years old and older is expected to increase among the population of 65 years old and older. Additionally, when people become 75 years old and older, an increasing number of people need some sort of support for daily living activities.

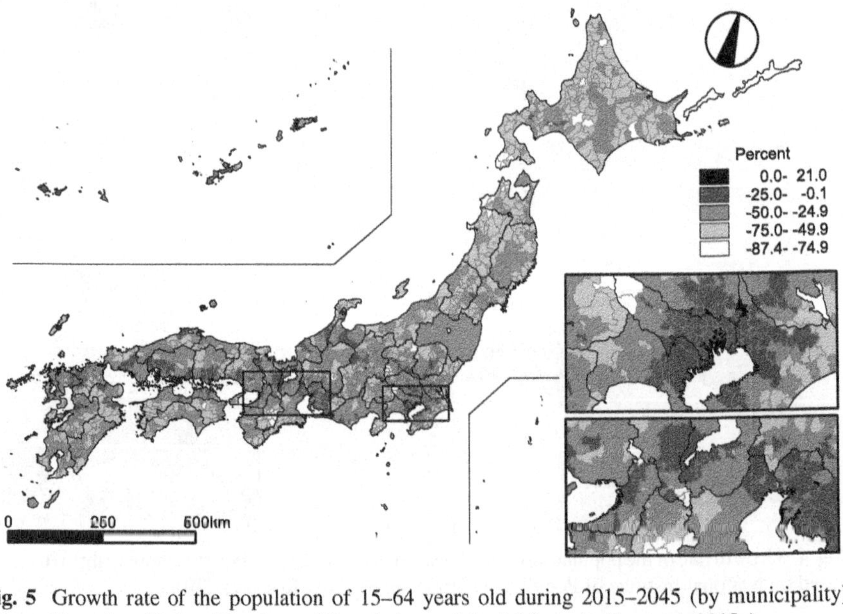

Fig. 5 Growth rate of the population of 15–64 years old during 2015–2045 (by municipality) (Created based on National Institute of Population and Social Security Research 2018a)

The population of 75 years old and older is expected to continue to increase in Japan as a whole. However, regarded by municipality, the growth rate of the population of 75 years old and older in the future varies widely. Figure 6 shows the growth rate of the population of 75 years old and older during 2015–2045. The number of municipalities for which the growth rate of the population of 75 years old and older is higher than 50% is 336, accounting for 20.0% of all municipalities, but 698 municipalities have a growth rate below 0%, accounting for 41.5% of all municipalities.

Geographical patterns indicate that the growth rate of the population of 75 years old and older during 2015–2045 is high in large cities and their surrounding cities and towns in the Metropolitan and Non-metropolitan areas. However, it is low in the remaining municipalities in the Non-metropolitan Area. These patterns are further noticeable compared to the population of 0–14-year-olds and those of 15–64-year-olds described above. These patterns reflect differences in the distribution of the population that is expected to reach old age: they reflect that the population expected to reach old age itself is large in large cities and their surrounding cities and towns in the Metropolitan and Non-metropolitan areas because past in-migrants reside in these municipalities.

The increase of elderly people and the expansion of their population share affect social systems supporting them. Here we specifically examine the population of 15–64-year-olds per person among the population 75 years old and older as an index to present the population balance between elderly people and young and middle-aged

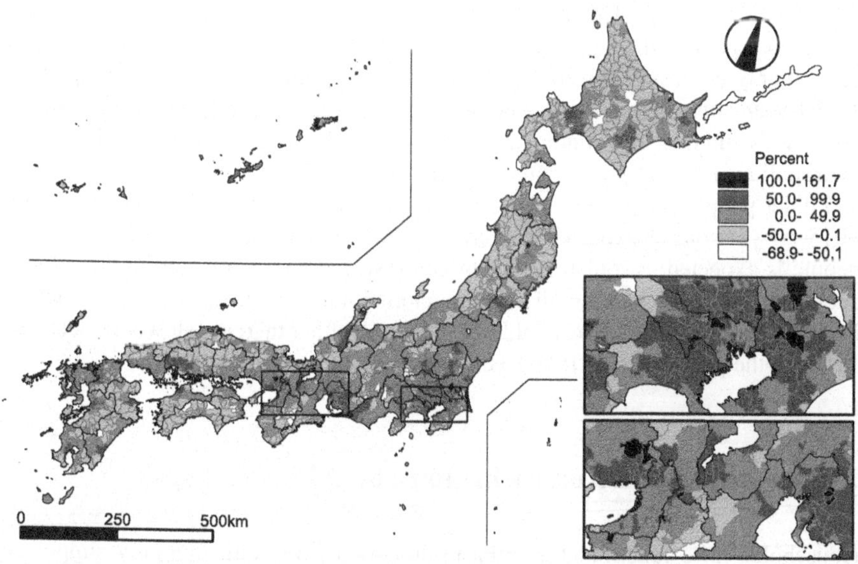

Fig. 6 Growth rate of the population of 75 years old and older during 2015–2045 (by municipality) (Created based on National Institute of Population and Social Security Research 2018a)

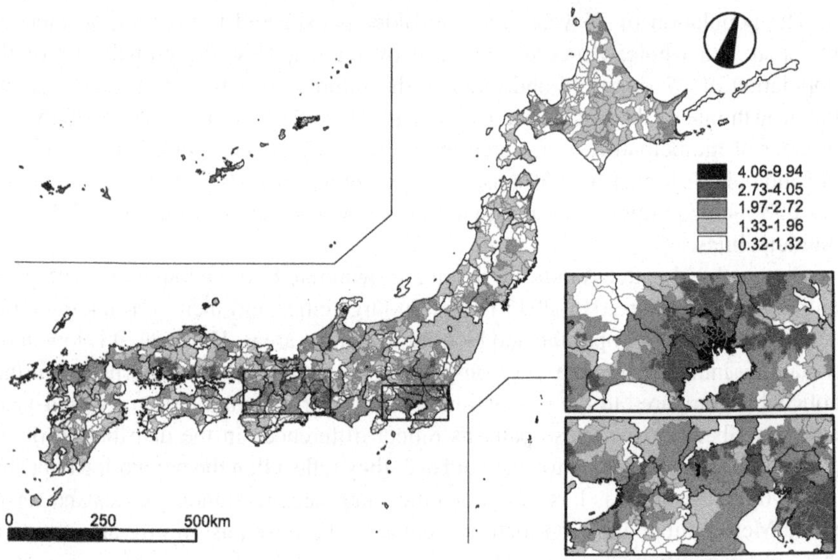

Fig. 7 Population of 15–64-year-olds per person among the population of 75 years old and older (in 2045 by municipality) (Created based on National Institute of Population and Social Security Research 2018a)

people. The mean value for municipalities is expected to decrease gradually from 3.8 in 2015, becoming 1.9 in 2045.

Because geographical patterns show little or no distinctive change after 2015, we present only the results for 2045 in Fig. 7. Municipalities with a high population of 15–64-year-olds per person in the population of 75 years old and older in 2045 will be large cities and their surrounding cities and towns in the Metropolitan and the Non-metropolitan areas. These are municipalities in which the population of 75-year-olds and older people is expected to increase rapidly. However, because the population of 15–64-year-olds is expected to decrease slowly, the population supporting elderly people is expected to remain large. In contrast, the population supporting elderly people will continue to be small in the remaining municipalities. In these municipalities, the population of 75 years old and older will either increase slowly or decrease, although the population of 15–64-year-olds will decrease remarkably.

3 Geographical Diversity in Household Family Types

Households have functioned as a basic unit for living, with mutually supporting members. In Japan, the government has developed social security systems on the premise of the function of households. However, in recent years, household family types are becoming diverse.

Japan had 53,332,000 private households in 2015, representing an increase even after the population started decreasing (Fig. 8) because the average number of housc-hold members is decreasing. However, because the influence of population decline will increase further, the number of private households is expected to start decreasing from 2019.

With the decreasing average number of household members, household family types are becoming increasingly diverse (Table 2). In the 1980s, married-couple households with children accounted for most (42.1%) of all households; other types of households that consisted mainly of three-generation family households accounted for second-most (19.9%). Subsequently however, the number of one-person households, married-couple-only households, and single-parent households with children increased. Consequently, one-person households were the most common in 2010, coming to account for 32.4% of all households. It is projected that household family types will continue to diversify and that the relative proportions of one-person households, married-couple-only households, and single-parent households with children will increase further.

Household family types are becoming increasingly diverse. Clear regional differences exist (Fig. 9). The proportion of one-person households is high in prefectures in the Metropolitan Area, and Hokkaido and southwestern Japan in the Non-metropolitan Area. The proportions of married-couple-only households are high in Hokkaido and southwestern Japan in thc Non-metropolitan Area. In southwestern Japan in the Non-metropolitan Area, successors of households were more likely to live closer to their parents after marriage than to live with them.

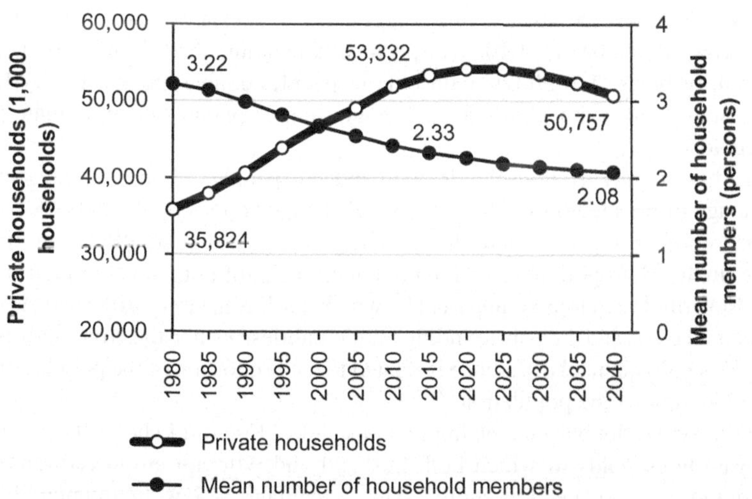

Fig. 8 Total number of private households and mean number of household menmbers in Japan (Created based on Population Census of Japan and National Institute of Population and Social Security Research 2018b)

Table 2 Household composition by family type in Japan (%)

Year	One-person households	Married-couple-only households	Married-couple households with children	Single-parent households with children	Other types of households
1980	19.8	12.5	42.1	5.7	19.9
1990	23.1	15.5	37.3	6.8	17.4
2000	27.6	18.9	31.9	7.6	14.0
2010	32.4	19.8	27.9	8.7	11.1
2020	35.7	20.5	26.1	9.3	8.3
2030	37.9	20.8	24.5	9.6	7.2
2040	39.3	21.1	23.3	9.7	6.6

Created based on Population Census of Japan and National Institute of Population and Social Security Research (2018b)

The proportion of married-couple households with children is high in prefectures in the Metropolitan Area. By contrast, the proportion of single-parent households with children is high in prefectures in the Non-metropolitan Area. The proportions of other types of households are high in prefectures in northeastern Japan and along the coast of the Sea of Japan in the Non-metropolitan Area. In these prefectures, successors of households exhibited a strong tendency to live with their parents after marriage.

On these bases, we assess the geographical diversity of household family types by municipality. The geographical units we use in this section are 1,756 municipalities (786 cities, 757 towns, 190 villages and 23 special wards of Tokyo) by the boundaries of October 1, 2010. We specifically examine the following four household family types: households with elderly people, double-income family households, households with children, and single-female-parent or single-male-parent households.

The first family is households with elderly people. Figure 10 presents the percentage of private households with members aged 65 years old and older among the total number of private households. The percentage is generally low in municipalities in the Metropolitan Area, but high in municipalities in the Non-metropolitan Area. Nevertheless, a trend is apparent by which it is low in large cities such as prefectural capital cities and their surrounding municipalities, even in the Non-metropolitan Area. This geographical difference is related to the proportion of the population aged 65 or older among the population.

The second is double-income family households. Figure 11 shows the percentage of private households in which both husband and wife are employed among the total number of private households. The percentage is low in municipalities in the Metropolitan Area, but high in municipalities in the Non-metropolitan Area, especially in municipalities along the coast of the Sea of Japan side in the Non-metropolitan Area. As explained above, clear regional differences exist in the labor force participation of married women.

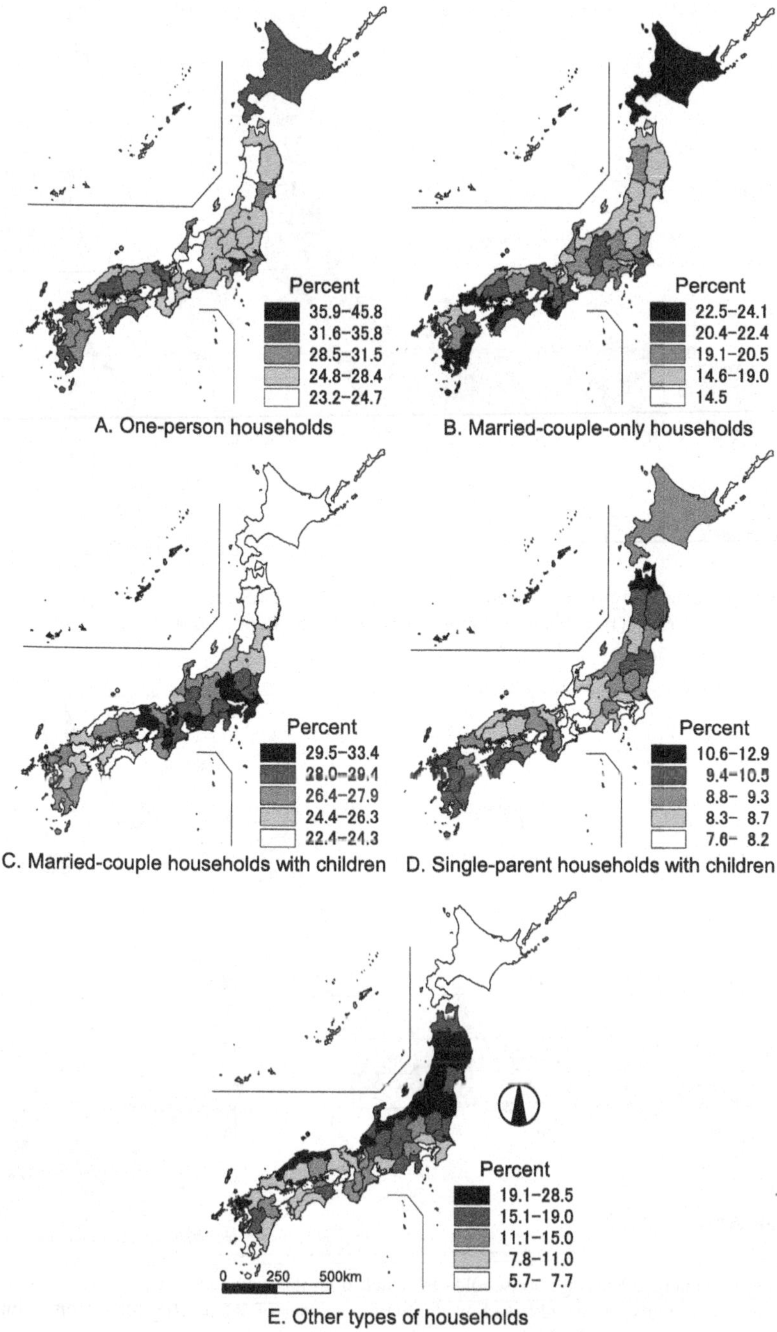

Fig. 9 Family types of households (2010) (Created based on Population Census of Japan. Reprinted from Yamauchi 2017 with permission of Akashi Shoten)

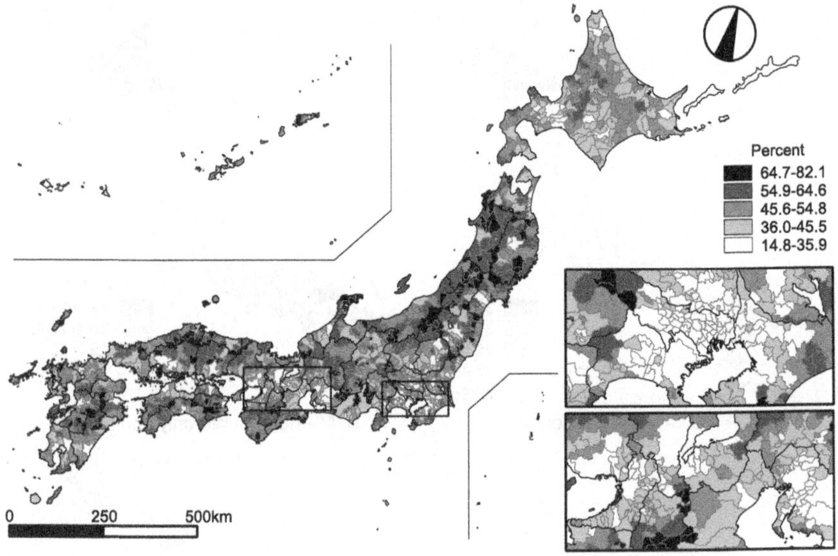

Fig. 10 Percentage of private households with members aged 65 years old and older (in 2010 by municipality) (Created based on Population Census of Japan. Reprinted from Yamauchi 2017 with permission of Akashi Shoten)

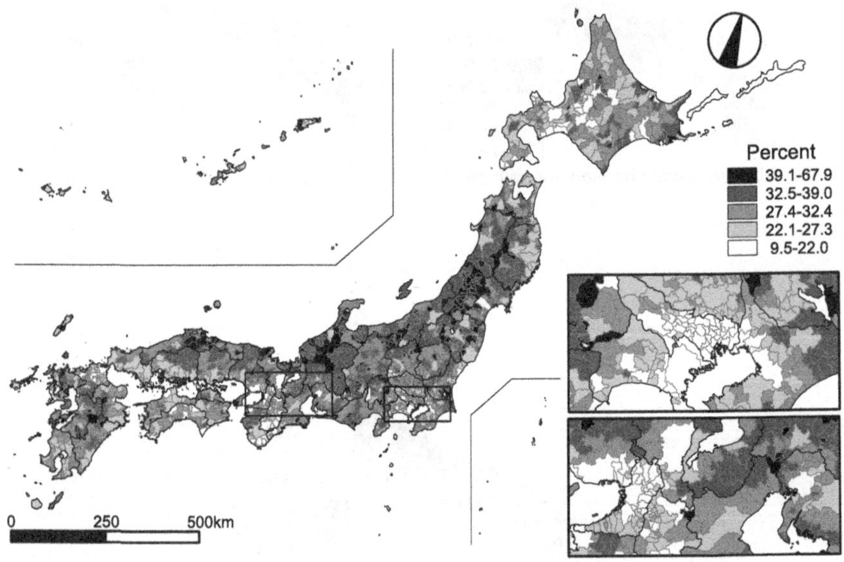

Fig. 11 Percentage of private households in which both the husband and wife are employed (in 2010 by municipality) (Created based on Population Census of Japan. Reprinted from Yamauchi 2017 with permission of Akashi Shoten)

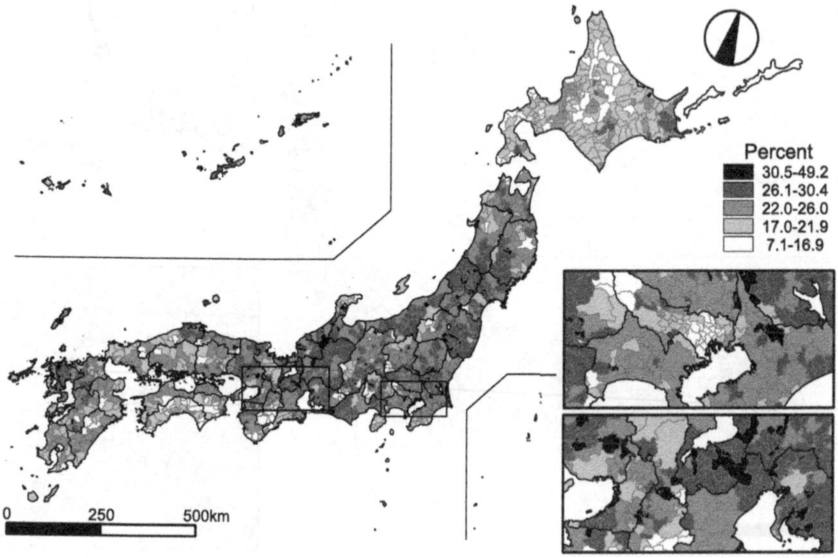

Fig. 12 Percentage of private households with members aged 18 years old and under (in 2010 by municipality) (Created based on Population Census of Japan. Reprinted from Yamauchi 2017 with permission of Akashi Shoten)

The third is households with children. Figure 12 shows that the percentage of private households with members aged 18 years old and younger accounted for the total number of private households. The percentage is high in municipalities in Tohoku, Chubu, and the western part of the Kinki region, and along the Sea of Ariake in the Kyushu region. It is low in municipalities in the regions of Hokkaido, the southern Kinki and Shikoku, and in mountainous and island areas of Kyushu. In municipalities for which the proportion of population of 65 years old and older among the population is low, the percentage of households with children tends to be high. However, such is not the case in the cities and towns in the Metropolitan Area because they have many one-person households. In municipalities in the Tohoku region, where the proportion of three-generation family households is high, the percentage of households with children is high even when the proportion of population of 65 years old and older among the population is high.

The fourth is single-female-parent or single-male-parent households. Figure 13 shows the percentage of single-male-parent or single-female-parent households with members aged 18 years and younger among the total number of private households. Although the value is low in municipalities along the coast of the Sea of Japan side from the Tohoku to Chugoku regions as well as in the Metropolitan Area, it is high in municipalities from Hokkaido to Aomori Prefecture and in the Shikoku and Kyushu regions.

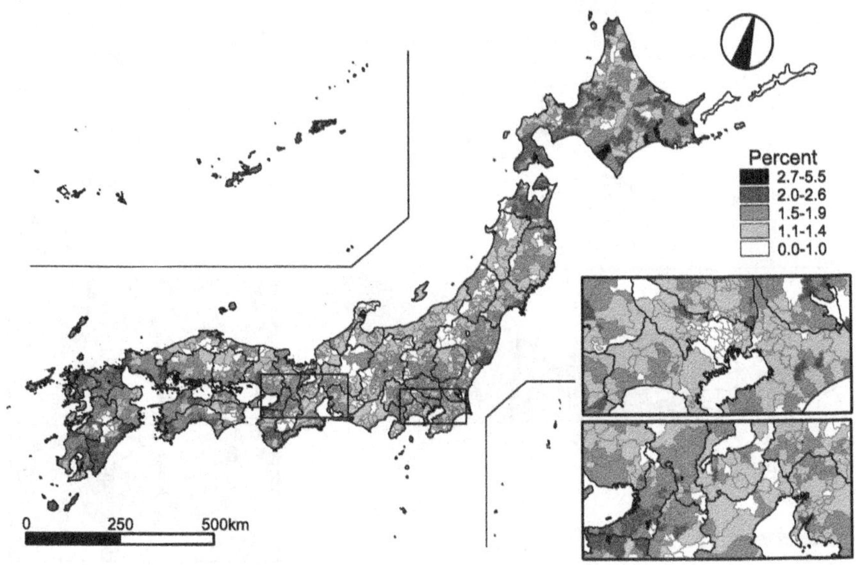

Fig. 13 Percentage of single-male-parent or single-female-parent households with children aged 18 years old and younger (in 2010 by municipality) (Created based on Population Census of Japan. Reprinted from Yamauchi 2017 with permission of Akashi Shoten)

4 Regional Differences in Health

4.1 Life Expectancy and Mortality Rate

Next we specifically examine regional differences in health in Japan. Regional differences exist in life expectancy, health conditions, and health behaviors at each scale, from a national level to a neighborhood level. These regional differences in health engender regional differences in the demand for social security outlays for services such as medical and nursing care.

Japan is among the countries with the highest average life expectancy in the world. However, in 1947, shortly after the war, the average life expectancy for Japanese people was about 50 years. Subsequently the average life expectancy continued to increase at a pace exceeding those of other countries. Japan soon overtook other economically developed countries and achieved the highest life expectancy in the world. As of 2017, it was 87.3 years for women and 81.1 years for men.

However, the examination of the average life expectancy within Japan reveals inter-regional differences by which the average life expectancy is high in some and low in others. We selected the top and bottom five ranked prefectures according to life expectancy based on The Prefectural Life Tables and arranged the values by sex as shown in Table 3 Prefectures with the highest average life expectancy in 2015 were Shiga (81.8 years) for men and Nagano (87.7 years) for women. The prefecture

Table 3 Average life expectancy (2015, Top and bottom 5 prefectures)

Rank	Men		Rank	Women	
	Prefecture	Average life expectancy		Prefecture	Average life expectancy
1	Shiga	81.8	1	Nagano	87.7
2	Nagano	81.8	2	Okayama	87.7
3	Kyoto	81.4	3	Shimane	87.6
4	Nara	81.4	4	Shiga	87.6
5	Kanagawa	81.3	5	Fukui	87.5
43	Kagoshima	80.0	43	Fukushima	86.4
44	Wakayama	79.9	44	Akita	86.4
45	Iwate	79.9	45	Ibaraki	86.3
46	Akita	79.5	46	Tochigi	86.2
47	Aomori	78.7	47	Aomori	85.9

Created based on Prefectural Life Tables

with the lowest average life expectancy was Aomori for both sexes, remaining at 78.7 years for men and 85.9 years for women. In the case of men, the difference from that of Shiga is greater than three years. It is often pointed out as a background that Aomori has high drinking and smoking rates as well as high salt intake.

Prefectures showing the greatest life expectancy have differed over time. Many people might still imagine Okinawa when considering prefectures associated with longevity. In fact, Okinawa recorded the highest average life expectancy in Japan in the past. Particularly, it maintained its number one position for women until 2005. Longevity in Okinawa has gained attention in the field of epidemiology and public health. It was hypothesized as attributable to the healthy local food and abundant social support in the region. However, its ranking fell to 7th out of 47 for women and 36th out of 47 for men in 2015. The mortality rate of younger generations is particularly high. Lifestyle-related diseases have come to be a serious problem in recent years, especially for people of younger generations. It has been pointed out that the spread of American-style eating habits, as well as a lack of exercise attributable to the widespread use of cars, lie in the background. They are both influenced by the history and geographical features of the region.

Here we also present the standardized mortality ratio (SMR) of malignant neoplasms (or cancer, Fig. 14), cardiac diseases (Fig. 15), and cerebrovascular diseases (Fig. 16) by prefecture as examples. A high SMR for cancer is found in prefectures from Hokkaido to Tohoku, in Kinki, and in the northern part of Kyushu. By contrast, the SMR is low in prefectures in Chubu, Shikoku, the southern part of Kyushu, and the Okinawa region. No great difference is apparent between men and women. Next, the SMR for cardiac diseases is high in the Tohoku and Kinki regions, but low in the Chubu region, which is similar to the SMR for cancer. However, different patterns are also apparent: a high SMR is successively observed from the Kanto to Tohoku regions; the SMR is low in the northern part of the Kyushu region

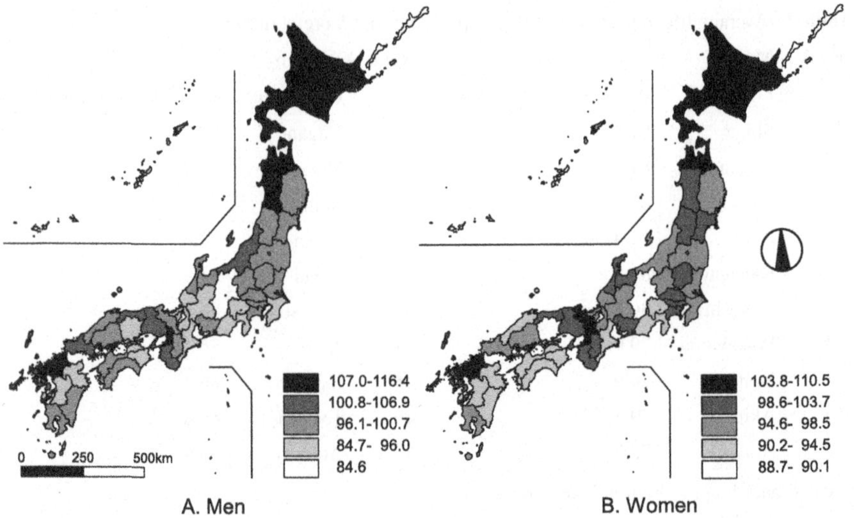

Fig. 14 Standardized mortality ratio for malignant neoplasms (cancer) (2008–2012) (Created based on Specified Report of Vital Statistics of Japan. Reprinted from Hanibuchi 2017a with permission of Akashi Shoten)

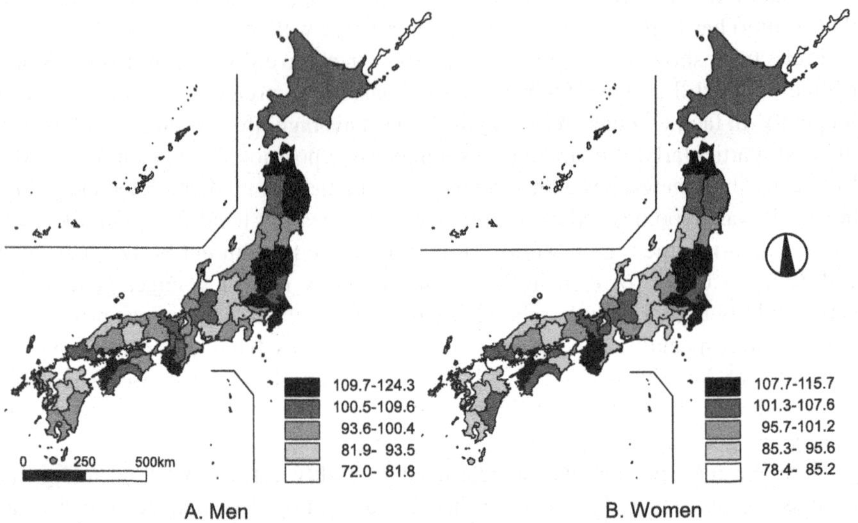

Fig. 15 Standardized mortality ratio for cardiac diseases (2008–2012) (Created based on Specified Report of Vital Statistics of Japan. Reprinted from Hanibuchi 2017a with permission of Akashi Shoten)

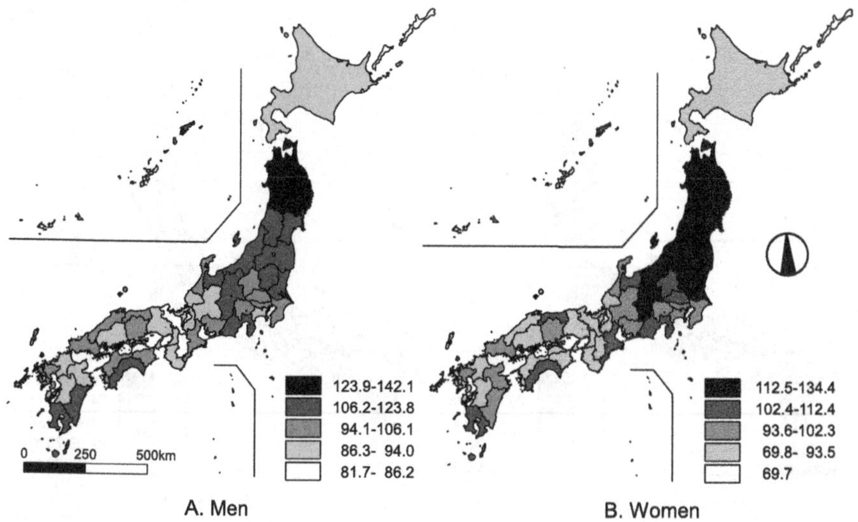

Fig. 16 Standardized mortality ratio for cerebrovascular diseases (2008–2012) (Created based on Specified Report of Vital Statistics of Japan. Reprinted from Hanibuchi 2017a with permission of Akashi Shoten)

but high in the Shikoku region. Furthermore, vast regional differences are found in SMR for cerebrovascular diseases. A considerably high SMR for the diseases is successively distributed in regions from Tohoku to northern Kanto. In the Kinki region, where the SMRs for cancer and cardiac diseases are high, the SMR for cerebrovascular diseases is considerably low in both men and women.

4.2 *Healthy Life Expectancy*

The period during which people can maintain good health and live independently is referred to as their "healthy life expectancy." In Japan, where longevity and aging have progressed, many elderly people must live with some sort of mental or physical difficulty. When conditions worsen, the number of people who are in need of support or long-term care (i.e., those who lost healthy life expectancy) is expected to increase. Given that background, it has become necessary to increase the healthy life expectancy and to reduce the gap between this and the average life expectancy.

Several means of calculating healthy life expectancy have been proposed. The Guidelines for Calculating Healthy Life Expectancy by a Ministry of Health, Labor and Welfare research group (2012) defined healthy life expectancy as the average of the period with no limitations on daily activities. The figures can be described as follows. For men, it was estimated as 69.40 years in 2001, 69.47 years in 2004, 70.33 years in 2007, and 70.42 years in 2010. For women, it was estimated as 72.65 years in 2001, 72.69 years in 2004, 73.36 years in 2007, and 73.62 years

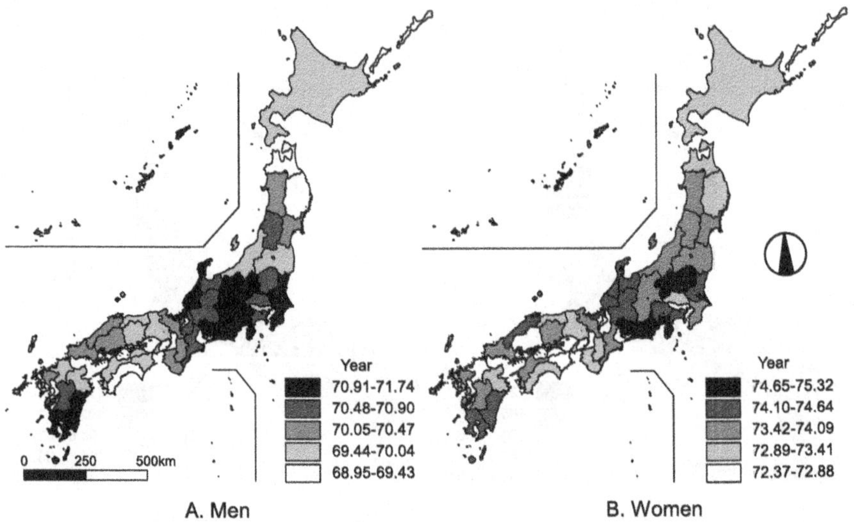

Fig. 17 Healthy life expectancy (2010) (Created based on Ministry of Health, Labor and Welfare research group 2012. Reprinted from Hanibuchi 2017b with permission of Akashi Shoten)

in 2010. Based on those results, Japanese people are regarded as having become able to maintain good health for a longer period of time.

Regional differences are apparent for healthy life expectancy. The maximum difference is about three years depending on the prefecture (Fig. 17). Prefectures with the highest healthy life expectancy are Aichi (71.74 years) for men and Shizuoka (75.32 years) for women. In contrast, the prefectures with the lowest healthy life expectancy are Aomori (68.95 years) for men and Shiga (72.37 years) for women. The maps show that regions with high healthy life expectancy are found successively from the Chubu to Kanto regions and that a similar distribution is found for southern Kyushu. Low healthy life expectancy is often found in the prefectures from Hokkaido to northern Tohoku and those in the Shikoku region. A distinctive feature is that healthy life expectancy in Tokyo and Osaka are conspicuously low compared to figures for surrounding areas. Although geographical characteristics in the distribution are largely common to men and women, a clearer pattern is observed for men.

4.3 Long-Term Care Need and Support Need

When losing healthy life expectancy because of illness or injury, people come to need nursing care or support to live everyday life. With the Long-Term Care Insurance System started in 2000, when people come to need support or nursing care, they can receive services by application for a Certification of Needed Long-Term Care to their insurers (mainly municipalities). The people certified as needing long-term care by this system are ever-increasing, from 2.56 million at the beginning (March 31, 2001) to 6.20 million at present (March 31, 2016). In Japan, the concern arises that when the elderly population is expected to continue to increase, the number of people known as "nursing-care refugees" is expected to increase because the supply of human resources might fail to meet the demand for nursing care.

Figure 18 is a map depicting regional differences in the certification rate of long-term care need by insurers as a unit. One can infer from the map that regions with a high certification rate of long-term care need are often distributed in the Chugoku and Shikoku regions, the southern part of the Kinki region, the Kyushu region, and the coastal areas of the Sea of Japan in the Tohoku regions. In contrast, regions with a low certification rate extend from the Kanto region to the Tokai region. The high or low certification rate of long-term care needs might reflect the health levels of local residents. In addition, the availability of family nursing care and the financial situation of insurers might have some effect.

However, it is not appropriate to consider the magnitude of the demand for nursing care based on this choropleth map because the map presents the "percentage" and does not present the "number": the quantity of demand itself. Considering this point, we present a cartogram in Fig. 19. Map A in Fig. 19 is shaded using the same data used in Fig. 18. The size of each insurer on the map has been converted in proportion to the number of people who were certified as being in need of long-term care or support. This makes it easier to distinguish the quantity of demand by region at first

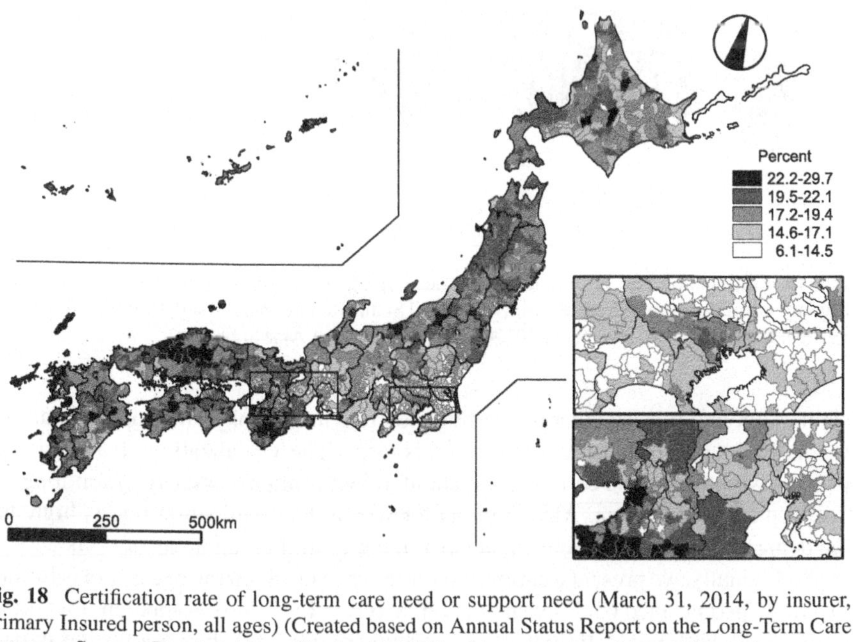

Fig. 18 Certification rate of long-term care need or support need (March 31, 2014, by insurer, Primary Insured person, all ages) (Created based on Annual Status Report on the Long-Term Care Insurance System. Reprinted from Hanibuchi 2017b with permission of Akashi Shoten)

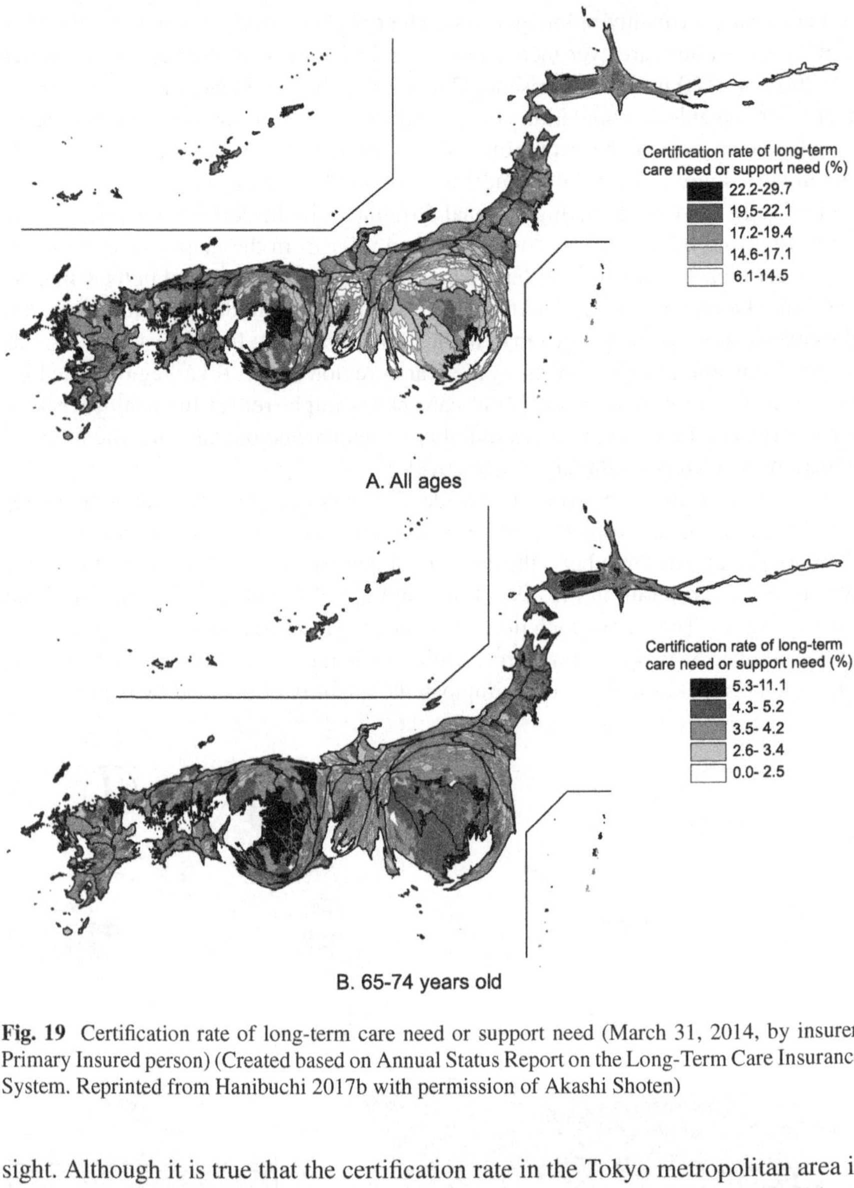

Fig. 19 Certification rate of long-term care need or support need (March 31, 2014, by insurer, Primary Insured person) (Created based on Annual Status Report on the Long-Term Care Insurance System. Reprinted from Hanibuchi 2017b with permission of Akashi Shoten)

sight. Although it is true that the certification rate in the Tokyo metropolitan area is not high, the number of people certified is large. The visualization also elucidates the following facts: The certification rate of Tokyo is not necessarily low among the Tokyo metropolitan area. Also, regions for which the certification rate is high and the demand for nursing care is great exist in the central urban areas of Osaka.

Additionally, we present a cartogram for the case in which the age group is limited to elderly people aged 65–74 years old on map B in Fig. 19. It is apparent that urban areas with a high population of the age group are depicted showing further expansion. Regional differences between the Chugoku and Shikoku regions in which the certification rate is high and metropolitan areas in which the rate is low are no longer clear. A large number of people in need of long-term care and the high certification rate in metropolitan regions stand out: the certification rate was high in rural areas (Map A), reflecting the population composition, which included many elderly people aged 75 and over who have a high certification rate. Taking demographic points into consideration, the high certification rate stands out rather in urban areas, particularly in Osaka.

Presuming that the number of people certified because of aging shall continue to increase at the same pace, Map B can be regarded as showing demand for nursing care in the near future. It is still fresh in our minds that rural migration was proposed as an option for addressing the increase of demand for nursing care in metropolitan regions, which provoked controversy (Masuda 2015). This cartogram portrays the "size" of the problem, indicating the difficulty of how to respond to increasing demand for nursing care in urban areas.

5 Regional Differences in Health-Related Behaviors

What produces regional differences in health levels? To ascertain the reasons, it is important to elucidate how health behaviors of residents differ among regions because, for present-day Japanese people, the strongest factor leading to death or long-term care is not infectious disease such as tuberculosis, gastroenteritis, and pneumonia, which ranked high among the causes of death for people in the past, but rather non communicable diseases such as malignant neoplasms (cancer), cardiac diseases, and cerebrovascular diseases. Non-communicable diseases are caused by an accumulation of unhealthy lifestyle habits. More specifically, smoking, unbalanced eating habits, and a lack of exercise are regarded as typical causes. For example, in Aomori, where the average life expectancy is the lowest, the high smoking rate and high-salt diet are often cited as reasons.

We look at smoking first. The smoking rate among Japanese men has declined sharply from 80% to about 30% during the past 50 years or so. However, regional differences are still observed in smoking rates. Figure 20 is a map showing the distribution of smoking rates (age-adjusted values) as estimated by prefecture. Differences exist in smoking rates depending on the prefecture, up to 1.3–1.4 times among men and surprisingly more than double that among women. For men, the prefecture with the highest smoking rate is Akita. That of the lowest is Tokyo. For women, the highest is Hokkaido. The lowest is Shimane. It is common to men and women that the smoking rate is high in Hokkaido and the Tohoku region and that it is low in many regions from Chubu through the Chugoku and Shikoku regions. However, a major difference is that the smoking rate among women is high in urban areas such as Tokyo and Osaka. Although prefectures with a low smoking rate stand out in the case of men in the Tokyo metropolitan area, the smoking rate is slightly higher in the case of women; contrasting results are apparent between men and women.

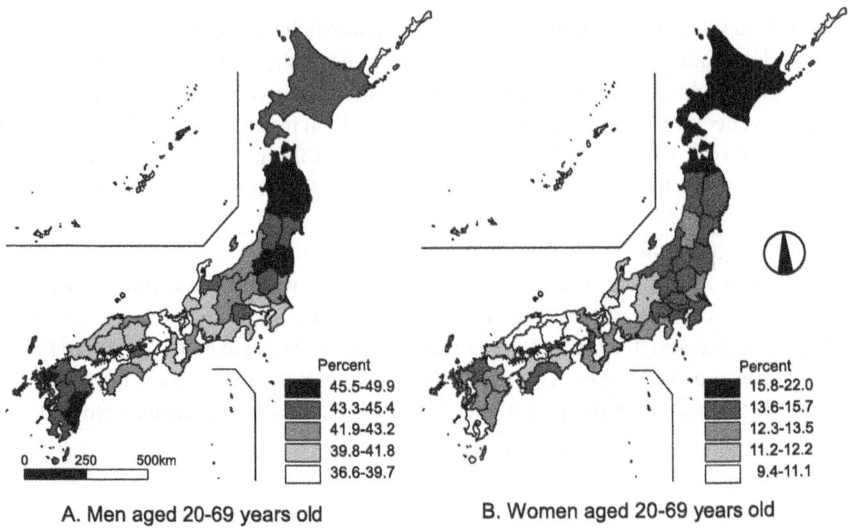

Fig. 20 Smoking rate for people 20–69 years old (in 2010, age-adjusted) (Created based on Ministry of Health, Labor and Welfare research group 2013. Reprinted from Hanibuchi 2017c with permission of Akashi Shoten)

Next, we confirm regional differences in eating habits by specific examination of the amount of vegetable intake. Because eating habits have been formed historically in accordance with the climate of the region to some extent, regional variations in eating habits are apparent within Japan. Figure 21 portrays maps that show the mean value of the amount of vegetable intake in the "National Health and Nutrition Survey" by prefecture. The amount of vegetable intake is often high in regions from the Koshinetsu to Tohoku region. It is low in regions from Tokai to Kinki. Although the trend is common to men and women, it is noticeable among men that the intake is low in the Shikoku and Kyushu regions. Looking at individual prefectures, the prefecture with the maximum amount is Nagano. That of the minimum amount is Aichi. It is noteworthy that the two adjacent prefectures show contrasting results related to the vegetable intake amounts. Nagano exhibited a high mortality rate attributable to cerebrovascular diseases in the past because people had food with much salt, such as pickles. Later, the prefecture launched a salt reduction campaign and recommended promotion of vegetable intake while promoting a modification of eating habits.

Finally, we examine regional differences in the amounts of physical activity. We specifically examine mean values of the numbers of steps taken daily by residents by prefecture based on the "National Health and Nutrition Survey" (Fig. 22). In general, prefectures with residents reporting numerous steps are in urban areas of the Kanto, Tokai, and Kinki regions; prefectures with residents reporting a small number are in surrounding areas in the country, such as Hokkaido, and the Tohoku and Kyushu regions. The prefectures with residents reporting the greatest number of steps are Hyogo (men) and Chiba (women) and those with residents reporting the fewest are

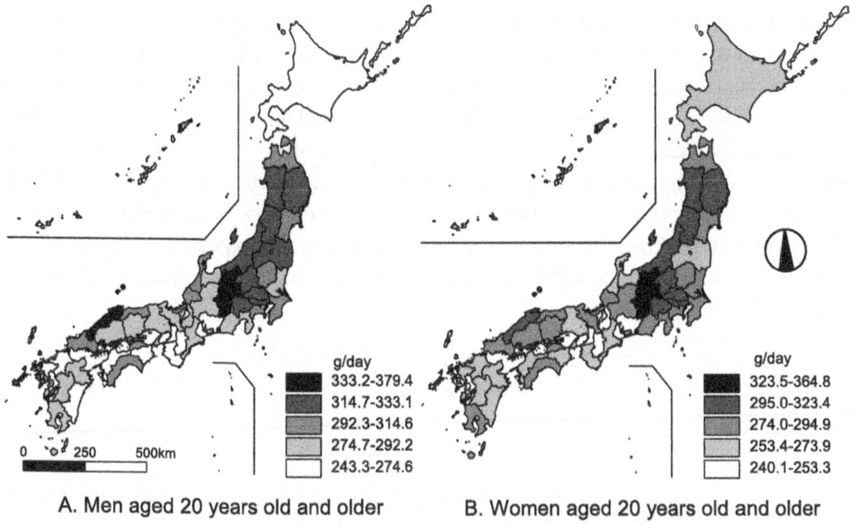

Fig. 21 Mean value of the amount of vegetable intake of 20 years old and older (in 2012, age-adjusted) (Created based on National Health and Nutrition Survey. Reprinted from Hanibuchi 2017c with permission of Akashi Shoten)

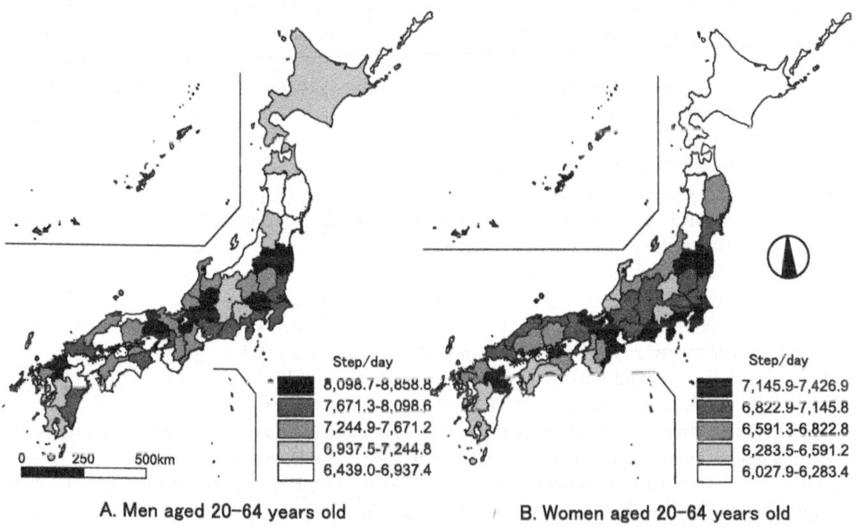

Fig. 22 Mean value of the number of steps of people 20–64 years old (in 2012, age-adjusted) (Created based on National Health and Nutrition Survey. Reprinted from Hanibuchi 2017c with permission of Akashi Shoten)

Miyagi (men) and Akita (women). The reason that the number of steps is large in urban regions is explainable by the idea of walkability. Walking is done not only for purposes of exercise; it also moves pedestrians to destinations such as work and shopping. If the destinations in everyday life are located within walking distance, then the place has an environment in which it is easy to walk (i.e., high walkability); the amount of walking is naturally greater. By contrast, in an environment where people use cars wherever they travel, they have a chance to walk only from the entrance of a building to a parking lot. Differences in walkability are expected to contribute to regional differences in the number of steps between urban and rural areas (Inoue and Nakaya 2015).

References

Hanibuchi T (2017a) Life expectancy and mortality. In: Miyazawa H (ed) Mapping health, medical care, and welfare in Japan. Akashi Shoten, Tokyo (in Japanese)

Hanibuchi T (2017b) Health status. In: Miyazawa H (ed) Mapping health, medical care, and welfare in Japan. Akashi Shoten, Tokyo (in Japanese)

Hanibuchi T (2017c) Health behavior. In: Miyazawa H (ed) Mapping health, medical care, and welfare in Japan. Akashi Shoten, Tokyo (in Japanese)

Inoue S, Nakaya T (2015) Urban environment and health. In: Kawakami N, Hashimoto H, Kondo N (eds) Society and health: an integrated approach to close health gap. University of Tokyo Press, Tokyo (in Japanese)

Masuda H (2015) The disappearance of Tokyo: nursing care breakdown and rural migration. Chuokoron-Shinsha, Tokyo (in Japanese)

Ministry of Health, Labor and Welfare Research Group (2012) Guidelines for calculating healthy life expectancy. http://toukei.umin.jp/kenkoujyumyou/syuyou/kenkoujyumyou_shishin.pdf. Accessed 6 October 2016 (in Japanese)

Ministry of Health, Labor and Welfare Research Group (2013) Empirical study of progress in methods for assessing cancer control using existing data: Promotion and assessment of policies to reduce smoking, hepatitis, and liver cancer. https://oici.jp/ocr/common/images/profile/2011report_tabuchi.pdf. Accessed 1 Sept 2019 (in Japanese)

National Institute of Population and Social Security Research (2018a) Regional population projections for Japan: 2015–2045. Population Research Series 340 (in Japanese)

National Institute of Population and Social Security Research (2018b) Household projections for Japan: 2015–2040. Population Research Series 339 (in Japanese)

Statistics Bureau Ministry of Internal Affairs and Communications Japan (2019) Annual Report on Internal Migration in Japan derived from the Basic Resident Registration 2018

United Nations, Department of Economic and Social Affairs, Population Division (2017) World population prospects: The 2017 revision. DVD edition

United Nations, Department of Economic and Social Affairs, Population Division (2018). World urbanization prospects: The 2018 revision. Online edition

Yamauchi M (2017) Number and type of household. In: Miyazawa H (ed) Mapping health, medical care, and welfare in Japan. Akashi Shoten, Tokyo (in Japanese)

Nursing Care Provision Systems for Elderly People and Geographical Distribution of Services in Japan

Teruo Hatakeyama, Shin'ichiro Sugiura, and Hitoshi Miyazawa

Abstract In 2000, with the rapid progress of population aging after the high economic growth period, Japan introduced a long-term care insurance system in which nursing care services are provided through the social insurance system. Under the long-term care insurance system, private business operators were encouraged to enter a quasi-market environment. As a consequence, their service centers came to be located mainly in metropolitan areas. Such uneven distribution of the services to metropolitan areas led to regional differences and disparities in the services. That trend persists even in community-based long-term care services where the authority to establish the services was transferred to municipalities. In fact, some municipalities have no established services.

Keywords Community-based long-term care services · Home-based care services · Institutional care services · Long-term care insurance premium · Long-term care insurance system

1 Nursing Care Provision Systems in Japan

Many economically developed countries have experienced population aging ahead of economically developing countries. Even in economically developing areas, some Asian countries, including the Republic of Korea and Singapore, are expected to experience extremely rapid population aging (United Nations, Department of Economic and Social Affairs, Population Division 2017). Those countries have put together

---

This chapter is a revised version of Hatakeyama (2017), Miyazawa (2017), and Sugiura (2017a, b).

T. Hatakeyama (✉)
Naruto University of Education, Naruto, Japan
e-mail: thatakeyama@naruto-u.ac.jp

S. Sugiura
Meijo University, Nagoya, Japan

H. Miyazawa
Ochanomizu University, Tokyo, Japan

H. Miyazawa and T. Hatakeyama (eds.), *Community-Based Integrated Care and the Inclusive Society*, International Perspectives in Geography 12,
https://doi.org/10.1007/978-981-33-4473-0_3

public care guarantee systems along with the progress of population aging. However, care guarantee systems differ among countries in terms of the following items: the ways and means of financial resources, the scope of persons to be guaranteed, the number of institutional care users, the availability of cash benefits, the body providing services, and the availability of copayments (Masuda 2014).

By devoting attention to the ways and means of financial resources in the care guarantee systems, we can divide the countries into those that adopt the social insurance system and those that provide nursing care services from general taxes. For example, in European countries where public care guarantee systems are the mainstream, Germany and the Netherlands have adopted a long-term care insurance system using a social insurance system, although Sweden and the United Kingdom have adopted systems that provide nursing care services from tax revenues.

Japan had publicly guaranteed nursing care. However, with the rapid progress of population aging, an assumption was made that it would become difficult to support welfare services for elderly people sustainably solely using public funds available from taxation. Given this, Japan introduced a long-term care insurance system using the social insurance system in 2000 under the idea that society supports elderly people as a whole. The long-term care insurance system in Japan is said to be modeled on the long-term care insurance system in Germany (Seon 2010). The following sections describe the long-term care insurance system, which is the core of the nursing care provision systems in Japan, with comparison to the long-term care insurance system in Germany, as is appropriate.

1.1 Characteristics of the Long-Term Care Insurance System in Japan

Insurers of the long-term care insurance systems in Japan are municipalities that have provided welfare services for elderly people as the basic local government closest to residents.[1] With long-term care insurance, 50% of the insurance benefits, except 10%[2] copayment by users, are financed by long-term care insurance premiums that persons aged 40 years and older pay; the remaining 50% are financed by taxes from the central government, prefectures, and municipalities. Although Germany, too, introduced a long-term care insurance system using the social insurance system, the financial resources are premiums paid by residents only; no copayment is made by users (Masuda 2014). In Japan, public involvement is secured in operating the long-term care insurance in addition to taxpayers' payments for it. For example, the central government develops laws and regulations and also revises the system. Prefectures undertake the designation and supervision of long-term care insurance service providers and support for municipalities (human resources development,

[1] In some cases, multiple municipalities set a wide-area insurer.

[2] Persons who earn more than a certain level of income were to pay 20% from August 2015. Persons who earned more were to pay 30% from August 2018.

explanation of legal systems, etc.). Municipalities manage a special account for long-term care insurance and the certification committee of needed long-term care.

In the long-term care insurance premiums in Japan, the payment method and calculation basis for premiums vary according to the age of the insured person. Insured persons aged 65 years and older are to be first-category insured persons. The long-term care insurance premiums to be paid by first-category insured persons are based on the base amount specified by the long-term care insurer; they differ according to a person's income.[3] The long-term care insurance premiums to be paid by first category insured persons are also calculated in the long-term care insurance planning that is revised every three years as one term based on the demand for nursing care services in each long-term care insurer. Insured persons aged 40–64 years are to be second-category insured persons, with premiums calculated differently from first-category insured persons. They paid the premiums, with half paid by their employers. In the long-term care insurance system in Germany, the insurer (Krankenkass[4]) for public medical insurance, in which 90% of the population participates, also serves as the insurer (Pflegekasse[5]). Insured persons under public medical insurance are also positioned as persons insured by long-term care insurance.

Figure 1 portrays the flow by which residents use long-term care insurance services in the long-term care insurance system in Japan. In long-term care insurance in Japan, benefit recipients are specified as insured persons aged 65 years and older, except those in an in-need-of-care state because of certain diseases. When an insured person[6] intends to use long-term care insurance services, the person must first apply for certification of a need for long-term care to a municipality. A committee certifying a need for long-term care informs the insured person of certification results based on the insured person's physical condition as determined by computation, results of a screening by a physician, and results of a home-visit interview by a municipal officer. The certification results are of three kinds: Independent, In need of assistance, and In need of care. Services to be received vary according to the condition of the insured person. Those In need of assistance comprise assistance level 1 and level 2. In need of care is set at five levels: levels 1–5. The number of services to be received also varies according to these levels. In the long-term care insurance system in Germany, medical service (MDK: Medizinischer Dienst der Krankenversicherung) established by Krankenkass undertakes the certification of needed long-term care

[3]In the 6th term (2015–2017), the Ministry of Health, Labour and Welfare presents Level 1 (0.5 times of the base amount) to Level 9 (1.7 times the base amount) as the standard model.

[4]The Krankenkass is a non-profit public corporation which is independent of government. An insured person can choose an affiliation among 116 (as of 2016) Krankenkass such as Allgemeine Ortskrankenkasse (AOK), Betriebskrankenkasse (BKK), Innungskrankenkasse (IKK) etc. Financial resources are covered only by insurance premiums. The insurance rates differ among Krankenkass. However, the upper limit of insurance rates is 15.5%; when it is insufficient to compensate for spending, additional premiums will be collected. Additionally, the remaining 10% of people subscribe to private insurance.

[5]The Pflegekasse belongs to Krankenkass. Although they are financially separate, the Krankenkass collect insurance premiums together. The subscribers are obligated to subscribe to Pflegekasse.

[6]A care management provider can submit an application on behalf of the person.

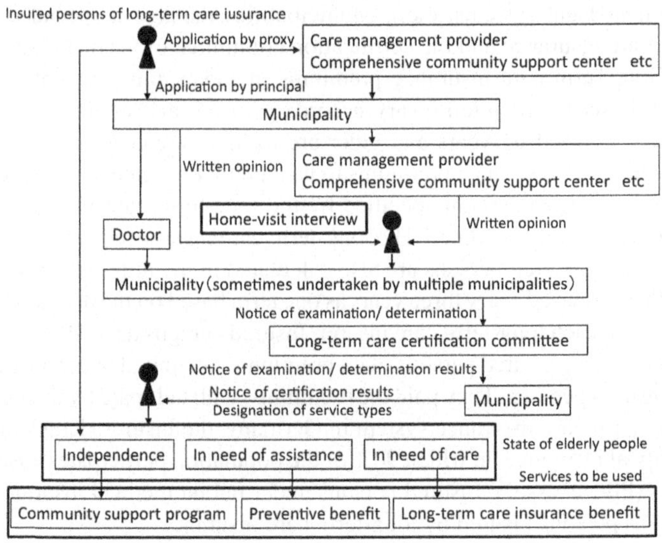

Fig. 1 Flow of use of services in long-term care insurance in Japan (Created based on Ido 2017)

similarly. However, the determination results for the classification of levels of care are more finely defined in Japan. In Germany, the examination is conducted in accordance with the standards by which persons who are certified as care level 2 or lower in Japan are excluded from insurance benefits.

The long-term care insurance system in Japan assigns importance based on user-centered care support. Consequently, users can receive various services from different providers based on their own choices. However, because the work is troublesome for insured persons themselves, care management providers or care managers in comprehensive community support centers usually create care plans. Users receive services based on the care plans. In Japan, to guarantee users to choose services and providers freely, a variety of providers, including private business operators, were allowed to enter service-providing businesses with the start of the long-term care insurance system. It had been expected that competitions between providers would improve service quality. As explained below, numerous private business operators actually entered home-based care services after long-term care insurance started in 2000. Care supply through the long-term care insurance system in Japan clearly has a quasi-market nature that introduces a market mechanism in care provision while securing control of governmental regulations and financial resources.

The introduction of long-term care insurance has encouraged private business operators to enter service-providing businesses in Germany, which is the same in Japan. Also in the long-term care insurance system in Germany, users and their families can select which services to use. For the requests, Pflegekasse offers options and makes decisions. It became possible from 2009 to obtain advice on that occasion from care counselors who undertake case management.

As described above, the long-term care insurance system in Japan has been developed using the long-term care insurance system in Germany as a reference. However, they are different in numerous ways because of differences in their respective historical backgrounds and socioeconomic conditions.

1.2 Types of Long-Term Care Insurance Services

Types of long-term care insurance services in Japan have increased through repeated revisions to the system. Table 1 shows long-term care insurance services as of 2018. The long-term care insurance system that was launched in 2000 provided only long-term care insurance benefits, for which most prefectures, ordinance-designated cities, and core cities[7] have the authority to designate and supervise. Services of the long-term care insurance benefits are intended for elderly people in need of care. Services of long-term care insurance benefits are classified broadly into "home-based care services", by which elderly people receive services while living at home, and "institutional care services", by which elderly people enter a facility and live there. Home-based care services are divided further into the following categories: "home services", by which a helper visits an elderly person's home and provides services; "outpatient services", by which an elder person visits a facility to receive services; and "short-term stay services", by which an elderly person enters a facility for a short period of time while living at home. The services also include "care management", which creates care plans for service users. Institutional care services of three types exist. Welfare facilities for elderly people are welfare residential facilities for elderly people who have difficulty living at home independently. A health facility for elderly people is a medical-care residential facility where elderly people aim at readjusting to life at home while undergoing rehabilitation. A designated long-term care hospital is also a medical-care residential facility affiliated with a medical institution to provide long-term care.

As described earlier, regarding long-term care insurance services in Japan, people can receive welfare services such as those provided at a welfare facility for elderly people as well as adult day care and home care, similarly to other countries. In addition, they can receive medical-care services such as the health facilities for elderly people as well as outpatient rehabilitation and home health. Long-term care insurance services in Japan are characterized by which receivable services of various types exist.

[7] In Japan, each municipality is included in a prefecture. Generally, a city has a population of 50,000 or greater. It is vested with the authority to provide more services than towns and villages. Among cities, local governments with a large population are given authority for numerous services from prefectures: they are ordinance-designated cities and core cities. An ordinance-designated city has a population of 700,000 or more. A core city has population of 200,000 or more, by and large. Ordinance-designated cities have the authority to provide more services than core cities. That authority is more or less equal to the authority that prefectures have.

Table 1 Long-term care insurance services in Japan (as of 2018)

	Services designated and supervised by prefectures, ordinance-designated cities, and core cities	Services designated and supervised by municipalities
Long-term care insurance benefit	• Home-based care services – Home services Home care In-home bathing services Home health Home-based rehabilitation Medical management – Outpatient services Adult day care Outpatient rehabilitation – Short-term stay services Short-term stay for personal care Short-term stay for health care • Institutional care services Welfare facility for the elderly Health facility for the elderly Designated long-term care hospital • Others Residential care for residents of long-term care facilities Reimbursement for purchasing adaptive equipment benefit Home modification benefit	• Community-based long-term care services Routine-visit and on-call home health care Night time home care Community-based day care services Day care services for people with dementia Small-scale multifunctional home-based care services Group home for people with dementia Community-based specified facility care Community-based residential care Combined senioir care services(Small-scale multi-functional nursing care services) • Care management

(continued)

Table 1 (continued)

	Services designated and supervised by prefectures, ordinance-designated cities, and core cities	Services designated and supervised by municipalities
Preventive benefit	• Preventive care services – Home services Preventive home bathing services Preventive home health Preventive day care rehabilitation – Outpatient services Preventive day care rehabilitation – Short-term stay services Short-term preventive stay for personal care Short-term preventive stay for health care – Others Preventive residential care for residents of long-term care facilities Disability prevention adaptive equipment rental services Preventive reimbursement for purchasing adaptive equipment benefit Preventive home modification benefit	• Community-based preventive care services Preventive day care services for people with dementia Preventive small-scale multifunctional home-based care services Preventive group home for people with dementia – Preventive care services

Created based on Ido (2017)

Revision of the long-term care insurance systems in 2006 raised awareness of building community-based integrated care systems, as discussed in Second Part. Community-based integrated care systems are aimed at providing housing and services of medical care, nursing care, preventive care, and livelihood support in the local community in an integrated manner. This system is for elderly people to be able to continue to live life in their familiar community and in their own way even if they become needful of care. Under the circumstances, community-based long-term care services provided from long-term care insurance benefits and preventive benefits were newly established. In terms of community-based long-term care services, the authority to designate and supervise providers was given to municipalities because prefectures had authority over designation in many services until that time, and facility development plans were not created in accordance with community needs (Hatakeyama 2009). Services of preventive benefits were newly established to prevent elderly people from declining to the point of needing care. The background is that long-term care insurance benefits and expenses soared after the long-term care insurance system was introduced because of the entry of numerous service providers. Furthermore, community-based preventive care services that combined the services explained above were introduced. Preventive benefit services are intended for elderly people who need assistance.

At around the same time, a community support program was established in which municipalities take the initiative in providing services,[8] although it is financed by long-term care insurance. This is a service for elderly people who are not deemed to be in need of care or in need of assistance to prevent them from declining to a point at which they would be in need of care or in need of assistance (Fig. 1). More specifically, home-visit type and day-care type preventive care programs have been undertaken in addition to public awareness-enhancing activities related to preventive care.

However, with long-term care insurance in Japan, people are not allowed to choose cash benefits as nursing care allowances, which is possible in Germany and Netherlands, and in other countries. They can receive services only. In Germany, when a person cares for a family member, the person can receive allowances from cash benefits. However, when the long-term care insurance system was introduced, Japan did not institutionalize cash benefits with the aim of outsourcing care provided by family members. This is a characteristic of Japan, where long-term care insurance was introduced to socialize care.

2 Increase of Service Providers and the Surge of Long-Term Care Insurance Premiums

Before the long-term care insurance system was introduced in Japan, the use of nursing care services depended on public financial resources from taxes; service

[8]Services can be entrusted to private business operators.

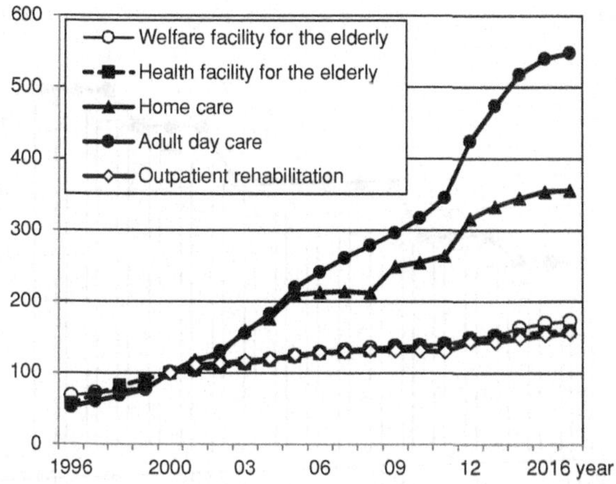

Fig. 2 Changes in the number of major long-term care insurance services centers in Japan (indexes equal to 100 in 2000). With regard to adult day care, small-scale facilities with a capacity of 18 or fewer people were positioned as community-based adult day care in community-based services in 2016. Consequently, the number of centers in 2016 is a combination of the number of adult day care centers and that of community-based adult day care centers (Created based on Survey of Institutions and Establishments for Long-term Care)

providers were limited to the public sector (local governments, social welfare corporations,[9] etc.), in principle. However, as described earlier, the introduction of the long-term care insurance system made it possible for various providers to enter service-providing businesses. As a result, numerous private business operators, mainly for-profit corporations and nonprofit corporations, entered markets of those related businesses. Figure 2 shows changes in the number of major providers of long-term care insurance services. The figure shows that providers for home-based care services such as adult day care and home care increased sharply, influenced by the increase of services run by private business operators. With that increase, users of long-term care insurance services also increased.

Consequently, long-term care insurance benefits and expenses surged. Figure 3 shows changes in the total cost of long-term care insurance benefit expenses in Japan. The total cost has tripled from 3.6 trillion yen in 2000 to 10.8 trillion yen in 2017. To make an international comparison here, Table 2 presents public care expenditures (including compulsory insurance systems) of nine major countries as a percentage of GDP, taking regional balance into consideration. Public care expenditures in Japan account for 2.0% of GDP. Although not as high as Scandinavian countries such as Denmark and Sweden, it is slightly above the OECD average and is ranked 7th among 26 countries. The growth rate of public care expenditures in Japan is 4.6%, which

[9]Non-profit and public interest corporations in Japan differ among corporation types according to the business domain. Social welfare corporations are private and non-profit corporations mainly engaged in social welfare business operations. Furthermore, medical corporations provide medical; and NPO corporations are private organizations engaged in social activities.

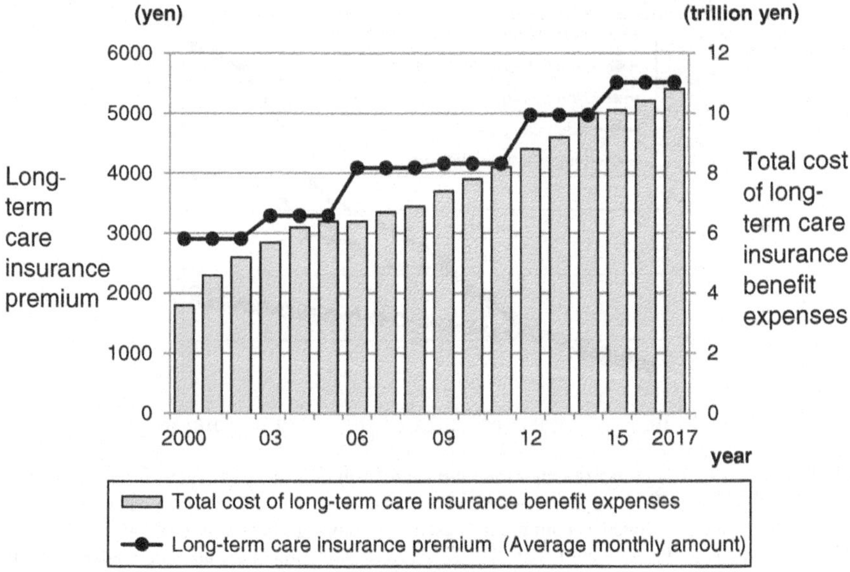

Fig. 3 Changes in the total cost of long-term care insurance benefit expenses and long-term care insurance premiums in Japan. Actual results are used until 2014 and the initial budget is used during 2015–2017 (Created by the Committee for Promoting the Integrated Economic and Fiscal Reforms Integrated Economic and Fiscal Reforms)

is the same as the OECD average, during 2005–2015. However, the growth rate is high compared to Scandinavian countries. It is ranked the 6th among 26 countries. It is readily apparent that public care expenditures in Japan are average-sized among economically developed countries and that they show an expanding trend in recent years.

With increasing long-term care insurance benefits and expenses, long-term care insurance premiums that insured persons pay are also soaring. Figure 3 shows the average of long-term care insurance premiums (a monthly amount) for each local government. The average of long-term care insurance premiums nearly doubled from 2,911 yen at the start of the long-term care insurance to 5,514 yen in the 6th program term (2015–2017).

However, long-term care insurance premiums shown in Fig. 3 are national averages. Long-term care insurance premiums are set by long-term care insurers. Therefore, the amount differs depending on the long-term care insurer because the distribution of care resources and the progress of population aging vary from one long-term care insurer to another. Figure 4 shows the base amount of monthly long-term care insurance premium in the 6th program term for each long-term care insurer. In general, a trend exists by which long-term care insurance premiums are low in metropolitan areas and in high in underpopulated areas. However, the premium is low overall in Hokkaido. In the program plan for the 6th term, the difference between the highest and the lowest premiums is tripled.

Table 2 International comparison of care expenditure (Unit:%)

Country	Long-term care expenditure(health and social components) by government and compulsory insurance schemes, as a share of GDP, 2015(or nearest year)	Government and compulsory insurance spending on long-term care(health) by mode of provision, 2015(or nearest year) (rate of inpatient long-term care)	Annual growth rate in expenditure on long-term care(health and social) by government and compulsory insurance schemes, in real terms, 2005-15(or nearest year)
Japan	**2.0**	**68**	**4.6**
Denmark	2.5	36	2.5
Sweden	3.2	64	2.0
Netherlands	3.7	86	2.9
Hungary	0.2	96	0.3
Poland	0.4	14	5.1
Canada	1.2	87	2.0
United states	0.5	–	1.8
Republic of Korea	0.8	85	32.1
OECD average	1.7[a]	65[b]	4.6[b]

[a]15 OECD countries reporting "health and social LTC"
[b]26 OECD countries (Created based on OECD Health Statistics)

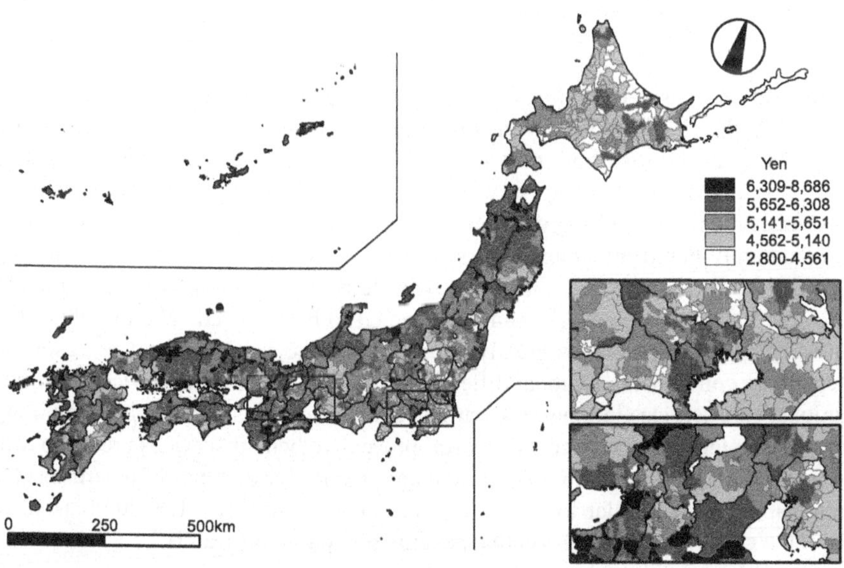

Fig. 4 Base amount of long-term care insurance premiums in the 6th long-term care insurance planning (Created based on the Ministry of Health, Labour and Welfare 2015. Reprinted from Sugiura 2017a with permission of Akashi Shoten)

A key factor for areas where long-term care insurance premiums are high is that because the degree of the improvement of the service infrastructure is high in relation to the size of the population of first-category insured persons, long-term care insurance benefits and expenses per person surged. As described later, a direct factor in the surge of long-term care insurance benefits and expenses is often derived from the location of institutional care services centers, including the welfare facility for elderly people. Table 2 also presents the percentage of hospitalization care services in public care expenditures in various countries. The value is 68% in Japan, which is slightly above the OECD average and is ranked 12th among 26 countries. Japan has shifted from institutional care services to services centering on home-based care by introducing the long-term care insurance system. Nevertheless, many areas still greatly need institutional care services. However, the level of the improvement of care infrastructure is often low in areas where the level of long-term care insurance premiums is low. Figure 4 shows that areas where the level of long-term care insurance premiums is low are often distant from the prefectural capital city and central cities, or are mountainous regions in prefectural border areas or isolated islands in each prefecture. Because the population size is small and because the demand for nursing care services is low in these areas, nursing care service providers are less willing to enter the services.

3 Regional Differences in Long-Term Care Insurance Services

This section considers regional differences in long-term care insurance services to clarify the distribution of care resources in Japan. If there are large regional differences in the distribution of nursing care services and providers, then users are prevented from choosing services and providers freely. Networking regional resources mainly by respective municipalities is regarded as important for building a community-based integrated care system. If regional differences occur in long-term care insurance services that are the key element for regional resources, then the creation of effective networks can be expected to be hindered. The distribution of nursing care services varies greatly depending on the service. For that reason, we separately consider the regional differences in home-based care services and institutional care services in long-term care insurance benefits and those in community-based long-term care services. However, preventive benefit services are often affiliated with home-based care services and community-based long-term care services centers. They are similar to the distribution of those services. For this reason, we omit the consideration of preventive benefits for space constraints.

3.1 *Home-Based Care Services (Long-Term Care Insurance Benefits)*

Home-based care services in long-term care insurance benefits consist of home services, outpatient services, and short-term stay services (Table 1). The distribution of resources depends on the service. For this reason, based on an analysis by Miyazawa (2017), we consider regional differences in the services using home care, adult day care, and short-term stay for personal care, respectively, as examples of home services, outpatient services, and short-term stay services.

The trend of the number of home-based care services centers is influenced by the entry behavior of service providers who consider conditions related to management. Service providers make a decision about entering service markets while considering regional characteristics. Consequently, the distribution of home-based care services has given rise to regional differences. Figures 5, 6 and 7 present the number of centers per 100,000 population of persons certified as being in need of care by long-term care insurers using home care, adult day care, and short-term stays for personal care as examples, respectively. The following trends are apparent from the figures.

Although many home care centers exist in metropolitan areas and in major cities in provincial areas, a small number of short-term stays for personal care centers exist in metropolitan areas. On the Sea of Japan coast, a small number of home care centers exist, whereas there are many short-term stay for personal care centers.

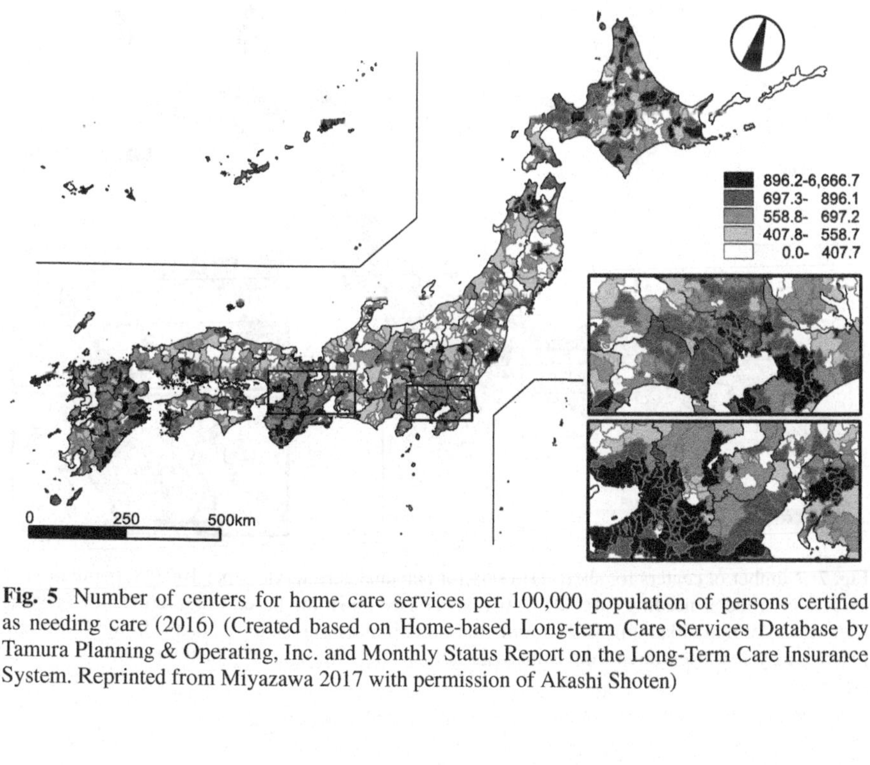

Fig. 5 Number of centers for home care services per 100,000 population of persons certified as needing care (2016) (Created based on Home-based Long-term Care Services Database by Tamura Planning & Operating, Inc. and Monthly Status Report on the Long-Term Care Insurance System. Reprinted from Miyazawa 2017 with permission of Akashi Shoten)

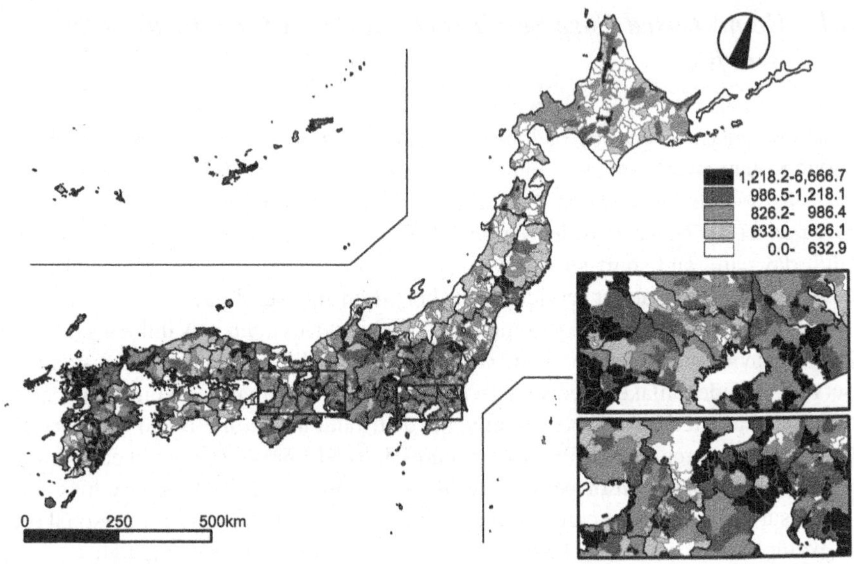

Fig. 6 Number of centers for adult day care services per 100,000 population of persons certified as needing care (2016) (Created based on Home-based Long-term Care Services Database by Tamura Planning & Operating, Inc. and Monthly Status Report on the Long-Term Care Insurance System. Reprinted from Miyazawa 2017 with permission of Akashi Shoten)

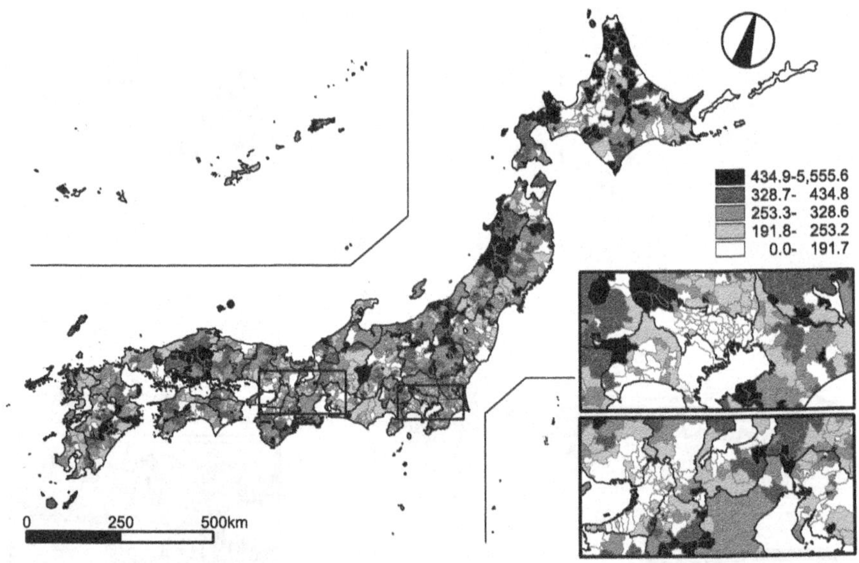

Fig. 7 Number of centers for short-term stay for personal care services per 100,000 population of persons certified as needing care (2016) (Created based on Home-based Long-term Care Services Database by Tamura Planning & Operating, Inc. and Monthly Status Report on the Long-Term Care Insurance System. Reprinted from Miyazawa 2017 with permission of Akashi Shoten)

No clear regional differences can be observed in adult day care centers. The distribution is intermediate between the other two services. Such regional differences in the distribution of centers are a consequence of which service providers selectively entered the areas. The result can be explained from differences in the forms of providing services and the conditions related to securing a space for business purposes (Miyazawa 2003).

Home care services require spaces for goods management, office work related to management and service provision, and consultation and for care staff to visit users' homes. With respect to the former, because no restriction exists on the location of offices, one can readily find a space, even in urban areas. Regarding the latter, because the travel time to make a visit is beyond the scope of long-term care insurance reimbursement,[10] saving as much care staff travel time as possible is necessary to improve profitability. That is, when considering the location of a home care center, areas with a high population density where the proximity to service users' homes will be close are advantageous. In contrast, hilly and mountainous areas where the population density is low and the Sea of Japan coast, where snow in the winter is an obstacle to travel,[11] are not amenable to home care, although regional additions to long-term care insurance reimbursement are made.

Services of short-term stay for personal care must have private rooms, a dining hall, a kitchen, and other facilities. For this reason, short-term stays for personal care centers are often affiliated with the welfare facility for elderly people that is institutional care services and use unoccupied beds and equipment in the facility. Therefore, as explained later, the distribution pattern of short-term stay for personal care centers is similar to institutional care services. To establish a short-term stay for a personal care center that requires space to provide services, non-urban areas where land prices are low and where land use can hardly compete with others present some advantages. In areas covered with snow, it becomes difficult for some people to live at home during the winter season. In rural areas, it becomes difficult for family members to give care during the busiest season for farmers. As described, the needs for short-term stay services change seasonally.

Adult day care requires a dining hall, a functional training room, and other facilities. However, no hindrance exists when the space is small compared to short-term stay services. In addition, although staff members need to travel to pick up and drop off users, travel is less frequent than that required for home care. Accordingly, the distribution pattern of adult day care centers is considered to be intermediate between that of home care and that of short-term stay for personal care.

---

[10]It is a consideration paid to service providers from the long-term care insurance finances when long-term care insurance services are used.

[11]Characteristics of the climate on the Sea of Japan side of Japan include high winter snowfall at low latitudes. The mechanism is the following. From Siberia to Japan in winter, cold seasonal winds blow from the northwest. When these winds cross the Sea of Japan, they absorb heat and moisture to form clouds, thereby becoming wet seasonal winds. These clouds bring snowfall on the Sea of Japan side of Japan. Particularly, west of Japan's central mountain range experiences heavy snowfalls.

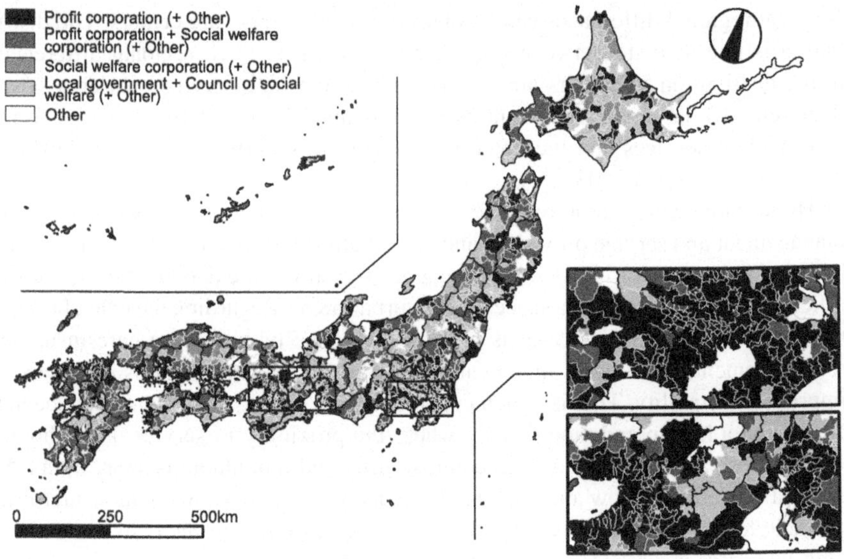

Fig. 8 Corporation types of service providers dominant in home care (2016). Insurers from whom data could not be obtained are included in Other (Created based on Database of Home-based Long-term Care Services by Tamura Planning & Operating, Inc. Reprinted from Miyazawa 2017 with permission of Akashi Shoten)

With the introduction of the long-term care insurance systems, many private business operators have entered home-based care services and regional differences have occurred also in entry trends. Figures 8, 9 and 10 respectively present the types of dominant corporations that established centers by long-term care insurers in terms of home care, adult day care, and short-term stays for personal care. Data show that many areas exist in which centers established by for-profit corporations dominate home care and adult day care. For-profit corporations are particularly dominant in metropolitan areas and major cities. Because those providers pursue profits, uneven distribution of services has occurred mainly in metropolitan areas (Miyazawa 2003). By contrast, centers established by public bodies such as local governments, councils of social welfare,[12] and social welfare corporations, and corporations with a high public benefit are dominant in small towns and villages.

In short-term stay for personal care, many centers have been established by social welfare corporations throughout the country. The reason is that only local governments and social welfare corporations can operate welfare facilities for elderly people

[12] A Council of Social Welfare is a private organization promoting local social welfare activities. However, many Councils of Social Welfare have a strong public character because welfare-related outsourcing projects are conducted by local governments. Based on social welfare laws, Councils of Social Welfare are organized in every municipality and prefecture. A national organization exists, supporting a federation structure. Furthermore, District Councils of Social Welfare are established as voluntary organizations of residents in small areas in a municipality.

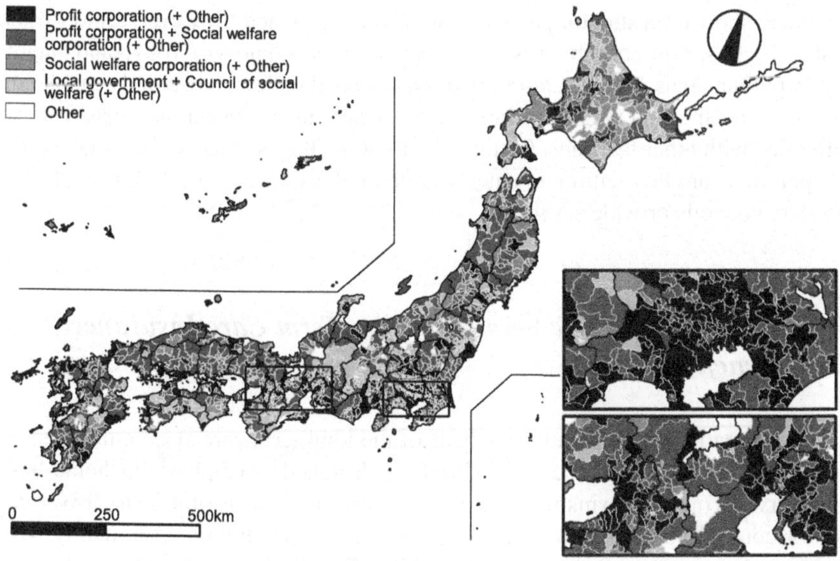

Fig. 9 Corporation types of service providers dominant in adult day care (2016). Insurers from whom data could not be obtained are included in Other (Created based on Database of Home-based Long-term Care Services by Tamura Planning & Operating, Inc. Reprinted from Miyazawa 2017 with permission of Akashi Shoten)

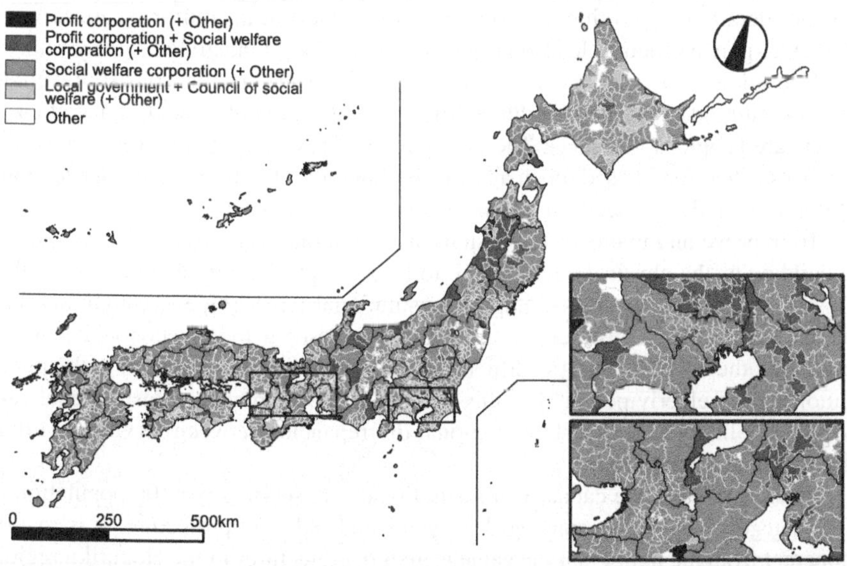

Fig. 10 Corporation types of service providers dominant in short-term stay for personal care (2016). Insurers from whom data could not be obtained are included in Other (Created based on Database of Home-based Long-term Care Services by Tamura Planning & Operating, Inc. Reprinted from Miyazawa 2017 with permission of Akashi Shoten)

to which short-term stay for personal care is often attached. However, centers established by for-profit corporations are dominant in the suburbs of Tokyo and Nagoya metropolitan areas, where centers are often affiliated with private elderly care homes and in some areas on the Sea of Japan coast, where many standalone centers are not affiliated with other facilities. In small towns and villages where even social welfare corporations are less willing to enter service markets, local governments and social welfare councils provide services.

3.2 Institutional Care Services (Long-Term Care Insurance Benefits)

In an international comparative analysis of the long-term care system in Japan, we ascertained that although the system has been designed by emphasizing home-based care services rather than institutional care services in Japan, institutional care services still account for a large percentage of public care expenditures. Community-based integrated care systems promoted since 2006 aim to watch over elderly people by various actors in areas where they live while further emphasizing home-based care services. However, as we have described, home-based care services are affiliated with institutional care services in many cases. This is true because the operations of nursing care service providers are becoming increasingly diversified according to considerations of profitability in a market mechanism. Institutional care services therefore play a salient role also in the community-based integrated care system.

Institutional care services in long-term care insurance benefits consist of welfare facilities for elderly people, health facilities for elderly people, and designated long-term care hospitals (Table 1). The distribution of resources differs for each of the services. Therefore, based on an analysis by Sugiura (2017b), we consider regional differences in the services, emphasizing welfare facilities for elderly people.

To improve and maintain the system of institutional care services, laws require a building as the physical environment to be a receptacle, with a generous number of staff. Consequently, costs to improve institutional care services are high. In urban areas, because the construction of facilities imposes a heavy economic burden, including the cost of land, it is difficult to develop facilities in relation to the population size of elderly people. For this reason, the development of institutional care services is likely to lead to large regional discrepancies between services in urban areas and non-urban areas.

Figure 11 shows the capacity of institutional care services per 100 population of first category insured persons aged 75 years and older by prefecture. It is readily apparent from the figure that the value is high in prefectures in the Hokuriku region, including Toyama Prefecture, which is the highest, Tokushima Prefecture, and Tottori Prefecture. The next level is often found in provincial areas, such as Akita and Kochi prefectures, distant from metropolitan areas, in addition to Ibaraki Prefecture.

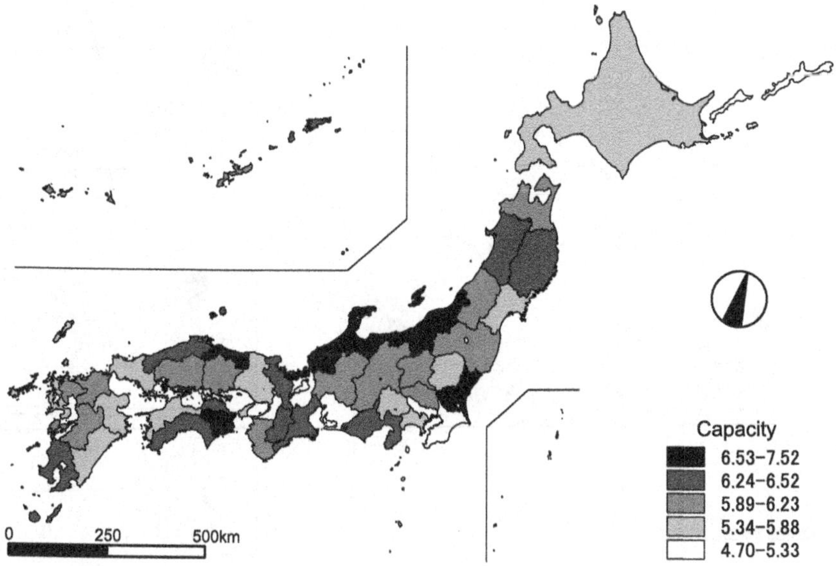

Fig. 11 Capacity of institutional care services per 100 population of first category insured persons aged 75 years and older (2014) (Created based on Survey of Institutions and Establishments for Long-term Care and Annual Status Report on the Long-Term Care Insurance System. Reprinted from Sugiura 2017b with permission of Akashi Shoten)

However, most areas with low capacity are dominated by prefectures such as Tokyo, Osaka, and Aichi, which serve as the core of metropolitan areas.

The welfare facility for elderly people accounts for 56.3% of facilities in institutional care services of three types. Many people are on waiting lists: they have applied but cannot enter welfare facilities for elderly people. This has persisted as an issue for the welfare facility for elderly people. Under the circumstances, Japan has emphasized the functions of the welfare facility for elderly people as facilities to support persons in need of care who have moderate or severe difficulty living at home. More specifically, users of welfare facilities for elderly people are limited to elderly people who are certified as care level 3, in principle, since 2015.

Figure 12 shows the capacity of the welfare facility for elderly people per the number of persons certified as care level 3 or higher by long-term care insurers. Compared to home-based care services, welfare facilities for elderly people cover a wide geographical range in which demand and supply of services are linked. Accordingly, not all residents are inhabitants of the municipality where the facility is located. However, people do not generally prefer to enter a facility in an area far away from the place where they have lived. They often choose a facility located in their municipalities. Therefore, for analyzing the location trends of facilities in this paper, we recognized the necessity of grasping them in the unit of each municipality. In Fig. 12, municipalities with a high level of capacity are noticeable in Hokkaido. Many areas have high capacity in mountainous regions in the Tohoku and Kyushu regions. In this

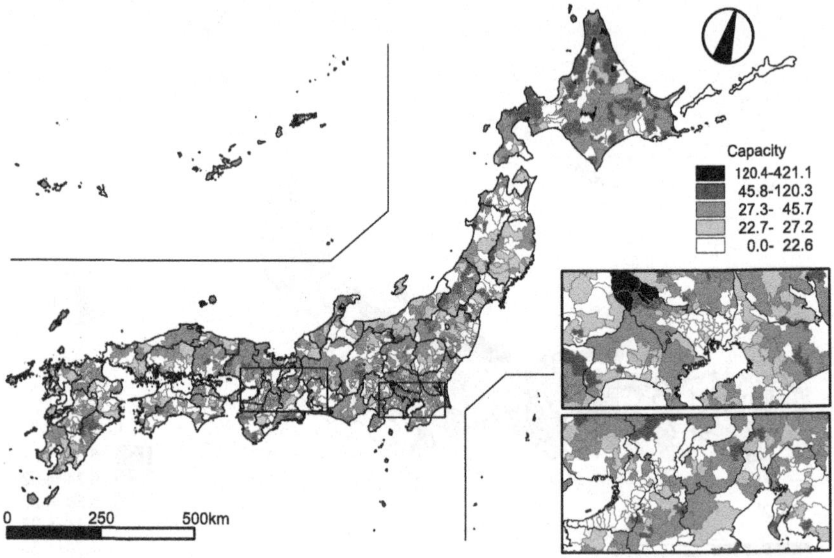

Fig. 12 Capacity of welfare facilities for elderly persons per 100 population of first category insured persons certified as care level 3 or higher (2014). Insurers from whom data could not be obtained are treated as 0 (Created based on Survey of Institutions and Establishments for Long-term Care and Annual Status Report on the Long-Term Care Insurance System. Reprinted from Sugiura 2017b with permission of Akashi Shoten)

way, areas with a high level of capacity are concentrated in provincial areas. However, it is characteristic of metropolitan areas that the Okutama region of western Tokyo shows a high value. Facilities have been actively developed in the Okutama region for a long time to service residents of urban areas, such as special wards of Tokyo. The reason is that it was difficult to develop facilities in urban areas because of high land prices.

It is readily apparent that many areas in Japan have no facility. Such areas are small towns and villages. The main reason that facilities have not been developed is that the demand size is insufficient to operate a welfare facility for elderly people as a business even when including neighboring municipalities. If one facility is located in a municipality in this manner, then it is reasonable to expect that a certain number of residents shall enter the facility. As a consequence, long-term care insurance benefits and expenses can be expected to rise, which might cause a surge of long-term care insurance premiums. For this reason, facilities are not developed in many cases.

3.3 Community-Based Long-Term Care Services

As of 2018, community-based long-term care services consist of nine types of long-term care insurance benefits and preventive benefits of three types (Table 1). These services present differences according to their geographical distribution. Based on an analysis conducted by Hatakeyama (2017), we consider regional differences in the services.

Community-based long-term care services are new services established by the revision of the long-term care insurance system in 2006. The purpose for which the service provision was newly established was to guarantee, as much as possible, that an increasing number of elderly people with dementia or those living alone would be able to continue to live in the community with which they are familiar. A characteristic of the services is that municipalities, which are closest to residents, have the authority to designate, instruct, and supervise providers. That is, the services are of importance in the community-based integrated care system which municipalities must build according to the initiative.

A Survey of Institutions and Establishments for Long-term Care revealed that, as of October 2016, the number of centers for community-based adult day care (21,063) is the highest among community-based services. Although community-based adult day care had been positioned as home-based care services, small-scale services with a capacity of 18 people or less were positioned as community-based services in 2016. The number is followed by the institution-based type of group home for people with dementia (13,069), home-based type of small-scale multifunctional home-based care services (5,125), and home-based type of adult day care for persons with dementia (4,239).

To clarify regional differences in community-based long-term care services, we calculated coefficients of variation from the number of centers per person in need of care for each long-term care insurer (Table 3). The service with the lowest coefficient of variation is the group home for people with dementia (0.73); other services had a

Table 3 Differences between insurers in the number of centers for community-based long-term care services per person certified as needing care

Service name	Coefficient of variation	Percentage of insurers who have not yet established the services
Night time home care	5.20	91.7
Small-scale multifunctional home-based care services	1.93	37.0
Group home for people with dementia	0.73	9.6
Community-based specified facility care	4.28	89.4
Day care services for people with dementia	1.41	41.7

Only five services, those from which detailed data by insurers were obtained, are compared. Data for 2015 are used for group homes for people with dementia. Data for 2016 are used for other services (Created based on Database of Housing for the Elderly and Database of Home-based Long-term Care Services by Tamura Planning & Operating, Inc. and Monthly Status Report on the Long-Term Care Insurance System. Reprinted from Hatakeyama 2017 with permission of Akashi Shoten) coefficient of variation exceeding 1. Particularly, the value exceeded 4 for night time home care and community-based specified facility care. As described, considerably numerous large regional differences in community-based long-term care services because many long-term care insurers do not institute community-based long-term care service centers. The percentage of insurers who have not yet established services is 9.6% for the group home for people with dementia, which is the lowest; it exceeds 30% for other services (Table 3). Approximately 90% of long-term care insurers of night time home care and community-based specified facility care have not yet established services. In this way, large differences prevail among services in terms of the percentage of long-term care insurers who have not yet established services.

Group homes for people with dementia and small-scale multifunctional home-based care services are on a steeply increasing trend among community-based long-term care services and have a low percentage of insurers who have not yet established the services. With regard to them, Figs. 13 and 14 present the number of centers per 100,000 population of persons certified as being in need of care by long-term care insurers. The group home for people with dementia is a service to provide specialist care for users with dementia. In the service, a small number of users live in a group home with specialists. Small-scale multifunctional home-based care services are a service that provides support for everyday life and functional training to enable users

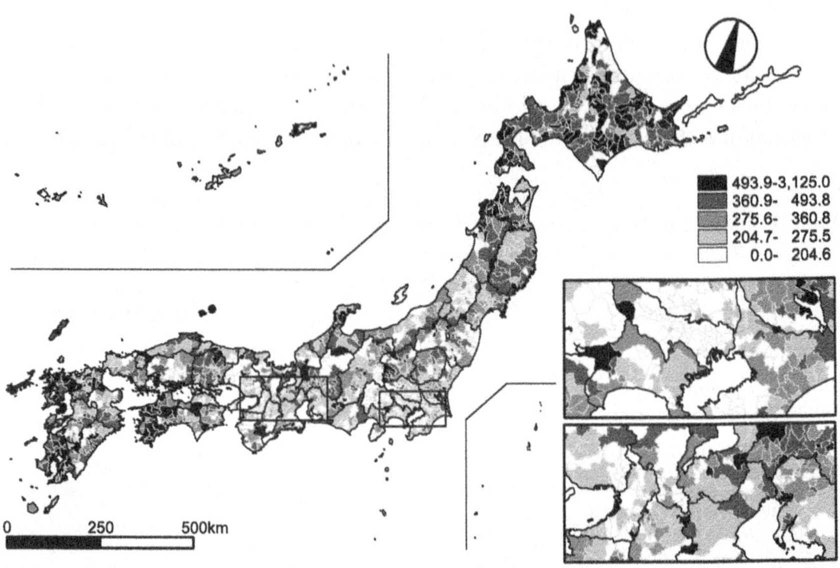

Fig. 13 Number of centers for the group home for people with dementia per 100,000 population of persons certified as needing care (2015). Insurers from whom data could not be obtained are treated as 0 (Created based on Database of Housing for the Elderly by Tamura Planning & Operating, Inc. and Monthly Status Report on the Long-Term Care Insurance System. Reprinted from Hatakeyama 2017 with permission of Akashi Shoten)

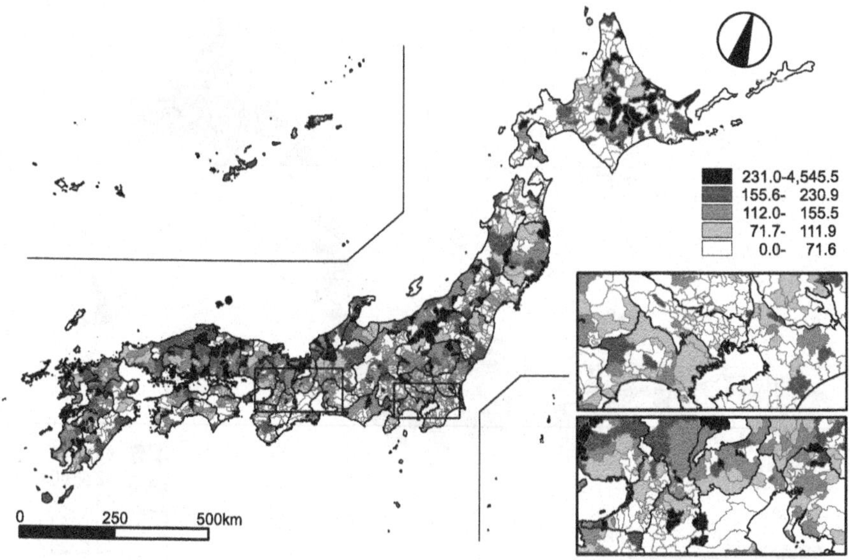

Fig. 14 Number of centers for small-scale multifunctional home-based care services per 100,000 population of persons certified as needing care (2015). Insurers from whom data could not be obtained are treated as 0 (Created based on Database of Home-based Long-term Care Services by Tamura Planning & Operating, Inc. and Monthly Status Report on the Long-Term Care Insurance System.Reprinted from Hatakeyama 2017 with permission of Akashi Shoten)

to live everyday life as independent as possible, combining outpatient services, home services, and short-term stay services according to the user's choice.

Figures 13 and 14 show that a few centers exist in the Tokyo metropolitan area and the Kyoto–Osaka–Kobe metropolitan area in both services, although many centers exist mainly in underpopulated areas in Hokkaido and the Tohoku, Chugoku, Shikoku, and Kyushu regions. However, in small-scale multifunctional home-based care services where the percentage of insurers who have not yet established the services is 37.0%, many insurers have not yet established services in underpopulated areas in Hokkaido, the Tohoku, Chubu, Shikoku, and Kyushu regions, and southern Kinki. That is, in underpopulated areas, many cases exist in which services have not been established. However, once they are established, the degree of satisfaction can be expected to be high compared to those for metropolitan areas because the population is small. As explained earlier, regional differences in community-based long-term care services are large between metropolitan areas and underpopulated areas in non-metropolitan areas and between insurers who have and those who have not yet established services in underpopulated areas.

Similarly to home-based care services, many private business operators have entered community-based long-term care services. The entry of for-profit corporations is particularly noticeable. As of 2015, the percentage of centers established

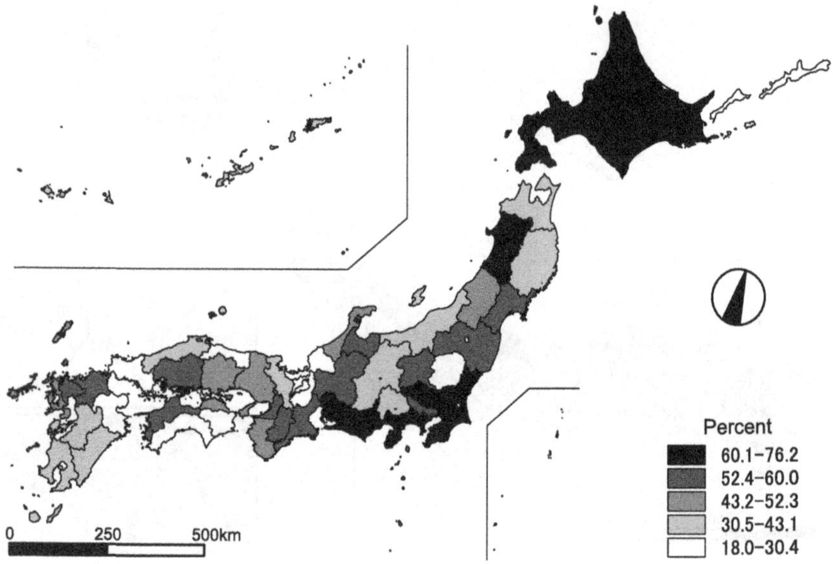

Fig. 15 Percentage of centers run by for-profit corporations in the group home for people with dementia (2015) (Created based on Survey of Institutions and Establishments for Long-term Care. Reprinted from Hatakeyama 2017 with permission of Akashi Shoten)

by for-profit corporations exceeded 40% in night time home care, small-scale multi-functional home-based care services, group homes for people with dementia, and community-based specified facility care.

Group homes for people with dementia have numerous centers in community-based long-term care services and a high percentage of for-profit corporations, accounting for 53.6% (2015). Figure 15 presents the percentage of for-profit corporations by prefecture with regard to group homes for people with dementia. From that figure, it is apparent that the values of prefectures located in metropolitan areas are generally high. In addition, a trend exists by which the values are high in eastern Japan and low in western Japan. The former is true because many for-profit corporations entered service in metropolitan areas with prospects of large-scale demand. Particularly, the percentage exceeds 70% in Saitama and Chiba prefectures. However, the latter is true because many medical corporations entered service mainly in the Chugoku, Shikoku, and Kyushu regions in western Japan. It is a characteristic of western Japan that medical corporations have a strong base. A similar trend was observed in the service of long-term care insurance benefits.

In community-based long-term care services, municipalities were expected to have the authority to designate providers, which made it possible to set the amount of service supply based on long-term care insurance planning by municipalities. However, regional differences in services are large. Furthermore, quite a few insurers have not yet established services. Two reasons exist for these discrepancies. One is the case in which a goal for the development of facilities was not set in long-term care insurance planning in the first place. The other is the case in which providers were less willing to enter because of the small demand, although having set a goal for the development of facilities.

In community-based long-term care services, although municipalities became able to designate providers with the enhanced authority, the effect has not been demonstrated. Particularly, different problems have arisen in areas on both ends of the spectrum: services have not been established in underpopulated areas; moreover, the services are in short supply in metropolitan areas.

References

Hatakeyama T (2009) The expansion of municipal authority and regional influence resulting from long-term care insurance system reform: The case of Fujisawa city, Kanagawa prefecture, Japan. Jpn J Hum Geogr 61:37–54 (in Japanese with English abstract)

Hatakeyama T (2017) Community-based long-term care services. In: Miyazawa H (ed) Mapping health, medical care, and welfare in Japan. Akashi Shoten, Tokyo (in Japanese)

Ido M (2017) A quick guide to revision of long-term care insurance. Nihon Jitsugyo Shuppansha, Tokyo (in Japanese)

Masuda M (ed) (2014) Long-term care systems in the World, 2nd edn. Horitsu Bunka Sha, Kyoto (in Japanese)

Ministry of Health, Labour and Welfare (2015) The sixth planning period - The first premium, expected service volume, and other matters of long-term care insurance of Heisei 37. https://www.mhlw.go.jp/stf/houdou/0000083954.html. Accessed 1 May 2019 (in Japanese)

Miyazawa H (2003) Uneven nursing care service opportunity and the behavior of service providers under the long-term care insurance system: A statistical analysis in the Kanto District. Geogr Rev Jpn 76:59–80 (in Japanese with English abstract)

Miyazawa H (2017) Home-based care services. In: Miyazawa H (ed) Mapping health, medical care, and welfare in Japan. Akashi Shoten, Tokyo (in Japanese)

Seon HK (2010) A comparative consideration of the long-term care insurance system in Japan, Germany and South Korea. J Kyoei Univ 8:1–18 (in Japanese)

Sugiura S (2017a) Long-term care insurance system. In: Miyazawa H (ed) Mapping health, medical care, and welfare in Japan. Akashi Shoten, Tokyo (in Japanese)

Sugiura S (2017b) Long-term care insurance facilities. In: Miyazawa H (ed) Mapping health, medical care, and welfare in Japan. Akashi Shoten, Tokyo (in Japanese)

United Nations, Department of Economic and Social Affairs, Population Division (2017) World population prospects: The 2017 revision

Medical Care Provision System and Geographical Distribution of Medical Resources in Japan

Tsutomu Nakamura, Kazumasa Hanaoka, and Hitoshi Miyazawa

Abstract The private sector-centered free medical practitioner system prevailing in Japan has contributed to quantitative expansion of medical institutions to a great degree. However, it has led to starkly uneven regional distribution because medical institutions have concentrated in urban areas. Regions with no doctors (no dentists) remain mainly in mountainous areas, where access to medical care is limited. Even core hospitals supporting medical services in remote areas cannot provide sufficient support because of a shortage of doctors. Similar regional differences in the location of pharmacies and in access to medical care are apparent in institutions implementing in-home medical care and emergency medical services. Existing medical care plans have specifically emphasized the setting of medical areas and regulation of the number of hospital beds. However, to prepare and implement the plans in accordance with actual conditions for each region, careful consideration is needed of medical and nursing provision systems, such as the development and cooperation of care facilities, in addition to the accurate assessment of demand for medical care, including in-home medical care.

Keywords Emergency medical service · In-home medical care · In-patient bed · Medical institution · Pharmacy

This chapter is a revised version of Hanaoka (2017), Miyazawa (2017), and Nakamura (2017a, b, c).

T. Nakamura (✉)
Ryutsu Keizai University, Ryugasaki, Japan
e-mail: nakamura-tsutomu@rku.ac.jp

K. Hanaoka
Ritsumeikan University, Kyoto, Japan

H. Miyazawa
Ochanomizu University, Tokyo, Japan

H. Miyazawa and T. Hatakeyama (eds.), *Community-Based Integrated Care and the Inclusive Society*, International Perspectives in Geography 12,
https://doi.org/10.1007/978-981-33-4473-0_4

1 Characteristics of Medical Care Provision System in Japan

1.1 International Comparison of Medical Care Systems

Medical care is a highly specialized human service that a person must buy when becoming injured or ill. For safe and healthy life of all people, medical care must be provided securely as a system in society. Speaking in economic terms, such a service can be supplied through markets. Nevertheless, perfectly free competitive markets do not necessarily secure either fair prices or supply to meet necessary demand: prices and supply are always unstable. With regard to medical care, which is highly specialized conduct, a large information gap persists, separating medical institutions, representing the supply side, and patients, representing demand. So-called information asymmetry exists in this market. Medical care is also a vital service to society. To develop a system in which anyone can receive medical care at a reasonable cost any time is necessary not only for individuals, but also for society as a whole. Therefore, through various historical circumstances, economically developed countries have developed systems in which people can receive proper medical care to the greatest extent possible when they need it because of illness or injury (Morita 2016).

Table 1 presents a comparison of the basic frameworks of medical care systems in six major countries. In Western countries, if medical care is supplied by the state, as in the United Kingdom and Sweden, then it is considered based on principles of universalism. If medical care is supplied through social insurance, as in Germany and France, then it is regarded as based on principles of selectivism (Mano 2012). In

Table 1 International comparison of medical care systems

	Supply	Finance
Japan	• Mainly private • Free access (Very weak gatekeeper function)	Public (Social insurance system)
Germany	• Mainly public • Weak gatekeeper function	Public (Social insurance system)
France	• Mainly public • Weak gatekeeper function	Public (Social insurance system)
U.K.	• Nearly all public • Very strong gatekeeper function	Public (Tax system)
Sweden	• Nearly all public • Not strong gatekeeper function	Public (Tax system)
U.S.A.	• Mainly private • Strong gatekeeper function in managed care type insurance	Public (exc. Medicare and Medicaid)

(Created based on Shimazaki 2011)

stark contrast, the United States is a unique country: it supplies medical care services and welfare services as private goods. Government involvement in medical care is regarded as hindering free economic activity.

The medical care provision system in Japan has three salient characteristics. First, it adopts a social insurance system and regulates only medical fees that the insurer pays uniformly, nationwide. The system controls the content and volume of medical care to be provided by adjusting medical fee points and calculation requirements (Morita 2016). Second, all residents have public medical care insurance based on the universal medical care insurance system. Third, the system is based on a free medical practitioner system, under which doctors can start practicing wherever they like as long as they satisfy facility standards. These characterize Japan as the only country with a different combination of public and private components in terms of the supply and finance of medical care. It has played a major role in the quantitative development of medical institutions (facilities providing medical care). However, as a result, most of them have been established by private bodies. The 2014 Static Survey of Medical Institutions shows that private medical institutions in Japan constitute 80.9% of hospitals[1] and 97.1% of medical and dental clinics (Nakamura 2017a).

In addition to the three points above, the medical care provision system in Japan is characterized by the adoption of free access by which anyone can visit a medical institution the person wants to consult, in principle. Being able to choose medical institutions freely irrespective of their symptoms has greatly improved patient access to medical care. A salient criticism is that moral hazard is likely to occur where people frequently visit large hospitals after relying upon word of mouth or reputation, despite minor symptoms. This point is particularly related to the fact that most private hospitals have developed as an extension of a clinic. For that reason, qualitative differences are slight among hospitals and clinics compared with those in Europe and the United States (Ikegami and Campbell 1996). From the perspective of streamlining medical care expenses, a shift toward a system that encourages patients to receive treatment in accordance with medical care functions is regarded as an important challenge for medical policy in Japan.

1.2 Revision of the Medical Care Act

The Ministry of Health and Welfare, presently the Ministry of Health, Labour and Welfare, revised the Medical Care Act in 1985, obligating prefectures to produce medical care plans to use medical resources effectively. The revision, called the first revision of the Medical Care Act, aimed at (1) appropriate allocation of medical resources and (2) building of a medical care supply system suitable for an aging

[1]The Medical Care Act restricts places to practice medicine to hospitals and clinics. The Act distinguishes those with 20 or more beds as hospitals and those with no bed or less than 19 beds as clinics. Of them, clinics that have beds are called clinics with beds and clinics that practice dentistry are called dental clinics.

society. The former is intended for hospital beds. Each prefecture was to formulate a medical care plan after specifying secondary medical areas as local units for the adjustment of general bed numbers. Later, long-term care beds, distinct from general beds, were added to them.[2] On that basis, the revision included calculation of the number of necessary beds (the name was changed to the standard number of beds in 2000) based on the population and the rates of estimated patients using the calculation method specified in the Ordinance for Enforcement of the Medical Care Act and setting the numerical target of hospital beds to be adjusted. For areas that are determined as surplus, where the existing number of hospital beds exceeds the standard number of beds, the prefectural governor can make recommendations related to the establishment and beds of private hospitals, which has not been subject to regulations. However, medical institutions are not obligated to reduce the number of hospital beds on the grounds of being in the surplus area. Rather, it caused a further increase in beds by "a last-minute increase in beds" (Inoue 2010; Shimazaki 2011). The establishment of hospitals that had been unrestricted until that time was halted (Yakuji Jiho-Sya 1999).

The second revision of the Medical Care Act made in 1992 institutionalized "special functioning hospitals" and "group of long-term care-type beds." The former were advanced medical institutions intended for main hospitals affiliated to universities, with the latter values of recuperation and care rather than treatment (Yakuji Jiho-Sya 1999). The third revision of the Medical Care Act enforced in 1998 newly added functional cooperation of medical facilities and the adjustment target for a "group of long-term care-type beds" according to the region, the system of emergency medical services, and the system of separation of dispensing and prescribing functions. It specified the methods of the medical care provision system more concretely in the medical care plan. In the fourth Medical Care Act revision enforced in March 2001, the calculation method was changed to a formula in which, if the length of stay decreases, the number of necessary beds will decrease in conjunction (Jiho 2001).

Subsequently, the fifth revision of the Medical Care Act was made in 2007. The medical care provision system was improved for each secondary medical area. However, the new medical care plan made a drastic shift to the cooperation system built for each major disease or project, such as cancer, stroke, acute myocardial infarction, and diabetes, or measures for emergency medical services for children, perinatal medical care, and remote area medical care. The revision makes it a principle to show the health and medical care provision systems in the region in an easy-to-understand way from residents' and patients' perspectives. This enables them to know what treatment they can get when contracting a disease and to receive seamless medical care

[2]Medical areas set by a prefecture consist of the "primary medical area," the "secondary medical area," and the "tertiary medical area." The "primary medical area" deals with general diseases, in which each municipality is a unit. The "secondary medical area" is cross-regional, providing highly specialized health and medical services. The "tertiary medical area" provides highly advanced specialized health and medical services with the entire prefecture as a unit. (Hokkaido has multiple tertiary medical areas.) As described, health and medical plans envision hierarchical medical areas according to the size of hospitals. Beds of the other types, except general beds and long-term care beds, are adjusted in units of the tertiary medical area (prefecture).

from the acute phase to the home-care phase (Jiho 2006). Furthermore, the regional medical care plan was directed to be originally made out in respective prefectures, considering health care finance. In addition, a large part of health and medical care administration was devolved from prefectures to municipalities.

The background by which these efforts are undertaken nationwide mainly by prefectures is that the Ministry of Health, Labour and Welfare intends to decrease the medical care expenses of elderly people, which is a key factor for the increase of medical care expenses, particularly inpatient expenses. The Ministry of Health, Labour and Welfare has an awareness of the difficulty that in Japan, the number of hospital beds is higher and the duration of hospitalization is longer than in other countries. A series of revisions of the Medical Care Act represents a clear intention of shortening the duration of stay by clarifying medical care functions such as the acute phase and the recovery phase, as well as shortening the total length of stay through strengthening cooperation among different functions.

In recent years, to respond to the increase of medical care expenses and to the diversification of medical care demand, the so-called community-based integrated care system has been built, in which family physicians and long-term care workers mutually cooperate with core hospitals based on their roles. The former addresses consultation and daily health care of patients in stable conditions. The latter undertakes specialized examinations and advanced medical care.

Following the "Outline of the Comprehensive Reform of the Social Security and Tax Systems" approved by the Cabinet in 2012, the sixth Medical Care Act was enforced from 2014. In response to which issues of mental health and dementia are becoming intensified, mental illness was newly added to diseases subject to the Act. Furthermore, to promote community-based integrated care, prefectures were required to state numerical targets to be achieved, as well as measures and projects, with regard to the building of in-home and other medical care systems, similar to other diseases or projects (Matsuda 2015). In the seventh Medical Care Act revision enforced in 2015, the recognition system for regional cooperative medical corporations was established to promote the sharing of functions between medical institutions and the cooperation of operations. This establishment was envisioned as an option for prefectures to estimate the medical care demand and the required amount of hospital beds in 2025 for each medical care function and to realize the created Community Health Care Vision to advance functional differentiation and cooperation of hospital beds toward 2025.[3]

Although existing medical care plans have emphasized the setting of medical areas and regulating of the number of hospital beds, the purpose of the plans must be confirmed (Ikegami 2017). Particularly, in-home medical care required for the fulfillment of improvement is greatly affected by geographical conditions. To prepare and implement plans in accordance with reality but which do not cause inconsistency between supply and demand of medical care after the reorganization of hospital beds, it is necessary to give careful consideration to medical and nursing provision

[3]This explanation is based on the website of the Ministry of Health, Labour and Welfare.

systems in each region, such as development and cooperation of care facilities. Ascertaining the demand for medical care accurately, including in-home medical care, is also necessary. Consequently, this chapter clarifies the features of the medical care provision system for each region based on the distribution of medical resources in Japan.

2 Distribution of Medical Institutions

2.1 Uneven Distribution of Medical Institutions

The 2014 Static Survey of Medical Institutions reports that there are 8,493 hospitals, 100,461 medical clinics, and 68,592 dental clinics in Japan. Hospitals account for only 4.8% of all facilities. Furthermore, the number of hospitals decreased by about 15% after peaking in 1990 (Fig. 1). Regarding the data by funding body, the primary causes of the decrease of private hospitals are expected to be a business failure and a shortage of successors. The primary causes of the decrease of public hospitals are expected to be the streamlining of facilities and the reorganization of functions with fiscal consolidation and a shortage of doctors. Nevertheless, medical and dental clinics by practitioners are likely to achieve stable management; the number is consistently increasing. However, clinics with beds have decreased considerably among medical clinics: data show that clinics with no beds have increased substantially.

Examination of the number of medical institutions per 100,000 population reveals that both hospitals and medical clinics show a trend by which the number is high in western Japan and low in eastern Japan (Fig. 2a, b). With regard to hospitals, many secondary medical areas show high values in Hokkaido, in addition to the Shikoku, Kyushu, and Chugoku regions. By contrast, the number of hospitals per 100,000

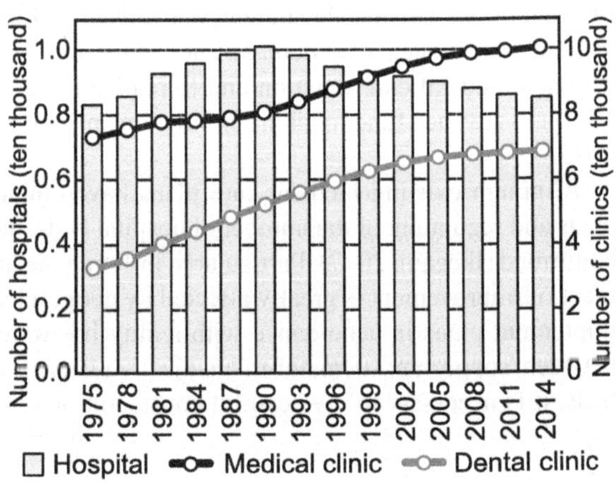

Fig. 1 Change in the number of medical institutions (nationwide) (Created based on Survey of Medical Institutions. Reprinted from Nakamura 2017a with permission of Akashi Shoten)

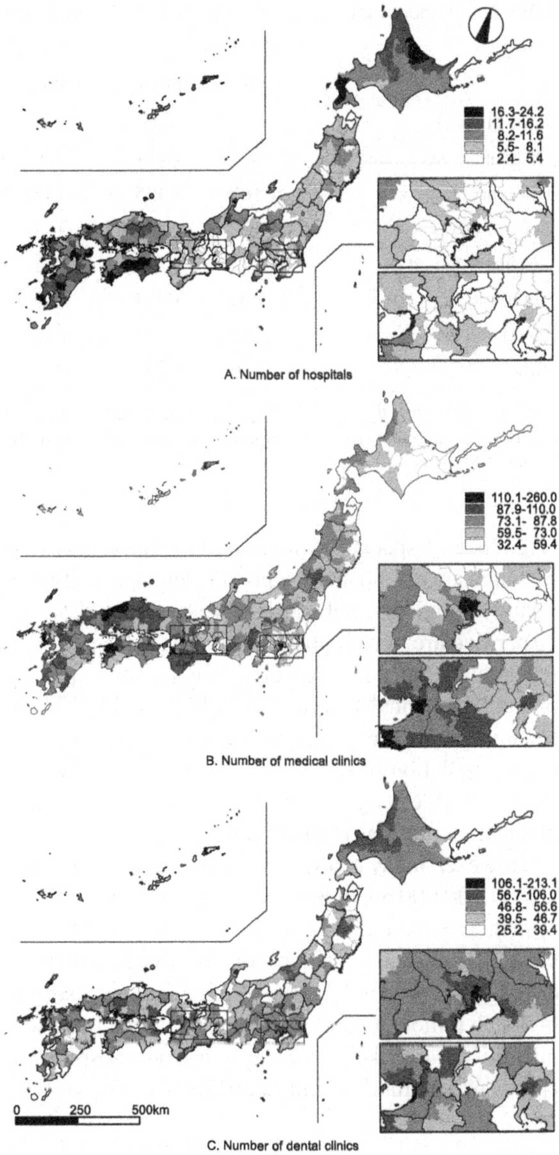

Fig. 2 Number of medical institutions per 100,000 population (by secondary medical area in 2014) (Created based on Survey of Medical Institutions and Basic Resident Register in Japan. Reprinted from Nakamura 2017a with permission of Akashi Shoten)

population is low from the Tohoku through the Kinki regions (Fig. 2a). In areas west of Kinki, the number of medical clinics per 100,000 population is high (Fig. 2b) (Nakamura 2017a).

Although the private sector-centered medical practitioner system has played a huge role in the qualitative expansion of medical institutions, it has led to a markedly uneven distribution among areas because medical institutions have concentrated in urban areas. This trend is particularly evident in figures for medical and dental clinics

Table 2 Number of medical institutions by founding body (2015)

Category	Hospital		Medical clinic		Dental clinic	
	Public	Private	Public	Private	Public	Private
Special ward	38	384	26	8,679	2	8,253
Ordinance-designated city	222	1,451	226	21,257	13	16,406
Core city	166	1,135	144	12,424	11	8,994
Special city	112	479	113	6,852	16	5,417
Medium-size city	241	1,385	304	15,600	39	12,643
Small city	492	1,564	829	15,249	61	12,137
Town and village	288	494	643	4,676	123	4,328
Total	1,559	6,892	2,285	84,737	265	68,178

Special cities are the special city at enforcement (Created based on SCUEL Data (medical institutions database) by Mecompany, Inc. Reprinted from Nakamura 2017a with permission of Akashi Shoten)

(Fig. 2b, c). Especially with regard to the number of dental clinics, the vast majority of which are established by private dental practitioners, per-capita figures are exceptionally high in special wards of Tokyo and large cities such as Osaka City. Regional disparities are very high in terms of access to dental consultation, which suggests that dental clinics in urban areas are in oversupply. By contrast, the component ratio of public medical institutions is high mainly in towns and villages. That is true because traffic conditions in towns and villages are unsatisfactory despite their scattered population of residents, making it difficult for private hospitals and clinics to expand business operations there (Table 2). Those areas were supported by public medical institutions (Nakamura 2017a).

However, many medical areas, including prefectural capital cities, have few hospitals per 100,000 population (Fig. 2a). Such areas seem to have experienced population growth above the rate of increase of the number of medical institutions throughout the high economic growth period. In contrast, mountainous areas and isolated islands have experienced a decrease in population. Among those areas and islands, regions in which remote area medical services have been secured mainly by public medical institutions are likely to have numerous hospitals and a high number of medical clinics per capita (Nakamura 2017a).

2.2 *Regions with No Doctors and Remote Area Medical Services*

It is apparently irrational that some regions and residents might not receive medical care services, despite being institutionally guaranteed by the universal medical care

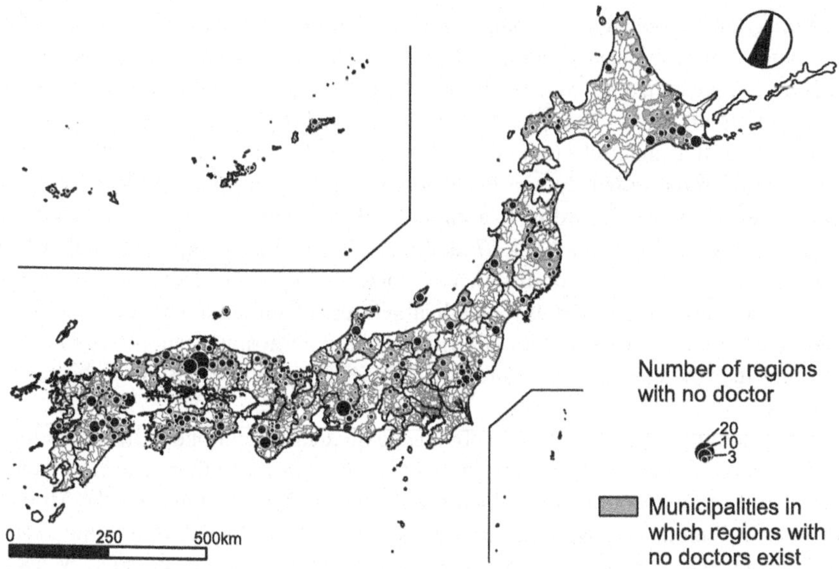

Fig. 3 Number of regions with no doctors (by municipality in 2014). Comparing 2009 to 2014, regions with no doctor increased in nine prefectures: Iwate, Miyagi, Tochigi, Toyama, Aichi, Wakayama, Shimane, Hiroshima, and Ehime. By contrast, they decreased in 26 prefectures 8 (Created based on Survey on No-doctor Districts. Reprinted from Nakamura 2017a with permission of Akashi Shoten)

insurance system. However, as of October 2014 in Japan, regions with no doctor[4] were 637: those with no dentists were 858. In some regions, medical institutions still do not exist. Prefectures with many regions with no doctors and their respective numbers of regions are as follows: Hokkaido, 89; Hiroshima, 54; and Kochi and Oita, each 38. Regarding results by municipality, such regions are more common in mountainous areas that have limited medical care access capable of responding to daily demands for medical care (Fig. 3).

Article 16 of the Act on Special Measures for Promotion for Independence of Underpopulated Areas under the jurisdiction of the Ministry of Internal Affairs and Communications requires the securing of doctors in regions with no doctors based on the Prefectural Plan. Clinics in remote areas[5] have been established by the prefecture and by the central government subsidizing one-half each for the municipalities that fall under the standards of that law. As of January 2015, 1,055 clinics have been

[4]A region with no doctors (no dentists) is a region in which there is no medical (dental) institution, in which 50 or more people live within about a 4 km-radius zone starting from the central place in the region, and for which it takes more than an hour each way by ordinary transportation to travel to the nearest medical institution.

[5]Standards for establishing clinics in remote areas are specified as follows: no other medical institution exists within about a 4-km-radius zone starting from the place where the clinic is to be established; 1,000 or more people reside there; and it takes more than half an hour by ordinary transportation to travel to the nearest medical institution from the planned site of the clinic.

developed in remote areas in Japan. Figure 4 portrays a map showing the distribution of remote clinics. Because northeastern Japan has low population density, many regions are presumed not to meet the standards for establishing the clinics in remote areas. In Japan, 302 facilities are designated as core hospitals for medical services in remote areas that undertake mobile clinics, dispatch doctors to clinics in remote areas, and dispatch replacement doctors when doctors working in clinics in remote areas are on holiday. However, according to a survey on the current state of medical services in remote areas, as of January 2014, 67 facilities (22.6%) did not implement any mobile clinic, dispatch of doctors, or dispatch of replacement doctors. Not all facilities have played the role of core hospitals for medical services in remote areas because of doctor shortages and other reasons. As described, even core hospitals supporting medical services in remote areas cannot provide sufficient support because of a shortage of doctors.

Traditionally, a person who has obtained a medical practitioner's license belongs to the doctor's office in the university hospital that has the authority over personnel matters of the university from which the doctor graduated. However, the Ministry of Health, Labour and Welfare introduced a new clinical resident training system in 2004 by which residents can choose a hospital at which to practice. Whereas residents strengthened the tendency to choose hospitals in urban areas, doctors' offices adversely affected by a shortage of staff took dispatched doctors back from affiliated hospitals in rural areas. This practice is said to have caused a decrease of dispatched doctors in regions such as remote areas, where there were insufficient

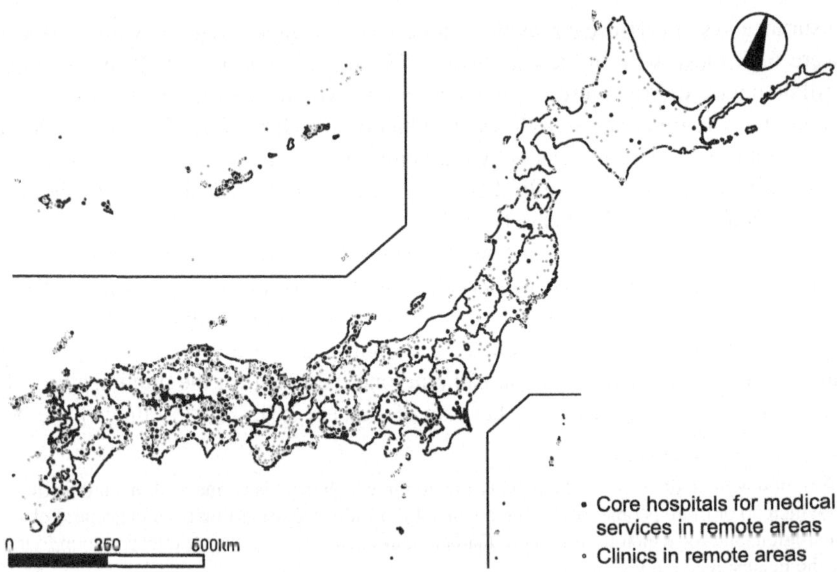

Fig. 4 Location of clinics in remote areas and core hospitals for medical services in remote areas (Created based on the website of Japan Association for Development of Community Medicine. Reprinted from Nakamura 2017a with permission of Akashi Shoten)

doctors in the first place. Support is necessary to reduce burdens on hospital doctors and to secure doctors.

3 Distribution of In-patient Beds

3.1 Regional Differences in the Number of Hospital Beds

Beds for hospitalization are classified into five types by the Medical Care Act: (1) general beds, (2) long-term care beds, (3) mental beds, (4) infectious disease beds, and (5) tuberculosis beds. The number of hospital beds in Japan shows a declining trend after reaching a peak in 1992. Regarding the trend by medical institution, with the increasing number of clinics with beds converting to clinics with no bed, the number of beds in medical clinics decreased from 285,000 beds in 1981, when it reached its peak, to 112,000 beds in 2014. The number of hospital beds also decreased from 1,681,000 beds in 1993, when it reached its peak, to 1,568,000 beds in 2014. However, assessing hospital beds by type, only long-term care beds show an increasing trend. They have decreased once since 2005, but are increasing again in recent years (Nakamura 2017b).

Although the number of hospital beds has decreased, the number of hospital beds per capita is greater in Japan than in other countries in Europe and in the United States. Coupled with a longer length of stay, as described later, that fact has been regarded as a primary factor for increasing medical care expenses. Cutting expenses has been an important challenge for medical policy.

Regional differences in hospital bed distribution are large (Fig. 5). The number of hospital beds per 100,000 population shows a trend of "west high, east low," except for Hokkaido. This tendency reflects the distribution of the number of hospitals and that of clinics with beds. Regional differences are also large between large city areas and rural areas. A three-fold difference exists between Kochi Prefecture (2,652 beds), the highest, and Kanagawa Prefecture (843 beds), the lowest. Regarding hospital beds by type, the number of long-term care beds and that of mental beds are high in western Japan, which indicates the contribution to a trend of "west high, east low" in terms of the distribution of the number of hospital beds.

3.2 Average Length of Stay and the Occupancy Rate of Hospital Beds

The average length of stay and the occupancy rate of hospital beds serve as indicators of the occupancy status of hospital beds. The average length of stay has been consistently shortening since 1981, except for long-term care beds that remain high since the 2000s. Particularly for general beds, it has greatly shortened from 39.7 days in

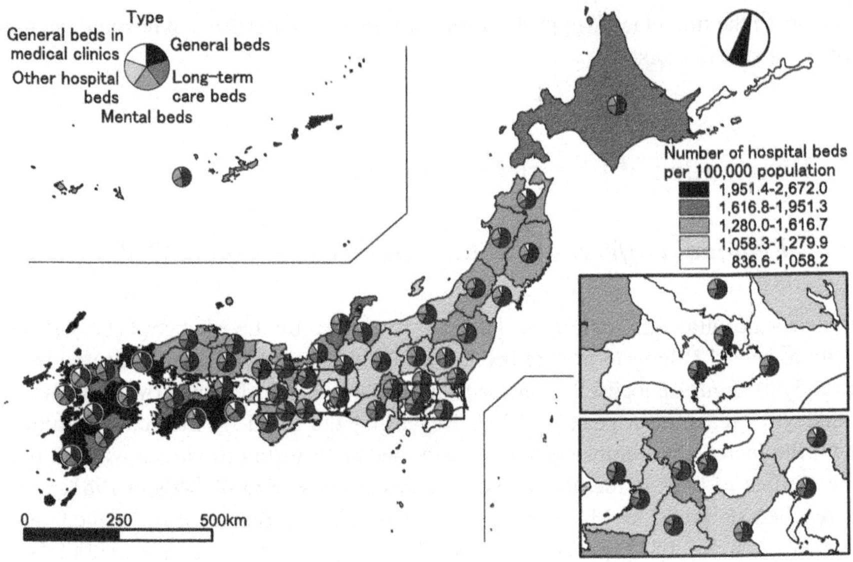

Fig. 5 Number of hospital beds per 100,000 population by medical institution and by type (2014) (Created based on Survey of Medical Institutions)

1981 to 16.8 days in 2014. The occupancy rate of hospital beds also shows a generally declining trend. The occupancy rate of general beds, among others, dropped by 70% in the late 2000s.

The average length of stay and the occupancy rate of hospital beds show a trend of "west high, east low," correlating mutually. Because the number of long-term care beds and mental beds associated with long duration of stay is high in the Shikoku and Kyushu regions, in addition to a trend of "west high, east low" in general beds, the average length of stay is longer in prefectures in the Shikoku and Kyushu regions, but shorter in metropolitan areas such as Kanagawa Prefecture and Tokyo. The occupancy rate of hospital beds is high in Okinawa and Saga prefectures, but it is low in Fukushima and Yamanashi prefectures.

Despite shortening the length of stay, the occupancy rate of hospital beds is declining. This trend suggests that the primary factor is that doctors are in chronically short supply and that bed management for smooth hospitalization and discharge is becoming difficult with shortening lengths of stay. To shorten the length of stay and to raise the occupancy rate of hospital beds simultaneously, it is necessary to coordinate patients' hospitalization and discharge, as well as to streamline hospital beds, in cooperation with other medical institutions.

3.3 *Challenges of Community Health Care Vision*

With the enactment of the Act on Promotion of Comprehensive Securing of Medical Nursing Care in 2014, the revised Medical Care Act came to require each prefecture to formulate a Community Health Care Vision as part of a medical care plan. In this plan, a prefecture estimates the medical care demand and the required number of hospital beds according to the categories of four functions in 2025 using the planned areas that are set by reference to the current secondary medical areas as a unit, and takes measures to secure and request facilities and equipment as well as medical care professionals. An expert panel established in the Cabinet Office estimated the number of necessary beds in 2025 considering the vision of future medical care, from which we can ascertain the direction of bed reorganization. In the estimation, the numbers of necessary beds in the acute phase and the chronic phase are estimated as fewer than today. This likely trend suggests that patients with minor symptoms using those hospital beds will be treated by entering care facilities, or by recuperating at home or in housing for the elderly using in-home medical care. In contrast, the number of necessary beds in the recovery phase is estimated as significantly greater, which reflects a policy to encourage returning home as early as possible by increasing hospital beds in the recovery phase. A shift to so-called "community-contained" medical care is regarded as reducing the number of hospital beds as a whole and cut down on increasing medical care expenses when population aging is progressing (Niki 2015).

Long-term care beds are regarded as a primary factor for medical care expenses per person remaining high. They account for most chronic phase hospital beds. With regard to long-term care beds, there is up to five times the regional difference in rates of estimated inpatients, including clinics with beds (number of patients per 100,000 population), even after adjusting for differences in gender and age composition between prefectures (Fig. 6). Consequently, correction of the differences is required, in particular.

4 Implementation of In-home Medical Care

In-home medical care, by which medical practice is performed in a patient's home, has been promoted. The primary political factor driving the trend is a shortage of care facilities, in addition to the shortened length of stay and bed reorganization, which were pointed out earlier. In-home medical care is becoming increasingly important when elderly people with chronic diseases or disorders are increasing because of population aging and epidemiologic transition.

In-home medical care is mainly conducted as home visit consultation in which a doctor visits a patient's home at regular intervals to perform the medical practice. It requires 24-hour support, including emergency home calls, in case the condition of the patient changes (Wada 2009). To support such in-home medical care in the

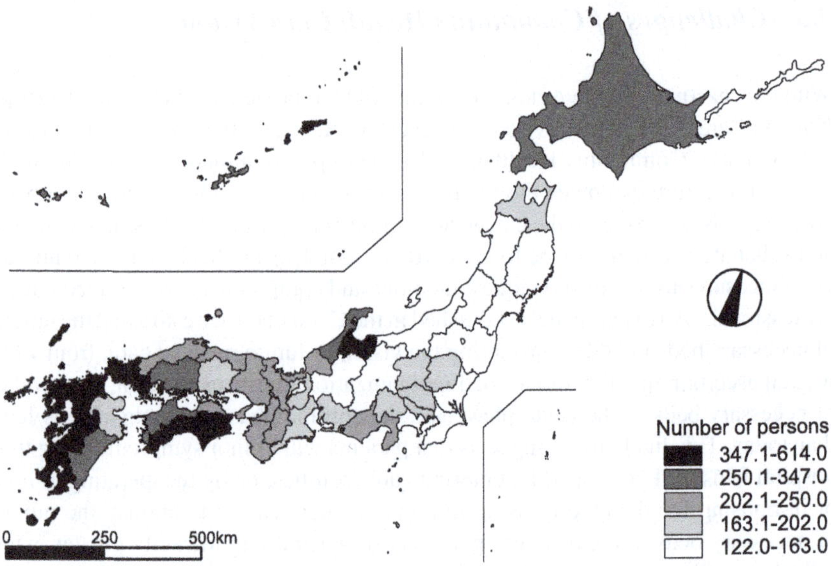

Fig. 6 Rates of estimated inpatients in long-term care beds adjusted by gender and age group (2011) (Created based on Hospital Report. Reprinted from Nakamura 2017b with permission of Akashi Shoten)

region with responsibility, home-care-supporting clinics and home care supporting hospitals have been established in Japan. Since 2012, home-care supporting clinics and hospitals that have met the following requirements[6] can be designated as the home-care-supporting clinic, or hospital, with enhanced functions[7]: to have three or more full-time doctors in charge of in-home medical care; and to have dealt with a certain number of emergency home calls and deathwatches during the prior year.

With in-home medical care, various medical care professionals provide services to patients at home under instructions from doctors. The professionals include: nurses from medical institutions and home-visit nursing stations, physical therapists and

[6]The home-care-supporting clinic is a system that was newly established in the medical fee scheme in 2006 and in which medical fees are set higher than those of other medical institutions. The conditions to be designated are several: (1) securing of a system for 24-hour communication, (2) 24-hour house calls and home-visit nursing, (3) securing of emergency admission beds, (4) provision of patient information to affiliated medical institutions and home-visit nursing stations, and (5) reporting of the number of deathwatch situations once a year. Actually, (2) and (3) can be dealt with by affiliated medical institutions or home-visit nursing stations. Requirements for home-care-supporting hospitals are the following: to be hospitals where the beds are fewer than 200 or for which no other clinic exists within 4 km; and to have functions equivalent to those of home-care-supporting clinics. However, only home-visit nursing can be dealt with by affiliated medical institutions. This system was newly established in 2008.

[7]Two types of home-care-supporting clinics and hospitals have enhanced functions: the independent type and the cooperative type. The latter can meet some requirements in cooperation with other medical institutions. Both of their medical fees are set higher than those of conventional home-care-supporting clinics and hospitals.

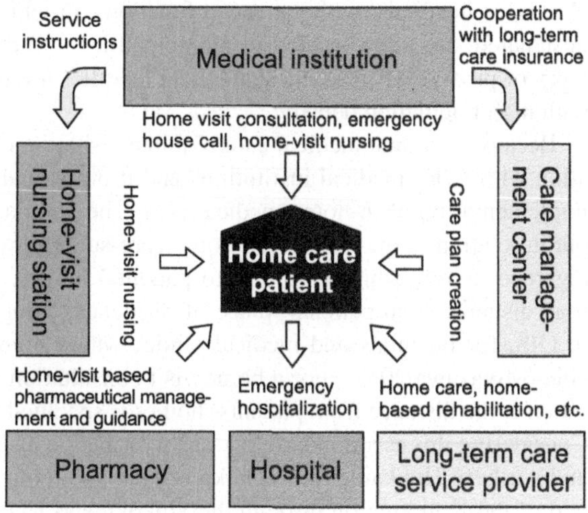

Fig. 7 System of in-home medical care (Reprinted from Miyazawa 2017 with permission of Akashi Shoten)

occupational therapists who support rehabilitation, and pharmacists from pharmacies who manage prescription and drug-taking. Furthermore, cooperation with nursing care professionals, such as helpers and care managers, is necessary in addition to cooperation with acute-phase hospitals in cases for which hospitalization is needed. Their interprofessional cooperation and networking of medical and nursing care resources in the region (Fig. 7) are necessary to promote in-home medical care, which is positioned as an element of the community-based integrated care system. The essence of in-home medical care is to support patients' quality of life. For that reason, the service system is obligated to secure the following: each element of home visit consultation, acute exacerbation admission beds, home-visit nursing, home-visit care, and housing of high quality. The elements mutually combine in an organic way, which supports patient life around the clock (Shimazaki 2013).

According to the 2014 Static Survey of Medical Institutions, as of September 2014, of 108,954 hospitals and medical clinics in Japan, about 40% implemented in-home medical care through medical care insurance and other insurance schemes, and about 10% implemented it through long-term care insurance. Home-care-supporting clinics were 14,188 facilities. Home-care-supporting hospitals were 1,016 facilities. Regarding dental clinics, 14,069 facilities out of 68,592 facilities implemented in-home medical care.

Regarding the implementation status of in-home medical care for each region, we specifically examine the number of home visit consultation implemented by hospitals and medical clinics, as well as the number of home-visit nursing implemented by home-visit nursing stations, per 1,000 population aged 65 years and over by prefecture. Both are more likely to be high in prefectures in western Japan and in prefectures with large metropolitan areas or large cities. A similar trend is apparent in home-visit dental consultation by dental clinics. Many home-care-supporting clinics and hospitals are located in municipalities in western Japan. Moreover, home-care-supporting clinics, or hospitals, with enhanced functions, in which high performance must be designated, are often located in municipalities in metropolitan areas and in urban areas of provincial areas: these medical facilities are likely to be located in regions with high population density.

Because in-home medical care requires visitation of patients' home and cooperation with other medical institutions and professionals, their proximity affects the implementation of in-home medical care. The population density that was pointed out in relation to the location of home-care-supporting clinics and hospitals can be regarded as reflecting proximity to patients' homes. Therefore, we measured the road distance from patients' place of residence, using a network analysis function in GIS, for hospitals and medical clinics where home visit consultation is available (Miyazawa 2017). For a home visit consultation to include in the medical fee scheme, the distance from patients' homes is specified as 16 km or less, in principle. Considering this as a standard, regions with a large distance are noticeable in eastern and northern Hokkaido, the Tohoku region, and mountainous areas in the southern part of the Kinki region through the Shikoku and Kyushu regions (Fig. 8). In practice, many medical institutions set the upper limit of the area of visit at 10 km each way, even home visits by car. It is especially difficult for small-scale clinics to make distant home visit consultation, in addition to outpatient consultation, and to respond 24 h a day because of human resource limitations. Consequently, cooperation among medical institutions is expected to deal with patients. However, in non-urban areas, few medical institutions are capable of dealing with home visit consultation within

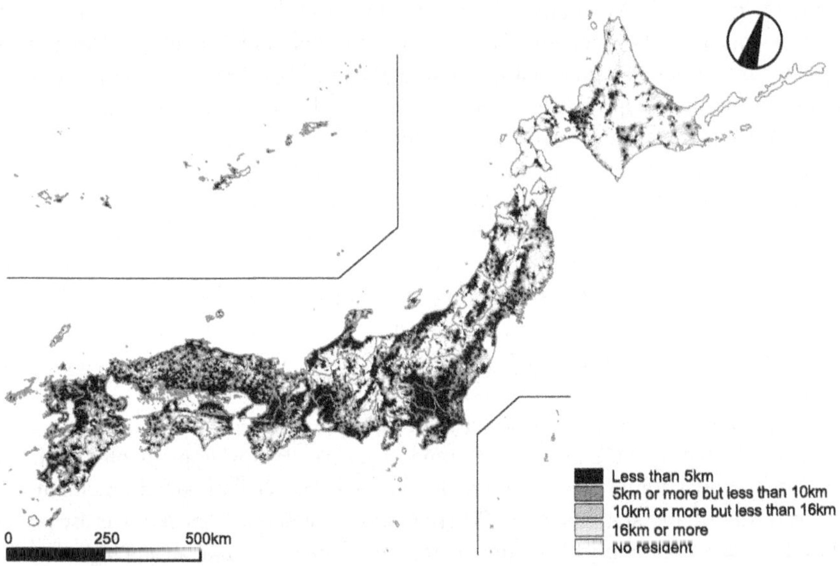

Fig. 8 Distance from home-visit consultation implementation medical facilities (Created based on SCUEL Data (medical institution database) by Mecompany Inc., Population Census of Japan, and Open Street Map. Reprinted from Miyazawa 2017 with permission of Akashi Shoten)

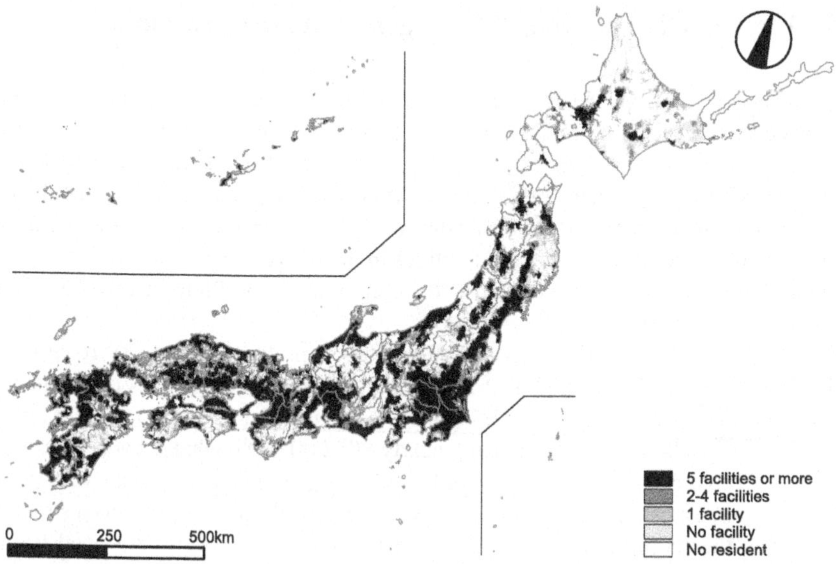

Fig. 9 Number of home-visit consultation implementation medical facilities located 10 km or less from a place of residence (Created based on SCUEL Data (medical institution database) by Mecompany Inc., Population Census of Japan, and Open Street Map. Reprinted from Miyazawa 2017 with permission of Akashi Shoten)

10 km of a patients' residence (Fig. 9). Furthermore, in cold regions, not only the deterioration of road conditions caused by snow during the winter constitutes an obstacle to visiting patients' home, but also users of home-visit nursing services decrease because patients admitted to the hospital or entering facilities increases seasonally, which engenders unstable management. The effects of the geographical conditions above are pointed out as the background by which in-home medical care is being only sluggishly pursued in eastern Japan and non-metropolitan areas.

The medical institutions implementing in-home medical care, especially home-care-supporting clinics and hospitals, have not increased as expected. Depopulated areas have difficulty in implementing in-home medical care against the background of geographical features, such as the remoteness and natural conditions. In urban areas, many medical institutions and professionals work on in-home medical care. However, the quantitative improvement of in-home medical care remains necessary because of the increase of elderly people and the absolute shortage of care facilities and housing facilities. Furthermore, in urban areas where there are many medical institutions and professionals involved in in-home medical care, when their policies and specialization mutually differ, time and labor are often required for the neces-sary coordination of cooperation among medical institutions and among different professionals.

5 Uneven Distribution of Emergency Medical Services

Medical care services that we have examined to date are premised upon services provided in normal conditions. However, in addition to those services, medical care includes emergency medical services that are provided in emergencies and during periods outside of ordinary office hours. After receiving acute phase medical care, sick and wounded patients often transfer to recovery phase rehabilitation wards, or move to affiliated hospitals or care facilities in the region. Consequently, emergency medical services are closely related to the community-based integrated care system (Nakamura et al. 2016).

According to "The 2015 Emergency Dispatches (preliminary report)" released by the Fire and Disaster Management Agency, the number of dispatches of ambulance cars (hereinafter designated as "ambulances") in Japan increased from 5,278,000 in 2005 to 6,051,000 in 2015, increasing nearly 800,000 during that decade. One point of background is an increasing demand for emergency medical services with population aging. The supply system for emergency medical services presents a structure similar to the central place system because this system has been developed from a perspective of how medical care services are supplied effectively within the limited space. However, because regional differences exist in the number of medical institutions as well as in organizational and financial strength of medical associations and local governments, the level of supply of emergency medical services is not uniform in spatial terms (Hayashi and Niimi 1998). Particularly, designated emergency hospitals per 100,000 population are few in metropolitan areas. However, a trend of "west high, east low" exists here as well. Inconsistent with the increase of emergency dispatches, an increasing number of hospitals and clinics are withdrawing from emergency services: more than 540 designated emergency hospitals have ceased supplying services the past decade or so.

Regarding annual emergency dispatches per 1,000 population by prefecture (Fig. 10), it is apparent that dispatches are frequent in the Tokyo metropolitan area and the Keihanshin (Kyoto–Osaka–Kobe) metropolitan area. Furthermore, a trend of "west high, east low" exists by which dispatches are infrequent in eastern Japan, particularly the Tohoku and Hokuriku regions, compared with western Japan. Based on the degree of the disease or injury severity of the person transported, it is readily apparent that nearly half of all persons transported have mild symptoms not requiring hospitalization. The percentage is high in metropolitan areas. Reasons for the frequent use of ambulance services are that the car ownership rate is low in large cities and because the proportions of elderly single and elderly couple households are high in the Kanto region through the west.

In ambulance transportation, the time from when a sick or wounded person is found and taken from one place to a hospital by ambulance to when the person is accepted by doctors greatly influences the person's survival rate and social reintegration rate. The national average of time required was 39.4 min. The longest was 51.8 min in Tokyo, which is extremely long compared with second-placed Saitama Prefecture: 45.5 min. In contrast, the shortest was 30.1 min in Fukuoka

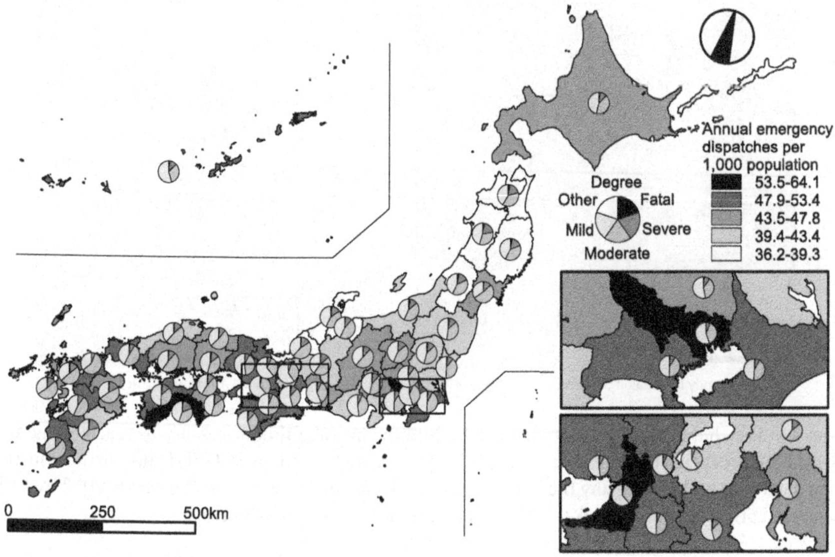

Fig. 10 Annual emergency dispatches and the proportions of persons transported by degree of disease or injury (2016) (Created based on Present State of Ambulances and Rescue)

Prefecture. Transportation time by ambulance is influenced by road traffic conditions and geographical distribution of fire stations and medical institutions in respective regions. We compared the average time between regions where a fire department covers a population of fewer than 50,000 people and where a fire department covers a population of 700,000 and more people. Results indicate that smaller populations have notably longer times before arriving at a medical institution. One reason is considered that although fire stations where ambulances are to be deployed are evenly distributed according to the population distribution, medical institutions are located only in the central parts of the regions where the population is small.

Then, using a network analysis function in GIS, we estimated reachable area by ambulance, which is defined by the range within which an ambulance can reach a "medical institution accepting emergency patients" within the specified time when moving on the road[8] (Hanaoka 2017). Here, we selected Tokyo as an example of large cities and Niigata as that of provincial cities, including rural areas (Fig. 11). Results revealed the following. In Tokyo, because medical institutions accepting emergency

[8]We originally define a "medical institution accepting emergency patients" as a medical institution capable of dealing with emergency patients requiring hospitalization based on the notification of the emergency medical service management addition and emergency admission addition. Assuming transportation by ambulance, the reachable area was estimated using the following as preconditions: the time zone is that when the volume of traffic is at peak hours on weekdays; the travelling speed set according to the type of roads is the same as that of regular vehicles; and there is no traffic regulation, such as one-way traffic and no turns, all the time. The time division in the reachable area is determined by the standard time at which the mortality rate exceeds 50% (3 min for cardiac arrest, 10 min for respiratory arrest) based on the Golden Hour Principle.

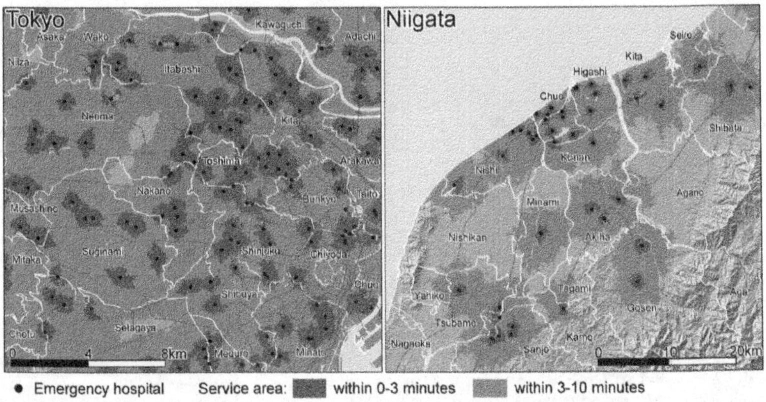

Fig. 11 Reachable areas by ambulance based on the average time necessary to reach a "medical institution accepting emergency patients" (2016) (Created based on SCUEL Data (medical institution database) by Mecompany Inc. and road network data in ArcGIS Data Collection 2012 by ESRI Japan. Reprinted from Hanaoka 2017 with permission of Akashi Shoten)

patients are densely distributed, the time is 10 min or less in almost all regions. However, doughnut-shaped "blank areas" exist in which the time required exceeds 10 min in some regions. In Niigata, except for the central part of Niigata City, it is more than 10 min in many regions. Comparing Tokyo with Niigata, one finds that distances that an ambulance can travel differ widely in the same 3 min: the former is about 500 m; the latter is about 1000 m. Those results indicate that the primary factor affecting the time required is road conditions in large cities, but it is the travel distance from medical institutions accepting emergency patients in provincial cities and rural areas.

Although ambulances have been improved across Japan to address increasing demand for emergency medical services, the decrease of emergency medical service institutions that are to be a receptor has been a major problem. Particularly, the ambulance transportation time must be shortened in large cities and provincial cities. In addition to maintaining the number of emergency medical service institutions, enhancing doctor cars and medical helicopters with doctors and nurses on board, for example, will engender shortening of the travel time, which is a bottleneck both in large cities and small cities.

6 Distribution of Pharmacies

6.1 Separation of Dispensing and Prescribing Functions and the Increase of Pharmacies

Traditionally in Japan, patients have received medicine dispensed inside medical institutions. In 1974, the prescription fee for doctors was raised. Consequently, issuing of legal prescriptions by medical institutions increased. Doctors issued legal prescriptions instead of giving medicine and pharmacists in health insurance pharmacies in the town came to hand medicine. Health insurance pharmacies are pharmacies that dispense medicine in accordance with prescriptions issued by doctors based on health insurance consultation.[9] Since the 1990s, the percentage of prescriptions for pharmacies accounting for the number of outpatient prescriptions (the separation rate of dispensing and prescribing functions) has been increasing gradually. Pharmacies that used to sell nonprescription medicine, cosmetics, and sanitary goods have changed the management style to a style that dispenses only medicine for medical care based on prescriptions as a receptor of legal prescriptions. Pharmacies are now securing the health and safety of residents through dispensing medications based on prescriptions, sales of nonprescription drugs, and efforts at in-home medical care.

Pharmacies have more than doubled in number, from 25,000 in 1972 to 58,000 in 2014. In recent years, retailers and wholesalers have entered dispensing pharmacy chains. The number of drugstores dealing with dispensing is increasing. Consequently, despite an increasing number of prescriptions, the number of prescriptions per pharmaceutical facility has been decreasing for several years. Currently, however, too many individual stores of small-scale operation exist. In actuality, the market shares of the top ten chain stores collectively account for only about 10% (Nakamura 2017c).

6.2 Regional Differences in Pharmacy Management Environments

The number of pharmacies per capita is high in the Chugoku, Shikoku, and Kyushu regions and on the Sea of Japan side of the Tohoku region (Fig. 12). The number of prescriptions per pharmacy is high in the regions of concentrated population in the Tokyo metropolitan area, where the number of pharmacies per capita is low, and in regions with a high separation rate of dispensing and prescribing functions. The

[9]In addition to pharmacies, drug sellers with a first-class license and household distributors exist as places selling medicines. Although both can sell nonprescription medicines, they cannot dispense medicine. Legally speaking, drugstores are divided into those with a license as a pharmacy and those with a license as a drug seller with a first-class license. Drug stores have a license as a drug seller with a first-class license.

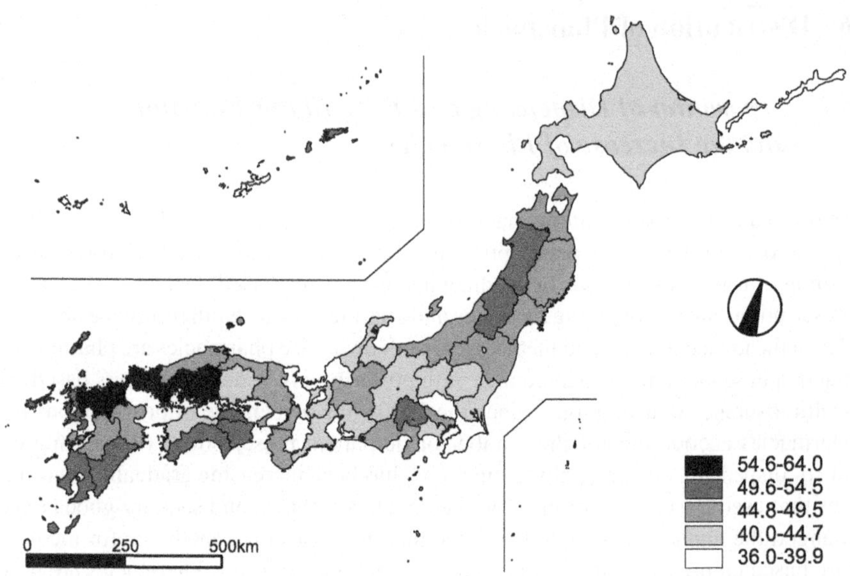

Fig. 12 Number of pharmacies per 100,000 population (2015) (Created based on Report on Public Health Administration and Services and Basic Resident Register in Japan. Reprinted from Nakamura 2017c with permission of Akashi Shoten)

management environments of pharmacies located in those regions are regarded as more beneficial than those of other regions.

Some municipalities in prefectures with isolated islands and remote areas only have medical institutions but have no pharmacy (Fig. 13). The number of municipalities with no pharmacy shows a declining trend. Some possible reasons include that the total number of pharmacies has increased; consequently, pharmacies have come to be located in municipalities with small populations. Moreover, by being merged into local governments with pharmacies, local governments that had been municipalities with no pharmacy were excluded from municipalities with no pharmacy. In municipalities with no pharmacy, in-house prescriptions in hospitals or clinics can be substituted for outpatients. Nevertheless, patients receiving in-home medical care have difficulty receiving guidance related to drug management from home-visiting pharmacists.

6.3 Change in the Positioning of Pharmacies and the Future Roles

When the main work is prescription dispensing, pharmacies try to obtain as many prescriptions as possible by being located next to large-scale medical institutions as pharmacies in front of hospitals. This location has made great contributions to stable

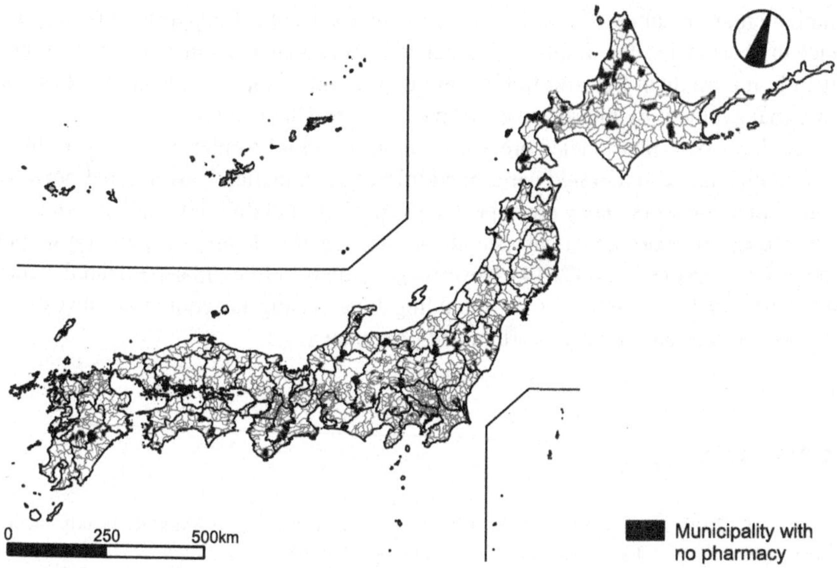

Fig. 13 Distribution of municipalities with no pharmacy (2015) (Created based on SCUEL Data (pharmacy database) by Mecompany Inc. Reprinted from Nakamura 2017c with permission of Akashi Shoten)

management. According to a pharmacy survey implemented by the Central Social Insurance Medical Council in 2015, pharmacies accepting prescriptions mainly from specific medical institutions account for approximately 70% of all.

With changes in social environments, such as the increase of population aging and the shift from acute diseases to chronic diseases, as a backdrop, the sites of medical care are shifting from wards and outpatient departments to home and care facilities. As a consequence, patients who do not, or who cannot, visit pharmacies are expected to increase. Therefore, the direction of conventional location strategy might change. For example, coupled with the increase of long-term medication, leftover medicine, such as drugs that patients forgot to take or which patients did not take as prescribed, are emerging in activities in pharmacists' in-home medical care.

The revision of medical fee in 2016 included the following as dispensing fee points: evaluation of family pharmacists who conduct patient compliance instruction after grasping patients' drug-taking status in a centralized and continuous manner in cooperation with prescribing doctors; evaluation of pharmacy systems and functions in which family pharmacists can play their role; the revision promoting proper evaluation of pharmaceutical management and in-home medical care according to the degree of contribution by family pharmacists and pharmacies, as well as proper use of medicines, including leftover medicine and double medication. However, the evaluation of large-scale pharmacies in front of hospitals was reconsidered. Chain store pharmacies with many large-scale pharmacies in front of hospitals are under pressure to make a transition from their existing business model. In fact, looking at the implementation status of home-visit based pharmaceutical management and guidance in pharmacies, the number of calculations related to Guidance for Management of In-home Medical Long-term Care in nursing care insurance has grown. In-home drug management by pharmacists has been progressing overall.

As described, pharmacies are increasingly required to play a role as a hub of community-based integrated care, in addition to conducting conventional prescription dispensing work. They are expected to check the condition of patients who used drugs based on pharmaceutical specialization, such as in-house recuperation support and primary care (Hazama 2014). Nevertheless, many hurdles must be cleared to meet standards for family pharmacies, including fund-raising for equipment investment and securing of chronically shorthanded pharmacists.

References

Hanaoka K (2017) Emergency medical services. In: Miyazawa H (ed) Mapping health, medical care, and welfare in Japan. Akashi Shoten, Tokyo (in Japanese)

Hayashi N, Niimi Y (1998) Spatial supply system of emergency medical service in Aichi Prefecture. Annals of the Japan Association of Economic Geographers 44:165–186 (in Japanese with English abstract)

Hazama K (2014) Pharmacy changes, regional medical care changes: In-home medical care innovation starting from the collaboration of doctors and pharmacists. Jiho, Tokyo (in Japanese)

Ikegami N (2017) Medical care and nursing care in Japan: history, structure, and the direction of reformation. Nikkei Publishing, Tokyo (in Japanese)

Ikegami N, Campbell JC (1996) Medical care in Japan: control and a sense of balance. Chuokoronsha, Tokyo (in Japanese)

Inoue Y (2010) Current meaning of regulation of numbers of hospital-beds: Consideration from the view points of competition policy among medical service suppliers and regional sovereignty. Yokohama Law Rev 18(3):1–26 (in Japanese)

Jiho (2001) Yakuji handbook 2001. Jiho, Tokyo (in Japanese)

Jiho (2006) Yakuji handbook 2006. Jiho, Tokyo (in Japanese)

Mano T (2012) Introduction to medical policies: Who determines, what is aimed for. Chuokoron-Shinsha, Tokyo (in Japanese)

Matsuda S (2015) How to create a community health care vision. Igaku-Shoin, Tokyo (in Japanese)

Miyazawa H (2017) In-home medical care. In: Miyazawa H (ed) Mapping health, medical care, and welfare in Japan. Akashi Shoten, Tokyo (in Japanese)

Morita A (2016) Politics of government councils III. Jigakusha Publishing, Tokyo (in Japanese)

Nakamura S, Miyake Y, Aruga T (2016) Community-based integrated care system and emergency medical care. J Jpn Council Traffic Sci 15(3):20–28 (in Japanese with English abstract)

Nakamura T (2017a) Medical institutions. In: Miyazawa H (ed) Mapping health, medical care, and welfare in Japan. Akashi Shoten, Tokyo (in Japanese)

Nakamura T (2017b) In-patient beds. In: Miyazawa H (ed) Mapping health, medical care, and welfare in Japan. Akashi Shoten, Tokyo (in Japanese)

Nakamura T (2017c) Phamacies. In: Miyazawa H (ed) Mapping health, medical care, and welfare in Japan. Akashi Shoten, Tokyo (in Japanese)

Niki R (2015) Community-based integrated care and regional medical care cooperation. Keiso Shob, Tokyo (in Japanese)

Shimazaki K (2011) Health care in Japan: institutions and policies. University of Tokyo Press, Tokyo (in Japanese)

Shimazaki K (2013) Current situations, ideas, and challenges of in-home medical care. In: Nishimura S, National Institute of Population and Social Security Research (eds) Community-based integrated care system: aiming at an "aging in a familiar community" society. Keio University Press, Tokyo (in Japanese)
Wada T (2009) Introduction to in-home medical care clinic. Nanzando, Tokyo (in Japanese)
Yakuji Jiho-Sya (1999) Yakuji handbook 1999. Yakuji Jiho-Sya, Tokyo (in Japanese)

Securing of Health, Medical, and Welfare Personnel and the Geographical Distribution in Japan

Hiroyasu Kamo and Akihito Nakajo

Abstract This chapter discusses the development and securing of personnel in the health, medical, and welfare fields in Japan and the characteristics of the related geographical distribution. The number of people employed in the fields has been increasing since 1985. Particularly, employment in the welfare field has been increasing rapidly since the Long-Term Care Insurance Act became effective in 2000. Nonetheless, a workforce that can meet increasing demands for labor is not sufficient: nursery teachers and caregivers are in particularly short supply in metropolitan areas. Furthermore, growing demand exists for labor quality to cope with the advancement of medical care and to address complicated and diverse needs. To this end, the duration of basic education for public health nurses has been extended. The total capacity of universities for training nurses is increasing mainly in metropolitan areas. Personnel from foreign countries have been accepted in nursing and long-term care. In 2008, Japan started accepting foreign candidates for nurses and certified care workers under an Economic Partnership Agreement. In 2017, Care Worker was added to the job categories in Technical Intern Training Program and the visa status "Care" was newly established for foreign students who have acquired qualifications as a certified care worker. Residents and their organizations in local communities other than professions are regarded as important as supporters in the welfare field. It has been pointed out that commissioned welfare volunteers, one category of supporter, has difficulties such as the failure to fill the quota and regional differences in the rate of filled vacancies. However, NPOs, which are the main actors of community welfare activities, are likely to concentrate in metropolitan areas and prefectural capital cities.

Keywords Community welfare activity · Human resources · Profession · Securing of personnel · Training

H. Kamo (✉)
Nihon Fukushi University, Tokai, Japan
e-mail: kamo@n-fukushi.ac.jp

A. Nakajo
Shizuoka University, Shizuoka, Japan

H. Miyazawa and T. Hatakeyama (eds.), *Community-Based Integrated Care and the Inclusive Society*, International Perspectives in Geography 12,
https://doi.org/10.1007/978-981-33-4473-0_5

1 Securing Personnel in Other Countries

When developing personnel in the health, medical, and welfare fields, Japan used systems in place in the United States and European countries as a reference. For example, it instituted the medical system in 1874 on the model of Germany and established medical education based on Western medicine (Sakai 2019: 508). After World War II, medical and nursing education was reformed under the direction of the GHQ (Shimazaki 2011: 52). The system of certified social workers was developed after the war using the theory and practice of social work in the United States and the United Kingdom (Kuga 2009: 57).

At present, the United States and European countries are confronting challenges of securing personnel in the health, medical, and welfare fields. Not only are facilities short-staffed in those countries: differences have also arisen in the distribution of human resources within the respective countries. Mori and Goto (2012: 185) report that the number of doctors increased during the 1990s and the early 2000s in the United States and the United Kingdom, which caused a gap separating urban areas and non-urban areas in the number of doctors per capita. In Germany, doctors are in short supply, particularly in rural areas in former East Germany. Doctors want to work in urban areas because of the convenience of life and a better educational environment for their children, an environment suitable for brushing up their specialty, and the presence of sophisticated medical equipment. To solve the uneven distribution of doctors, Norway and Sweden implement a policy that assigns priority to students from the provinces to enter the faculty of medicine, whereas the United States, Canada, and Australia operate a scholarship system on the condition of acceptance of work in remote areas (Mori and Goto 2012: 186). Germany and the United Kingdom are reviewing the division of roles, such as reducing doctors' burdens by extending the authority of nurses (Shirase 2015: 313).

Developed countries including the United States, Canada, and New Zealand, are short of nurses because the number of young people who desire to become nurses is decreasing and because of other reasons. Some countries that are short of nurses choose to employ foreign nurses as a countermeasure. Nurses' movements between member states and those from Africa have been confirmed in EU states and those from the Philippines in the United States (Kingma 2006). Time, energy, and money to spend on retaking a qualification or mastering a language in a foreign country are obstacles for nurses to emigrate. Even in cases where the mother tongue is the same, the accents or colloquial expressions might differ. Consequently, nurses are required to understand the patient's language in cases of life and death.

In the United States and European countries with advanced professionalization of welfare ahead of others, the social worker is positioned as a welfare professional. The United Kingdom enacted the Local Authority Social Service Act in 1970 and developed a qualification system for social workers (Editorial Committee of Social Welfare from Basics 2009: 167). By contrast, a care worker is a person employed by healthcare, medical care, or welfare providers without a specialized occupational qualification. The nurse's aide, care aide, home helper, and nursery assistant teacher fall under this category. Mitomi (2005) remarks that although it is difficult to arrive at an exact value because the scope of care workers is unclear, the percentage of care workers accounting for the total number of employees in 1999 was 4.8% in the United Kingdom and 3.4% in the United States. In the United States, 80–91% of "nurse's aides, etc." and 95–100% of "nursery teachers" are women, and 43–51% of "nurse's aides, etc." and 37% of "nursery teachers" are persons of races other than white. The country has a high turnover rate for care workers and a low rate of filled vacancies for care workers. Reasons include the low wage level for the occupation and the heavy physical and mental burden. In the case of the United States, regulations on occupational qualifications and support for education and training differ from one state to another. Reports describe that the turnover rate of care workers dropped in Wisconsin and Massachusetts as a result of subsidized education and training (Mitomi 2005).

Personnel shortages in the health, medical, and welfare fields are not limited to the United States and European countries. In some Asian countries, labor demand in related fields is growing with increasing population aging and women's participation in society. China began to address personnel development in the welfare field after the late 1980s when the former Soviet Union became less influential. It was after 2002, when the National Occupational Standards for Aged Care Workers was established, that the development of care workers went into full swing. Furthermore, it was after 2006, when laws and regulations related to the National Qualification Test for Social Welfare Professionals were promulgated, that the development of social workers went into high gear. Shimizu (2015) argues that although the establishment of the "Elderly Service and Management major" in higher vocational schools to develop managers at care sites was confirmed in 17 schools in 2013 throughout China, they are unevenly located: three are in Beijing; another three are in Nanjing. At present, most helpers with the qualification of aged care helper are migrant workers from rural areas and unemployed persons. They form the main force at care sites (Shen 2014: 198). In care centers in China run by Japanese providers, Chinese staff have been educated on Japanese care technology (Kaku 2018).

Based on the explanation above, we expound how personnel are developed in Japan and what measures are taken to secure necessary personnel, as well as what characteristics are found in the distribution of human resources. This chapter takes up doctors, nurses, public health nurses, certified social workers, care staff, and nursery teachers as professions and commissioned welfare volunteer, NPO, and voluntary organizations as actors other than professions.

2 Changes in Systems Related to Human Resources in Health and Medical Care and Welfare in Japan

2.1 Professions in the Health and Medical Care Field

After the war, the medical system in Japan was reformed under the control of the GHQ. As a result, Japan achieved universal health insurance coverage in 1961. Changes occurred after the achievement of universal health insurance coverage, including an increase in the consultation rate, free medical care for elderly people, and a rise in medical fees. This caused medical expenses in Japan to shoot through the roof. In addition, because of the end of high economic growth, population aging, and other reasons, Japan was compelled to reform the medical system. In 1983, the health service system for elderly people was enacted, which accelerated the correction of unnecessary long-term hospitalization and the shift of medical care to community and home care. Since the mid-1990s, with the progress of population aging and the deterioration in the finances of medical insurance systems, the medical system has been reformed frequently to date.

Changes in systems related to human resources are linked to the movement. Requirements for obtaining a medical practitioner's license and qualifications for taking a National Examination for Medical Practitioners were determined in 1946 (Nichigai Associates 2013:119; Shimazaki 2011: 73). In the revision of the Act on Public Health Nurses, Midwives, and Nurses in 1951, Assistant Nurses were introduced to make up for a shortage of nurses. The period of education was set at two years after graduating from junior high school. Since the 1960s, every time a shortage of nurses has become an issue, a prospective supply of nurses[1] has been presented and measures have been taken (Nomura 2015). In 2000, the Long-Term Care Insurance Act was enacted and the Medical Care Act was revised, in which the nursing staffing standards were raised. After the revision of medical fees in 2006, a movement was underway to increase the number of nursing staff members at acute care hospitals.

The development of physical and occupational therapists as rehabilitation professions was advanced with the promulgation of the Physical Therapists and Occupational Therapists Act in 1965 as a turning point. Since the 2000s, the places of employment have expanded from medical facilities to health services facilities for elderly people, day care service facilities, home-visit nursing stations, and sports-related facilities.

In 1982, the Health and Medical Service Law for the Elderly was established. Visits of nursing professionals to home care patients were established under a law for the first time. Subsequently, in line with the revision of the law in 1991, a home health system for elderly people was established. A home-visit nursing station that provides services is to place 2.5 or more nursing staff (full-time equivalent) and can place physical, occupational, and speech therapists and other professions. According to a "Survey of Institutions and Establishments for Long-term Care" conducted by the

[1] "Nursing staff" refers to public health nurses, midwives, nurses, and assistant nurses.

Ministry of Health, Labour and Welfare, the number of regular-worker-equivalent workers at home-visit nursing stations was nearly tripled from 22,302 in 2000 to 66,060 in 2017. However, some home-visit nursing stations suspend or abolish their operations because they cannot secure an adequate number of staff members (Kiyosaki 2018: 17).

The placement of municipal public health nurses started with the establishment of 39 health centers based on the enactment of the Health Center Law in 1937. Furthermore, the Community Health Act that revised the Health Center Law was enacted in 1994 leading to the creation of a new system for community health measures in which municipalities provide familiar and frequent health and welfare services while health centers provide specialized health services, including measures for infectious diseases, mental health, and intractable diseases as well as prevention of child abuse. In the operations of the comprehensive community support center introduced in 2006, public health nurses are instituted as the main provider in preventive care management (see Chapter "Regional Variation in the Community-based Integrated Care Systems in Japan"). In the comprehensive support center for childrearing generation that has been established in municipalities since 2017 with the revision of the Maternal and Child Health Act, public health nurses and other professions have been placed. They are regarded as actors supporting and managing pregnancy, childbirth, and parenting who understand the user's perspective while using their expertise (see Chapter "Establishing Community-Based Integrated Support Systems for Pregnancy, Childbirth, and Childcare in Japan: Focusing on Regional Differences").

With the revision of various systems, the number of staff members is increasing. However, emphasis is increasingly put on the advancement of medical care, medical safety measures, and the improvement of service quality. In fact, securing personnel who can advance these areas of competence has become an important issue.

2.2 *Professions in the Welfare Field*

The present social welfare professional system was developed during the post-war period when full-scale social welfare-related legislation came into effect. The requirements for qualifications of child welfare officers and child guidance center's directors were determined in the Child Welfare Act of 1947. Those of day nurse, child counselor, housemother, and other professions were determined in the Minimum Standards for Child Welfare Institutions of 1949. Consequently, these professions began to develop. The Act on Appointment of Social Welfare Officers was enacted in 1950. The development of social welfare officers also started. The social welfare officer qualification is necessary for those engaged in duties, such as public assistance, in welfare offices. Even today, the appointment of social welfare officer is prescribed in the Social Welfare Act.

After the period of high economic growth, Japanese society changed. The extended family is becoming a nuclear family. The population is becoming older. These changes have made people's difficulties in life more complicated and have promoted the development of social welfare facilities, such as nursery centers and homes for elderly people. As a consequence, a higher level of specialization has come to be required of persons engaged in the work (Kuga 2009: 25). To guarantee human resources as a national qualification for social welfare professions systematically, the Certified Social Worker and Certified Care Worker Act was promulgated in 1987. The Gold Plan formulated in 1989 presented specific goals in welfare measures for elderly people over the next 10 years, including great increases in the capacity of intensive nursing homes for elderly people and to increase the number of home helpers. Following that, the New Gold Plan, the Angel Plan, and the Disabled Plan were formulated, which further promoted the development and securing of personnel involved in social welfare. For example, the national qualification for nursery teachers was introduced.

In the field of welfare, certified social worker, psychiatric social worker, certified care worker, and nursery teacher are legislated professional qualifications. Professional qualifications equivalent to the above include long-term care support specialist, social welfare officer that is a qualification to be appointed, and child welfare officer. Long-term care support specialist is a professional qualification for care managers who provide consultation with elderly people in need of care or assistance and their families, assist the use of nursing care services, and manage long-term care benefits. When a person who holds a qualification as a doctor, nurse, certified social worker, or certified care worker and also has practical experience of the profession for at least five years has passed the Long-Term Care Support Specialist Examination and completed the training program, the person obtains a qualification as a long-term care support specialist. Furthermore, when a person who has practical experience of at least five years as a long-term care support specialist has completed the prescribed training, the person obtains the qualification of senior long-term care support specialist. Senior long-term care support specialists are placed in comprehensive community support centers to build a cooperative system with medical institutions, operate care management centers, and support long-term care support specialists.

2.3 Roles of the Resident Sector in the Welfare Field

In the field of welfare, specific emphasis is placed on residents and their organizations as supporters other than professionals in local communities. It has been pointed out that points of fragility exist in the specialization and continuity because they are not included in the framework of public welfare systems. However, the primary response to sudden illness and injury depends strongly on local residents. A function exists in a local community that prevents circumstances from worsening by finding a welfare issue when it is still small and by dealing with it at an early stage. Some advantages are pointed out for cases in which residents provide welfare support by forming an organization for welfare activities. They can develop support based on urgency and awareness. Emotional stability and social solidarity of residents' activities are achieved through the activities. Responses through residents' participation, such as the giving and receiving of support among residents, are effective. Those who play a central role in developing ties among residents in a local community are commissioned welfare volunteers and also community welfare activities in which residents participate.

A commissioned welfare volunteer is called a "commissioned welfare volunteer or child welfare volunteer" or a "commissioned child welfare volunteer" based on the Commissioned Welfare Volunteers Act, concurrently serving as a child welfare volunteer (hereinafter, a "commissioned welfare volunteer"). The idea and work of commissioned welfare volunteers are, as stipulated in the Commissioned Welfare Volunteers Act, to carry out overall activities related to welfare. Commissioned welfare volunteers have played a role in investigating the livelihood or poverty status of local residents and have supported them since the pre-war *Saisei komon seido* (social reform advisory system) period. After the war, in the course of which laws related to social welfare were being developed, changes in industrial structure caused a drift of population to large cities, which exacerbated the trend toward the nuclear family. Consequently, the caring and nurturing capacity of the local community has waned, leading to the emergence of changes in welfare needs and eligible persons. Commissioned welfare volunteers have worked in a position to support those changes as a member of local residents.

Residents participating in community welfare activities have attracted attention since the beginning of the twenty-first century as actors for welfare in local communities. Community welfare activities set a goal to solve welfare issues without people disrupting their rhythm of daily life and leaving the community where they have lived for a long time. This is regarded as important also in community-based integrated care systems, which are required to find and address welfare issues and difficulties that arise in the community: the resident sector is expected to play a role in grasping the information and giving daily support. Residents participating in community welfare activities have tried to stabilize and vitalize activities with the enactment of the Act on Promotion of Specified Non-profit Activities (the Act on NPOs) in 1998 and the Long-Term Care Insurance Act in 2000 as a turning point. For example, they acquire NPO corporate status and enter the long-term care insurance business. In urban areas, residents participating in community welfare activities have developed more in suburban areas where population decline and aging have progressed to a greater degree than in the city center as of now. The implemented activities include providing a place for elderly people and watching over them as well as supporting mothers as they undertake parenting. The characteristics are that they conduct activities to complement social relationships between local residents who create support in areas with a trend toward the nuclear family and the special dispersion of the family have progressed and community functions are weak. Furthermore, in hilly and mountainous areas, community welfare activities such as these are incorporated as important functions of the "region management organization (RMO)" that has been promoted in governmental measures for overcoming population decline and vitalizing the local economy in Japan. With increasingly diverse residents' needs, supplying effective and efficient services solely by the public sector, such as administration is difficult. Consequently, various bodies, including administration and local residents' organizations, must take on the responsibility in a collaborative manner.

3 Health and Medical Care-Related and Welfare-Related Employed Persons and Employment

According to a Population Census of Japan, the number of persons employed in occupations related to health and medical care and welfare is on an increasing trend. The total number of people engaged in health and medical care and those in social welfare professions doubled, from 1,980,000 in 1985 to 3,820,000 in 2015. Persons employed in the field of health and medical care, such as doctors, public health nurses, and nurses steadily increased in the period of 1985–2015 (Fig. 1). After 2000 when the Long-Term Care Insurance Act was enacted, the increase in employed persons is noticeable in the welfare field. In the period between 2000 and 2015, persons employed in other social welfare professions increased from 210,000 to 470,000 (2.2 times) and care staff (in medical and welfare facilities, etc.) increased from 360,000 to 1,260,000 (3.5 times).

One difficulty associated with healthcare, medical care, and welfare service industries is a shortage of labor. Employment methods most often used in the service industries are the following: employment placement through public employment security offices; free employment placement for welfare-related professions in welfare labor centers and welfare labor banks run by the social welfare council in each region; and

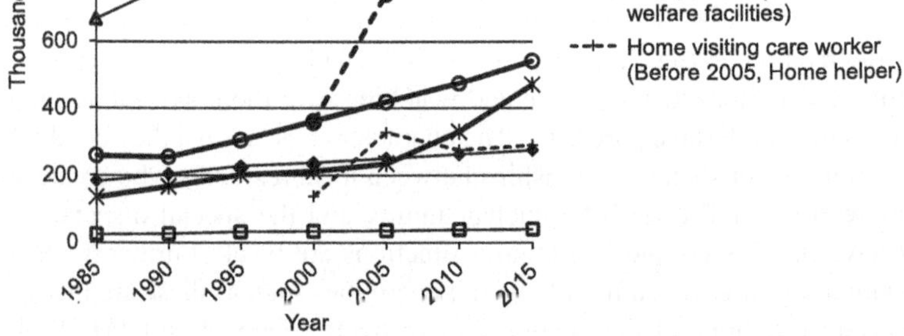

Fig. 1 Changes in the number of employed persons by profession (Created based onPopulation Census)

fee-charging employment placement. Looking at the status of employment placement by public employment security offices using a "Report on Employment Service" released by the Ministry of Health, Labour and Welfare, the active job openings to applicants ratio greatly exceeds 1 for all occupations in healthcare, medical care, and welfare. The active job openings to applicants ratio for "doctors, dentists, veterinarians, and pharmacists" was 5.54 in 2017. That for "public health nurses, midwives, and nurses" was 2.32. That for "professions in social welfare" was 2.65. That for "occupations in care services" was 3.67. Also in welfare labor centers and welfare labor banks, the active job openings to applicants ratio for all occupations in the welfare field (the total number) in 2016 is 4.32, which is very high.[2]

4 Geographical Distribution of Professions in the Health and Medical Care Field

4.1 Doctor

Since 2000, the number of doctors has been increasing in Japan. A Survey of Physicians, Dentists and Pharmacists shows that doctors working in medical facilities increased from 243,201 to 304,759 during 2000–2016. The number of doctors increased even in the pediatrics department and the obstetrics and gynecology department, where there is concern about a shortage of doctors. The number of doctors working in medical facilities increased from 14,156 to 16,937 in the department of pediatrics and from 11,059 to 11,349 in the department of obstetrics and gynecology (including obstetrics) in the same period. However, the rate of increase varies depending on the department; 25% for the total number of doctors, 20% for the pediatrics department, and 3% for the obstetrics and gynecology department. Some hospitals cannot maintain a pediatrics department, which has high demand for consultation after office hours, or an obstetrics department, which has high risks in birthing because of a shortage of doctors (Tsuji 2008: 142).

Wide gaps exist in the numbers of doctors among regions. A trend exists by which the number of doctors per 100,000 population is high in cities where the faculties of medicine and medical colleges are located and prefectures in western Japan; the number is quite low in suburbs in metropolitan areas and prefectures in eastern Japan (Kamiya 2017: 170). Prefectures with a low number of doctors working in medical facilities per 100,000 population are Saitama (160.1 people), Ibaraki (180.4 people), Chiba (189.9 people), and prefectures in the Tokyo metropolitan area, in ascending order (Fig. 2). However, the rate of increase in doctors (based on the place of work) during 2000 and 2016 was 44% for Saitama, 30% for Ibaraki, and 46% for Chiba.

[2] According to a Performance Report on Employment Placement (Japan National Council of Social Welfare National Center for Social Service Human Resources 2018), new job openings were 314,202 in 2016. The number accounts for 39% of the 799,234 new job openings for social insurance, social welfare, and care service in public employment security offices.

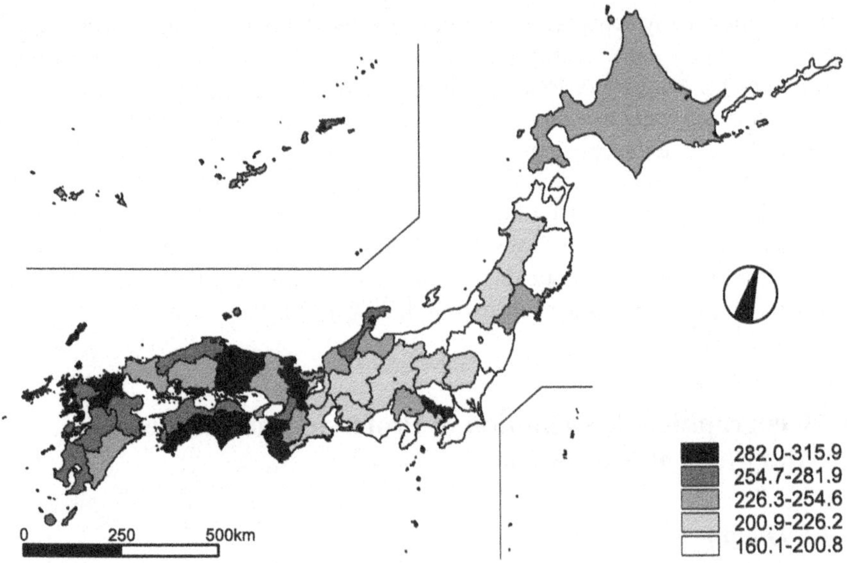

Fig. 2 Number of doctors working in medical facilities per 100,000 population (based on the main place of work, by prefecture in 2016) (Created based on Survey of Physicians, Dentists and Pharmacists)

They all exceed the national average of 25% and show a trend of approaching the national average for the number of doctors per 100,000 population. Mori and Goto (2012: 38) point out the reason for regional differences in the distribution of doctors as follows. In western Japan, where many faculties of medicine were established in the early postwar years, the supply of doctors is high. However, in prefectures neighboring Tokyo, where not many faculties of medicine were established, the number of doctors per capita is small because the population increased rapidly after the period of high economic growth. One point of the background is the difficulty in establishing the faculty of medicine flexibly according to demographic changes.

Although doctors in the obstetrics and gynecology department increased by 3% in Japan during 2000–2016, the degrees of change differ among regions. According to a Survey of Physicians, Dentists and Pharmacists, the number of doctors working in obstetrics and gynecology department in medical facilities increased in 15 prefectures and decreased in 32 prefectures in the period. Prefectures in which the number of doctors in the obstetrics and gynecology department increased were mainly those in metropolitan areas; the number decreased in many of the prefectures located in non-metropolitan areas. Prefectures with a high rate of decline in the number of doctors working in obstetrics and gynecology departments in medical facilities are Oita, Kochi, and Fukushima in descending order and declined from 110 to 80, from 67 to 49, and from 158 to 119, respectively, during that period.

4.2 Nurse

The environment surrounding nurses is changing toward high-quality nursing services capable of dealing with an aging population with a declining birthrate and the advancement of medical care. Consequently, changes have occurred in the composition of training schools for nurses. One change is the increase in the number of universities training nurses. A "Survey on Admissions into Nurse Schools and Work Statuses of Graduates" conducted by the Ministry of Health, Labour and Welfare reported that the number of universities training nurses in Japan increased from 158 to 267 during 2007–2017. With the increase in the number of universities, the percentage of university students accounting for the number of people who have passed the National Nursing Examination is also growing. In 2017, university students accounted for 34.9% of new graduates who passed the examination. An analysis by prefecture shows that the total capacity of universities to train nurses is expanding mainly in metropolitan areas. According to a "Survey on Admissions into Nurse Schools and Work Statuses of Graduates" by the Ministry of Health, Labour and Welfare, during 2007–2017, the total capacity expanded from 320 to 1,300 (4.1 times) in Nara Prefecture, from 280 to 1,040 (3.7 times) in Tokushima Prefecture, from 1,120 to 3,900 (3.5 times) in Saitama Prefecture, and from 1,600 to 5,200 (3.3 times) in Osaka Prefecture. However, prefectures with a low rate of increase are located in Kyushu and Okinawa prefectures. All eight prefectures have a rate below the national average rate of increase (1.8 times).

Another change is the decrease of assistant nurse training schools. The number of assistant nurse training schools in Japan reached a record high of 776 in 1970, but it has been declining since, falling to 231 in 2017. Of 47 prefectures, Fukui and Okinawa have no assistant nurse training school. According to Fig. 3, showing the percentage of assistant nurses, prefectures with a low percentage of assistant nurses are Tokyo (11.4%), Shiga (12.0%), and Kanagawa (12.5%) in ascending order. This order implies that the downsizing of assistant nurse training is occurring in metropolitan areas ahead of others. In contrast, prefectures with a high percentage of assistant nurses are Miyazaki (32.5%), Kumamoto (31.2%), and Saga (31.0%) in descending order. Seven prefectures in Kyushu have not only a high percentage of assistant nurses; they also have numerous assistant nurse training facilities. The number is 53 in 2017, accounting for 22.9% of the total number in Japan ("Survey on Admissions into Nurse Schools and Work Statuses of Graduates" by the Ministry of Health, Labour and Welfare). The Compilation Committee for the 60-year History of the Act on Public Health Nurses, Midwives, and Nurses (2009) points out part of the background of the shift from assistant nurse to nurse as follows. With an increasing percentage of women continuing to higher education and a growing orientation toward professions as well as advancing medical care, expectations have been raised for highly skilled nursing professions. In addition, the long-standing movement for abolishing the assistant nurse system in the nursing community influences the shift. In 1996, the Ministry of Health and Welfare proposed in the "Report of the Consensus Conference for Issues Regarding the Assistant Nurse System" (Ministry

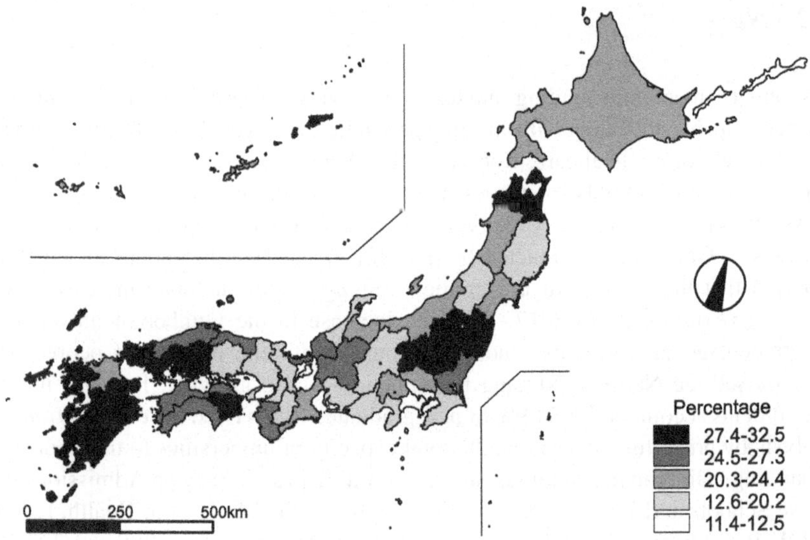

Fig. 3 Percentage of assistant nurses (by prefecture in 2016). These are percentages of assistant nurses accounting for the total number of nurses and assistant nurses (Created based on Report on Public Health Administration and Services)

of Health and Welfare 1996) that action should be taken toward the integration of nurse training systems at an early stage of the twenty-first century. In fact, the route from assistant nurse to nurse has expanded through correspondence courses for assistant nurses with practical experience, transfer from technical college to university, and with integrated education at high school. At the same time, there are opinions against abolishment of the system. Therefore, assistant nurse training schools continue to exist as of 2018.

4.3 Retirement of Nursing Staff and Preventive Measures

One reason that a shortage of nursing staff occurs is the high rate of turnover. The Japanese Nursing Association (2018) reports that the national average turnover rate for full-time nursing staff was 10.9% in 2016. In a prefecture-by-prefecture analysis, the turnover rate is low in the Tohoku and Hokuriku regions, but it is high in metropolitan areas such as Kanagawa Prefecture (14.7%), Tokyo Metropolis (13.8%), and Osaka Prefecture (13.4%) (Fig. 4). According to the Ministry of Health, Labour and Welfare (2011), the main reasons for leaving a job are childbirth and childcare, marriage, interest in other facilities, personal relationships in the workplace, and dissatisfaction with working hours, holidays, and wages. The survey shows that there are issues of the working environment and harsh conditions at work underlying the reasons. The former include the following: the burden of working night shifts

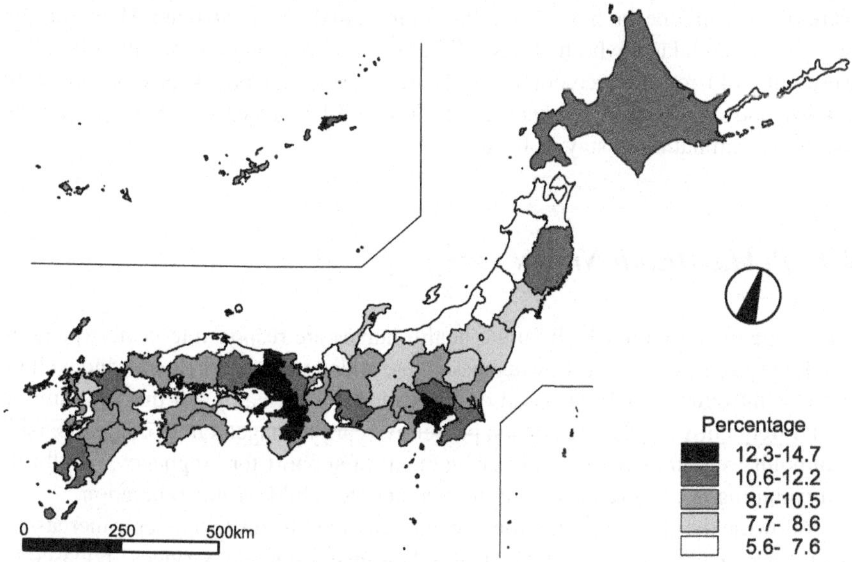

Fig. 4 Turnover rate of full-time nursing staff (by prefecture in 2016). These are percentages of assistant nurses accounting for the total number of nurses and assistant nurses (Created based on Japanese Nursing Association 2017)

is heavy; there is too much overtime work; and it is difficult to balance work with parenting and housework. The latter include the following: it is difficult to take sufficient rest under irregular shift work; and it is impossible to leave a child in someone's care. Dissatisfaction with wages is widespread, mainly among young nursing staff. According to the Japanese Nursing Association (2010), the percentage of nursing staff who answered that they had thought about leaving their job because of the low wages is 61.2% overall and 70.3% for those in their late 20s. In areas where the job opening to applicant ratio for nursing professions is high, if the working conditions are unsuitable for themselves, then it is easy for the nursing staff to leave the job and take a job at another medical institution. Han (2012) reports the following. Implementation of 7:1 nursing staffing standards in 2006 created a growing demand for nurses. At that time, large-scale hospitals in urban areas where working conditions were good actively acquired nurses, which made it difficult for small and midsize hospitals and hospitals in rural areas to secure nurses.

The acceptance of foreign candidates for nurses and certified care workers under an Economic Partnership Agreement (EPA) started in 2008 to strengthen bilateral economic partnership. In the acceptance program for EPA foreign nurse candidates, accepting organizations can employ foreign workers as employees or trainees during the period until they take the National Nursing Examination and can employ them as nurses after they have passed the national examination. As of 2018, three countries have sent candidates. The acceptance started in 2008 in Indonesia, 2009 in the Philippines, and 2014 in Vietnam. The total numbers of candidates accepted in 2017 were 622 for Indonesia, 506 for the Philippines, and 75 for Vietnam. However, the number of candidates who had passed the National Nursing Examination by 2017 were only 344 for all three countries. Therefore, the challenges have been how to employ and train candidates and how to induce EPA nurses who have passed the national examination to stay in Japan.

4.4 *Public Health Nurses*

The scope of work for which public health nurses are responsible is an expanding trend. In recent years, the following tasks were added to the work: health and welfare services introduced in 1994 based on the Community Health Act that are familiar to and frequently used by residents; preventive care management at comprehensive community support centers; and management of support for pregnancy, childbirth, and parenting at comprehensive support centers for childrearing generation.

Following the changing situation, the number of public health nurses is increasing. According to a "Report on Public Health Administration and Services" released by the Ministry of Health, Labour and Welfare, employed public health nurses increased by 39.4%, from 34,468 to 51,280, during 2000–2016. Regarding results by place of employment, the number of public health nurses belonging to municipalities is the highest, accounting for 55.6% of the total (Fig. 5). Although the number of staff decreased because of the reduction in the number of public employees resulting from the merger of municipalities, the number of public health nurses in municipalities increased. The number of public health nurses employed by municipalities was 20,646 in 2000, which increased to 28,509 in 2016.

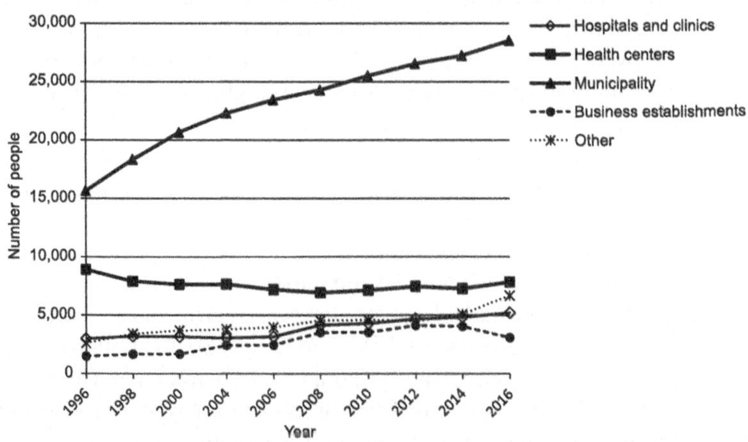

Fig. 5 Changes in the number of employed public health nurses by place of employment (Created based on Report on Public Health Administration and Services)

Prefectural health centers were realigned after enactment of the Community Health Act. In addition, local governments shifting toward establishing health centers increased by the merger of municipalities. As a consequence, the health centers in prefectures have been decreasing in number. The number of health centers was 594 nationwide in 2000, which declined to 480 in 2016 (according to research by the Health Service Bureau, Ministry of Health, Labour and Welfare). A "Report on Public Health Administration and Services" released by the Ministry of Health, Labour and Welfare describes that the number of public health nurses employed at health centers inched up 3.4%, from 7,570 to 7,829, during 2000–2016.

To provide health and medical care services of good quality, public health nurses must also be secured. The Act on Public Health Nurses, Midwives, and Nurses, which includes changes in qualifications for the national examinations, and the Act for Partial Revision of the Act on Assurance of Work Forces of Nurses and Other Medical Experts came into effect in 2010. This extended the period of study in basic education for public health nurses from six months to one year. With the increase of nursing universities, more than 90% of public health nurses are trained at four-year universities; approximately 90% of those who pass the National Public Health Nursing Examination are graduates from four-year universities (Ii and Arakida 2013: 13).

5 Distribution of Professions in the Welfare Field

5.1 Certified Social Workers as a Social Welfare Profession

Certified social worker is defined in paragraph (1), Article 2 of the Certified Social Worker and Certified Care Worker Act as "a person with expert knowledge and skills and uses the appellation "certified social worker" to provide advice, guidance, or welfare services in consultations about the welfare of persons with physical disabilities or mental disorders and intellectual disabilities or persons facing difficulty in leading a normal life because of environmental factors, and a person engaged in the business of communicating and coordinating with and providing other assistance to doctors, other health and medical service providers, and other related parties." It is clear from the definition that certified social workers are expected to be active in various fields including health and medical care and welfare. According to the Social Welfare Promotion and National Examination Center (2015), main places of employment for certified social workers as of November 2015 are elderly welfare-related (43.7%), disability welfare-related (17.3%), medical care-related (14.7%), community welfare-related (7.4%), child welfare and maternal and child welfare-related (4.8%), and administrative counseling centers (3.4%). Although certified social workers are employed in widely various fields, nearly half of them work in places related to welfare for elderly people.

The Ministry of Health, Labour and Welfare (2018a) reports that as of 2016, comprehensive community support centers accounted for 8.3% of all places where certified social workers worked. Certified social workers were chosen as one profession to be placed in comprehensive community support centers because they were deemed as the actor most appropriate for the consulting work and rights protection program (Takayama 2005). However, some studies point out that challenges lie in the performance of duties. Takayama (2015) gives high evaluation to consultation and support provided by certified social workers in the centers, but gave a low evaluation to regional networking. The latter is work to build a system that can respond smoothly during consultation through integrated management of information about elderly people's use of medical care, nursing care, and welfare services. Certified social workers have been engaged only slightly in the work. For this reason, the Expert Committee on the Securing of Welfare Personnel, Sectional Meeting on Welfare, Social Security Council (2018) presents an opinion that education, including a training curriculum, must be revised for certified social workers to play their anticipated role.

Qualifications for the National Certified Social Worker Examination can be acquired by three routes: (1) graduating from a welfare university after having completed prescribed subjects; (2) studying at least six months in a short-term training facility after completing basic subjects for social welfare and graduating from a welfare university; and (3) studying at least one year in a general training facility after graduating from a general or other university or after having been engaged in consultation and support work at least four years. Those who have passed the National Examination can use the title of certified social worker by registering with the Social Welfare Promotion and National Examination Center. According to the center, the number of certified social workers registered is 221,251 throughout the country as of March 2018 (Social Welfare Promotion and National Examination Center 2018). The number of registrations per 100,000 population is high in prefectures in the Hokuriku, Chugoku, and Kyushu regions, such as 251 in Niigata and 225 in Okayama, although it is low in prefectures in eastern Japan, such as 126 in Ibaraki and 134 in Fukushima (Fig. 6). The Japanese Association for Social Work Education (2018) reports that, as of 2018, welfare universities, two-year welfare universities, and training facilities are located in all prefectures except Tottori. In addition, prescribed subjects can be studied through correspondence courses. Consequently, regional differences in opportunities to take examinations are small.

5.2 *Demand for Nursing Care Professions and Acceptance of Foreign Workers*

With regard to labor demand in nursing care service industries, the Ministry of Health, Labour and Welfare (2018b) suggests that about 550,000, annually about 60,000, nursing care personnel must be secured by the end of 2025 in addition to

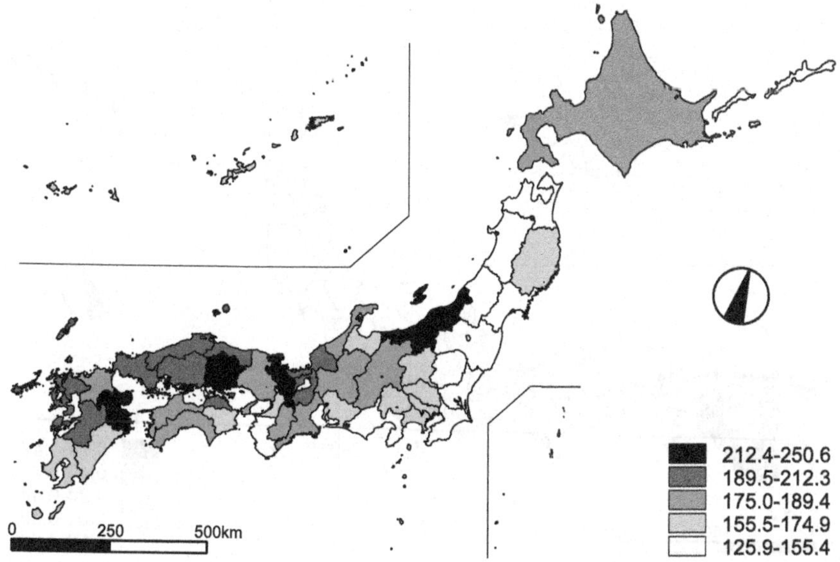

Fig. 6 Number of certified social workers registered per 100,000 population (by prefecture in 2018). Population based on the population estimate as of October 2, 2017 (Created based on the "Registration Status" by the Social Welfare Promotion and National Examination Center. http://www.sssc.or.jp/touroku/pdf/pdf_t04_3.pdf. Accessed 1 May 2019)

about 1,900,000 in 2016. However, a shortage of care labor will not occur uniformly throughout all parts of the country, but will progress with accompanying regional differences. A "Report on Employment Service" released by the Ministry of Health, Labour and Welfare shows that the active job openings to applicant ratio for care service professions was 5.0 or over in Tokyo Metropolis and Aichi Prefecture as of June 2017. Despite an increasing demand for care labor, demand has not been fulfilled in metropolitan areas because the labor to be supplied is limited not only in metropolitan areas, but also in non-metropolitan areas that have supplied labor to metropolitan areas. Among non-metropolitan areas, the Kyushu region is an important area for care facilities in metropolitan areas to obtain labor because welfare high schools are predominantly located there. According to a Basic Survey on Schools conducted by the Ministry of Education, Culture, Sports, Science and Technology, although the percentage of students attending welfare-related departments was 0.3% nationwide for the total of students in 2017, it was 1.5% in Oita Prefecture, 1.4% in Miyazaki Prefecture, and 1.3% in Kagoshima Prefecture (Fig. 7). However, the number of students attending school to train certified care workers, such as welfare high school, has been declining sharply. The number of high school students attending the welfare department declined from 10,697 to 8,769. Also, technical college students attending the welfare department declined from 19,468 to 9,537 throughout Japan during 2007–2017. Therefore, the numbers of students dropped by 18% and by 51%, respectively, during that period.

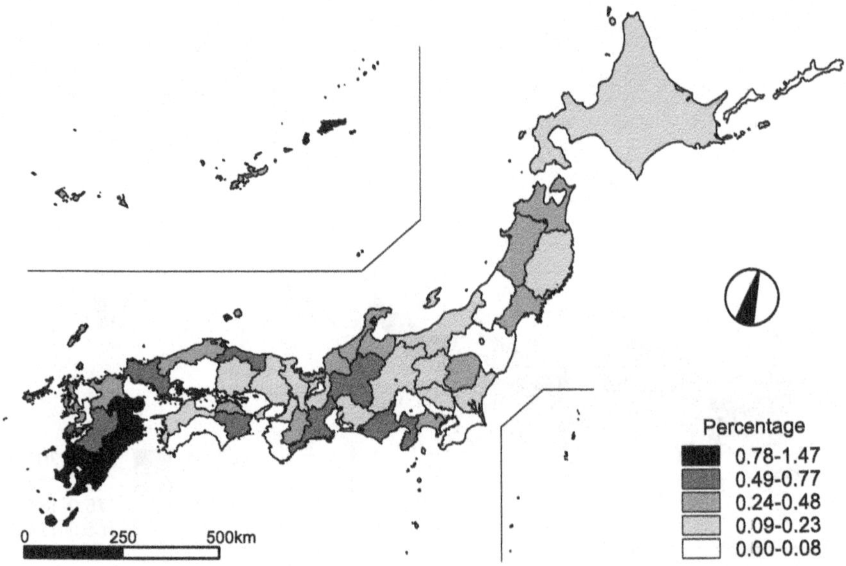

Fig. 7 Percentage of students in the welfare department accounting for the total number of high school students (by prefecture in 2017) (Created based on Basic Survey on Schools)

The acceptance of EPA foreign certified care worker candidates is a system to strengthen economic partnership between two countries. However, in nursing care service industries, it is expected to be one solution to labor shortage problems. For this reason, care facilities with difficulty filling vacancies with Japanese workers see this as an option to secure personnel. The total number of candidates accepted by 2017 is 1,494 from Indonesia, 1,400 from the Philippines (37 entering school), and 598 from Vietnam. Because the annual number to be accepted is set at a maximum of 300 people for each country, considering the influence on the labor market in Japan, the acceptance was slow until approximately 2012. Nevertheless, it has increased, reflecting a shortage of care labor since then. Nearly the maximum number of candidates were accepted in Japan in 2017 (Fig. 8).

Furthermore, because Care Worker was added to the job categories in the Technical Intern Training Program in 2017 and the visa status "Care" was newly established for foreign students who have acquired the qualification of a certified care worker, opportunities have increased for foreign workers to undertake care services in Japan. Compared to the EPA system, the number of countries sending laborers increased and the number of accepting facilities expanded in the two new systems. Judging from the situation in which problems of a care labor shortage are becoming increasingly severe, more foreign laborers than ever before will be accepted in nursing care service industries in Japan.

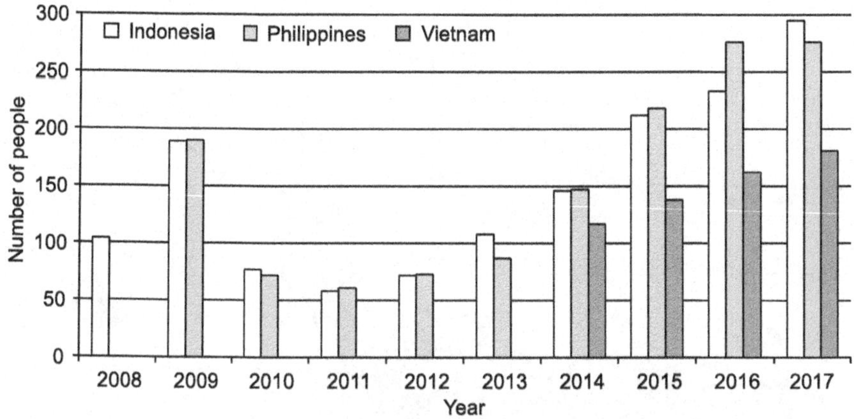

Fig. 8 Acceptance of EPA foreign certified care workers candidates (Created based on "Acceptance of foreign nurse and certified care worker candidates from Indonesia, the Philippines, and Vietnam" by the Ministry of Health, Labour and Welfare. https://www.mhlw.go.jp/stf/seisakunitsuite/bunya/koyou_roudou/koyou/gaikokujin/other22/index.html. Accessed 1 May 2019)

5.3 Shortage of Nursery Teachers and Potential Nursery Teachers

To reduce the number of children on waiting lists for nursery schools to zero, the securing of nursery teachers is an unavoidable issue. Areas in which it is difficult to secure nursery teachers are metropolitan areas such as the Tokyo Metropolis, which is the same as the issue of children on the waiting list. According to a "Report on Employment Service" released by the Ministry of Health, Labour and Welfare, as of November 2016, prefectures with a high ratio of active job openings to applicants for nursery teachers were Tokyo (5.68), Hiroshima (3.80), and Saitama (3.73) (Fig. 9). The ratio exceeded 1 in all prefectures except Yamanashi (0.95).

The qualification of nursery teacher is one qualification that the staff are expected to acquire not only for after-school children's clubs and children's recreational facilities, but also in the following services: consultation and guidance related to childcare and nursing in community-based parenting support center program; and support for children with disabilities in the inclusive service that was created with the revision of the long-term care insurance system and the welfare system for disabled persons in 2017. One reason for a shortage of nursery teachers is that the demand for nursery teachers is expanding to a wide spectrum of welfare services. As a countermeasure against it, the government is applying effort into supporting re-employment of those with qualifications but who are not working: so-called potential nursery teachers. The Ministry of Health, Labour and Welfare (2013) reports that of job seekers who have a nursery teacher qualification, 48.5% did not desire to work as a nursery teachers as of March 2013. The reasons include the following: wages do not meet expectations (47.5%); interest in other professions (43.1%); and the weight of responsibility and

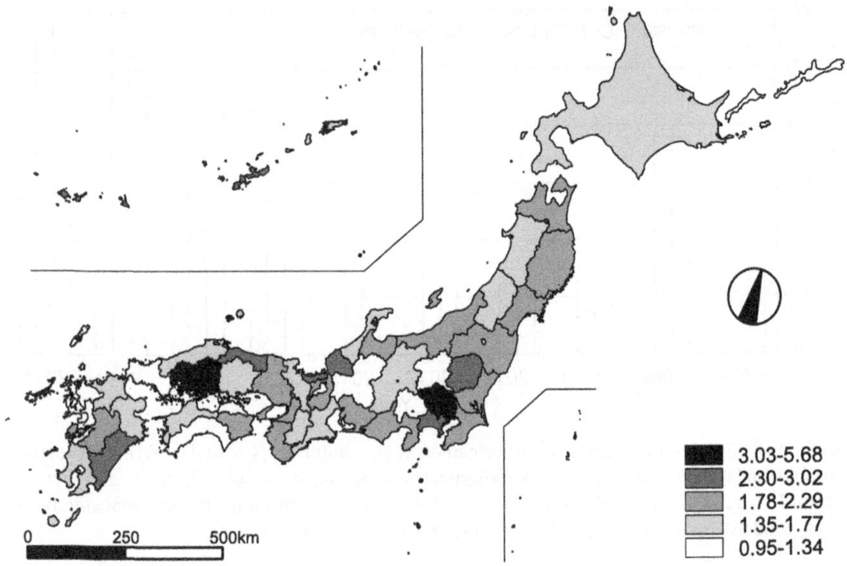

Fig. 9 Active job openings to applicants ratio for nursery teachers (by prefecture in November 2016) (Created base on Report on Employment Service)

the anxiety about accidents (40.0%). Comparing the wages of nursery teachers with other occupations based on a Basic Survey on Wage Structure, contractual cash earnings (total employees) in 2017 were 353,000 yen (41.7 years) for all industries and 230,000 yen (35.8 years) for nursery teachers. Contractual cash earnings of nursery teachers are at a low level of 270,000 yen (35.9 years) even in the highest prefecture, Kyoto. They are below the national average for all occupations.

The "Month for Intensively Addressing Measures for Promotion of Employment of Nursery Teachers," which is one measure taken by the Ministry of Health, Labour and Welfare, sets Tokyo Metropolis and Saitama, Kanagawa, and Osaka prefectures as areas to address the measure intensively and promote employment of potential nursery teachers (Ministry of Health, Labour and Welfare 2015). However, the working conditions of nursery teachers do not always match the conditions that they desire as conditions of re-employment. For instance, because the wages at re-employment are low and fewer nursery centers have introduced a regular pay hike, the wage structure is not one that they can work at throughout life; because most job openings are for full-time non-regular employees, it is not an easy option for women in the middle of parenting.

6 Regional Characteristics of the Resident Sector in the Welfare Field

6.1 Roles of Commissioned and Child Welfare Volunteers and Regional Differences in the Rate of Filled Vacancies

The Commissioned Welfare Volunteers Act was revised in 2000 for the first time in 50 years. The duties were specified as follows: (1) Proper understanding of residents' living conditions; (2) Support for persons requiring assistance, such as consultation and advice; (3) Provision of information about service use; (4) Collaboration and cooperation with welfare offices and social welfare-related organizations; and (5) Activities for the promotion of residents' welfare. Regarding the activities more specifically, the number of cases in which commissioned welfare volunteers dealt with consultation and support was about 6,460,000 and that of other activities was about 27,120,000 in 2014. The number of visits reached about 38,650,000. Examining the results by fields of activities, "Matters related to elderly people" accounted for 56.0%, "Matters related to children" were 20.6%, and "Matters related to persons with disabilities" were 5.1%. Commissioned welfare volunteers develop widely diverse activities, including the following: conducting surveys and grasping the situations of persons requiring support, such as elderly people and persons with disabilities, and households with children and mother–child households; offering advice; and participating and cooperating in various events as well as voluntarily participating in community welfare activities. In recent years, they have been actively addressing regional challenges such as mapping of persons requiring support in times of disaster. They have become the main actor in community welfare.

The standards for the placement of commissioned welfare volunteers are specified by law. To be more specific, one commissioned welfare volunteers must be placed for every 220–440 households in special wards of Tokyo and ordinance-designated cities, one for every 170–360 households in core cities and cities with a population of 100,000 and more, one for every 120–280 households in cities with a population less than 100,000, and one for every 70–220 households in towns and villages. An election is held every three years for all members. In the election held in 2014, of the quota of 236,296 members nationwide, 231,339 were elected as of the end of 2014, which increased by 1,178 compared to the previous year (0.5% increase year-on-year). The breakdown is 91,538 men, increasing by 111 from the previous year (0.1% increase year-on-year), and 139,741 women, increasing by 1,067 from the previous year (0.7% increase year-on-year).

As shown there, the number of commissioned welfare volunteers is on an increasing trend. However, looking at the trend by prefecture, the placement of commissioned welfare volunteers for the quota is not always filled. Although the rate of filled vacancies is growing, regional differences exist throughout Japan, which has become a great challenge in recent years. The number of vacancies nationwide was 4,957, decreasing by 398 from 2010 (the previous en masse election) in which the number was 5,355, and the rate of filled vacancies improved to 97.9%. Regarded by prefecture, all prefectures except Toyama failed to fill the quota in the 2014 en masse election. The percentage was particularly low in Okinawa (89.7%), followed by Tokyo (94.6%), and Miyagi and Osaka (97.0%).

It is increasingly difficult for commissioned welfare volunteers to carry out their duties because of the following reasons: progressing trends toward the nuclear family; weakening of relationships with neighbors; and increase of the younger generation taking a stance that refuses to have a relationship with public officers. Such current situations are pointed out as a key factor for unfilled vacancies. Furthermore, in addition to ageing of commissioned welfare volunteers, requests from administration are increasing. The requests ask to address child and elderly abuse, which has become a social issue in recent years, solitary deaths among elderly people, suicide issues, bank transfer scams, the adult guardianship system, and the long-term care insurance system. A growing sense is related to the burden in the activities.

6.2 *Residents Participating in Community Welfare Activities and the Geographical Characteristics*

Main actors in community welfare activities are NPOs and voluntary organizations. First, we look at the geographical characteristics using NPOs as a lead. NPOs that undertake "activities to promote healthcare, medical care, and welfare" account for 58.2% of the total, which makes the activities the largest item in those undertaken by NPOs. Figure 10 shows a significant concentration of NPOs in metropolitan areas: the highest is 4,772 in the Tokyo Metropolis (16.1% of the national total) followed by 2,014 in Osaka Prefecture (6.9% of the national total) and 1,909 in Kanagawa Prefecture (6.45% of the national total). Although the proportion of NPOs in rural areas for the total number is small because the NPOs are few, the item accounts for the largest proportion of the NPO activity items in each prefecture. Emphasis is placed on activities responding to population aging. The NPOs that undertake "activities to promote sound development of children" account for 44.6% of the total, which is lower than the former activities. Regarded by prefecture, the highest is 3,939 in Tokyo (17.3% of the national total), followed by 1,277 in Osaka (5.1% of the national total) and 1,171 in Kanagawa (5.1% of the national total). These are not as high as the former item. Specifically examining the situation of each prefecture, it is apparent that NPOs are most likely to gather in the prefectural capital city, probably because the largest populations gather in prefectural capital cities and because the demand for their activities is high. However, it is one characteristic that welfare NPOs are found not only in urban areas but also in hilly and mountainous areas.

In addition to NPOs, we confirm the activity state of welfare volunteers by prefecture (Fig. 11). Looking at it according to the activity content, it is apparent that the activities on which the most number of days were spent in a year were "activities for elderly people," where the national average is 31.4 days. These are a response to the

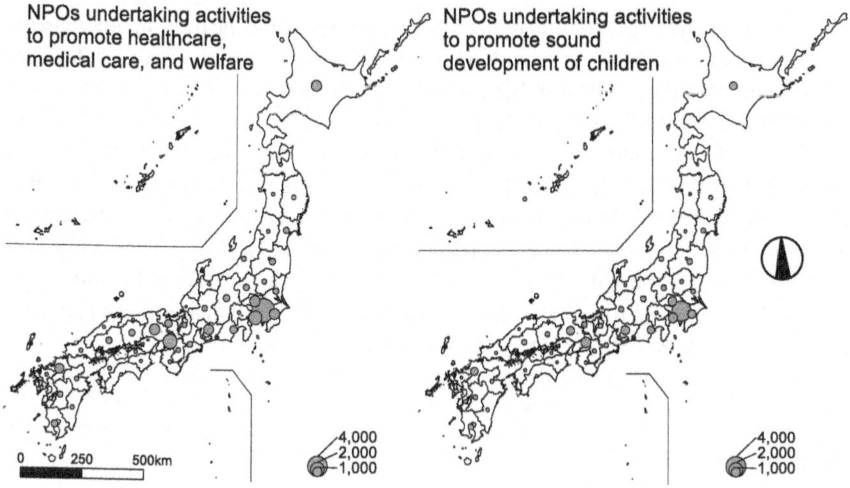

Fig. 10 Distribution of welfare NPOs (by prefecture in 2015) (Created based on the information about NPOs provided in the Cabinet Office "NPO Home Page." https://www.npo-homepage.go.jp/. Accessed 1 September 2015)

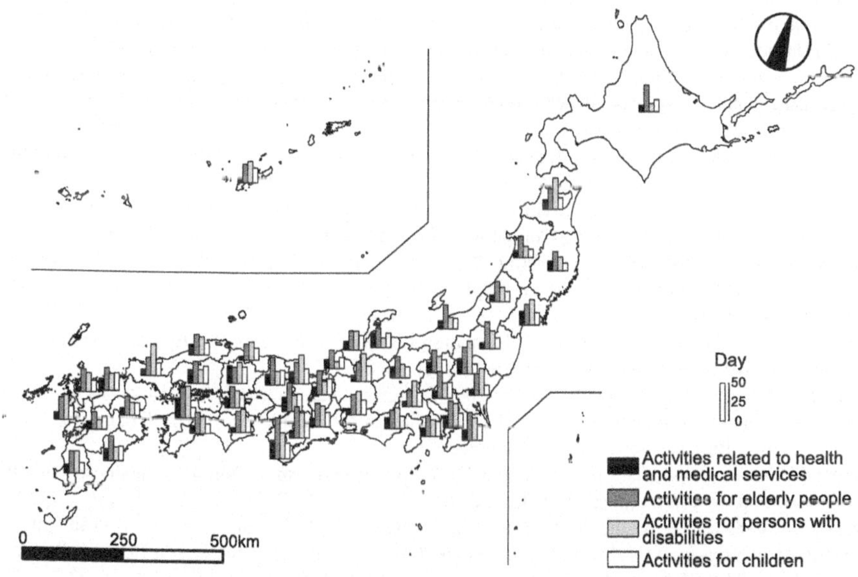

Fig. 11 Annual average number of days spending on welfare volunteer activities (by prefecture in 2016) (Created based on Survey on Time Use and Leisure Activities)

increase in the rate of population aging. Prefectures spending the most number of days on the activities are Wakayama with 54.4 days, followed by Nara with 41.5 days and Ehime with 40.1 days. The number exceeds the national average not only in provincial areas where population aging is progressing, but also in metropolitan areas such as Aichi Prefecture and Tokyo Metropolis. Following activities for elderly people, the activities on which the second most number of days were spent were "activities for persons with disabilities," where the national average was 26.1 days. Prefectures spending the most days on the activities are Tochigi with 44.1 days, Yamaguchi with 42.7 days, and Ehime with 42.3 days. It is noteworthy that Ehime holds third place in the activities in addition to the former activities. We think that in these activities, how to secure actors will influence the liveliness of the activities.

References

Compilation Committee for 60-year History of the Act on Public Health Nurses, Midwives, and Nurses (ed) (2009) The 60-year history of the Act on Public Health Nurses, Midwives, and Nurses: The years of nursing administration and development of nursing. Japanese Nursing Association Publishing Company, Tokyo (in Japanese)

Editorial Committee of Social Welfare from Basics (ed) (2009) Series: Social welfare from basics (1) Introduction to social welfare (second edition). Minerva Shobo, Kyoto (in Japanese)

Expert Committee on the Securing of Welfare Personnel, Sectional Meeting on Welfare, Social Security Council (2018) Roles required for certified social worker that is a social work profession (in Japanese)

Han H (2012) The study on the actual conditions of the nurse shortage in Japan. J East Asian Stud (Yamaguchi University) 10:1–24 (in Japanese)

Ii K, Arakida K (eds) (2013) New edition: handbook for public health nurses (3rd ed.). Japanese Nursing Association Publishing Company, Tokyo (in Japanese)

Japan National Council of Social Welfare National Center for Social Service Human Resources (2018) 2016 Performance report on employment placement. https://www.fukushi-work.jp/toukei/index_3.html. Accessed 28 Sept 2018 (in Japanese)

Japanese Association for Social Work Education (2018) List of schools. http://jaswe.jp/jascsw_members_list.html Accessed 1 May 2019 (in Japanese)

Japanese Nursing Association (2010) The 2009 Survey on Nursing Staff (in Japanese)

Japanese Nursing Association (2017) The 2016 Survey of Hospital Nursing. Japanese Nursing Association Research report 91 (in Japanese)

Japanese Nursing Association (2018) The 2017 Survey of Hospital Nursing. Japanese Nursing Association Research report 94 (in Japanese)

Kaku H (2018) Exploring the characteristics of "Japanese-style nursing care" entering in China's nursing market-analysis through case studies. Hyoron Shakaikagaku (Social Science Review) 124:107–124 (in Japanese)

Kamiya H (2017) Persons employed in the field of social security. In: Miyazawa H (ed) Mapping health, medical care, and welfare in Japan. Akashi Shoten, Tokyo (in Japanese)

Kingma M (2006) Nurses on the move: Migration and the global health care economy. Cornell University Press, Ithaca

Kiyosaki Y (2018) Home-visit nursing management from tomorrow: The main points to be grasped. Medicus Shuppan, Publishers Co., Ltd, Osaka (in Japanese)

Kuga H (2009) Introduction to social work: The basis of consultation support and professions. Minerva Shobo, Kyoto (in Japanese)

Ministry of Health and Welfare (1996) Report of the consensus conference for issues regarding the assistant nurse system assistant nurse. Accessed 23 Sept 2018 (in Japanese)
Ministry of Health, Labour and Welfare (2011) The 2010 Survey on Employment Status of Nursing Staff
Ministry of Health, Labour and Welfare (2013) Comprehensive efforts toward securing nursery teachers supporting childcare (in Japanese)
Ministry of Health, Labour and Welfare (2015) Month for intensively addressing measures for promotion of employment of nursery teachers (in Japanese)
Ministry of Health, Labour and Welfare (2018a) Current status of certified social workers (in Japanese)
Ministry of Health, Labour and Welfare (2018b) Required number of care personnel based on the seventh insured long-term care service plan (in Japanese)
Mitomi K (2005) MINERVA social welfare series (14) Western care workers: People forgotten in welfare states. Minerva Shobo, Kyoto (in Japanese)
Mori T, Goto R (2012) A study of doctors in Japan. Toyo Keizai Inc, Tokyo (in Japanese)
Nichigai Associates (2013) A cyclopedic chronological table of medicine in Japan 1722–2012. Nichigai Associates, Tokyo (in Japanese)
Nomura Y (2015) Nursing system and policy. Hosei University Press, Tokyo (in Japanese)
Sakai T (2019) The history of medicine with numerous illustrations. Igaku-Shoin, Tokyo (in Japanese)
Shen J (2014) MINERVA social welfare series (46) What social welfare reform in China is trying to achieve: harmonizing socialism with capitalism. Minerva Shobo, Kyoto (in Japanese)
Shimazaki K (2011) Health care in Japan: institutions and policies. University of Tokyo Press, Tokyo (in Japanese)
Shimizu Y (2015) Recent trends of vocational education of elderly service in China. Waseda Rev Soc-Sci 21:62–76 (in Japanese)
Shirase Y (2015) Medical personnel. In: Matsumoto K (ed) Medical system reform: a comparative analysis of Germany, France, and the United Kingdom and suggestions to Japan. Junposha Co., Ltd, Tokyo (in Japanese)
Social Welfare Promotion and National Examination Center (2015) Results of survey on employment status of certified social workers and certified care workers (in Japanese)
Social Welfare Promotion and National Examination Center (2018) Status of registration (in Japanese)
Takayama Y (2005) The Current trends of community-based comprehensive support centers associated with the reform of the long-term care insurance system: The feature and roles of certified social workers. Bull Jpn Lutheran Coll Theol Semin: Theologia-diakonia 39:51–67 (in Japanese with English abstract)
Takayama Y (2015) The Trend of dissertations and studies on social workers' practices at community-based comprehensive support centers. Bull Jpn Lutheran Coll Theol Semin: Theologia-diakonia 49:13–25 (in Japanese with English abstract)
Tsuji T (2008) What medical system in Japan is aiming at. JIJI Press Publication Service Inc, Tokyo (in Japanese)

[illegible] of Health, Medical, and Welfare Personnel.

Ministry of Health and Welfare [illegible] 2015 [illegible]
Ministry of Health, Labour and Welfare [illegible]
[illegible]
[illegible] Assessment [illegible]
[illegible]
[illegible]
[illegible]
[illegible]
[illegible]
[illegible]

Community-Based Integrated Care Systems in Japan

Regional Variation in the Community-Based Integrated Care Systems in Japan

Teruo Hatakeyama and Hitoshi Miyazawa

Abstract Significant regional differences exist in Japan's community-based integrated care system. The reason is that the community-based integrated care system in Japan must be built voluntarily in line with regional characteristics and with the leadership of local governments. This requirement naturally creates regional variation. More importantly, we must identify specific factors that create the regional variation and explain characteristics and issues associated with the respective differences in connection with regional characteristics. Accordingly, this chapter categorizes the types of community-based integrated care systems by municipality and identifies the characteristics of the respective categories. The chapter also presents analyses of the causes of regional variation in community-based integrated care systems based on their relations with regional characteristics. The results reveal a considerably strong of municipalities' population sizes on regional variation in community-based integrated care systems. Results also indicate the effects of local governments' financial capacity and historical, economic, and social conditions such as municipal consolidation and business distribution among regions.

Keywords Community-based integrated care system · Community care conference · Comprehensive community support center · Regional differences · Sphere of daily life

---

This chapter is a revised version of Hatakeyama et al. (2018).

T. Hatakeyama (✉)
Naruto University of Education, Naruto, Japan
e-mail: thatakeyama@naruto-u.ac.jp

H. Miyazawa
Ochanomizu University, Tokyo, Japan

H. Miyazawa and T. Hatakeyama (eds.), *Community-Based Integrated Care and the Inclusive Society*, International Perspectives in Geography 12,
https://doi.org/10.1007/978-981-33-4473-0_6

1 Introduction

Japan has an urgent need to build community-based integrated care systems. The Ministry of Health, Labor, and Welfare, which oversees health care, long-term care, and other care-giving services in Japan, aims at establishing community-based integrated care systems by 2025, when baby boomers will have become 75 years old or older. A community-based integrated care system is a system providing elderly people with housing, medical and long-term care services, preventive medicine, and disability support services in a unified manner to allow them to continue living in the communities where they are used to living in ways they prefer to live, even when becoming in need of care. The development of the community-based integrated care system is urgently needed for many reasons, including a rapid increase in elderly people requiring medical and long-term care services because the population is aging faster in Japan than in the U.S. and Europe. Moreover, there is a need for expansion of disability support services because of an increase in the number of elderly people living alone. Improvement of preventive care services to prevent elderly people from becoming in need of care is also demanded for the development of a community-based integrated care system.

Community-based integrated care systems in Japan are built based on the sphere of daily life. Spheres of daily life are areas determined by local governments. The area sizes are approximately equivalent to junior high school districts, which are assumed to be a range in which a supporter is able to rush to provide any necessary help. This system is intended to provide elderly people with monitoring and care in units of small areas: spheres of daily life. Values in the practice of community welfare are characterized by consciousness directed to the government, which is strong in Europe; in contrast, consciousness directed to local residents and municipal councils of social welfare is a stronger ideal in eastern Asian countries such as Japan and Korea (Uenoya and Saito ed. 2015: 232). Given such a background, Japan is particularly addressing organization of networks of professionals providing medical, healthcare, welfare, and other services to develop a foundation for elderly people to continue living in areas where they are used to living with a sense of security and networks of informal service providers, volunteer groups, community groups, and other helpers to monitor local residents. Spheres of daily life are the basic units for developing such networks.

The Comprehensive Community Support Centers placed in these spheres of daily life become the bases of Community-based Integrated Care Systems. Municipalities came to be required to establish the Comprehensive Community Support Centers after the revision of the long-term care insurance system in fiscal 2006. Key operations of the Comprehensive Community Support Centers include the treatment of the cognitive impairment of elderly people, monitoring of elderly people, and preventive care management. These are important duties in the development of community-based integrated care systems. Management of community care conferences is another fundamentally important task of the comprehensive community support centers. Community care conferences are meeting bodies established to help resolve difficulties and other issues[1] confronting local residents. Furthermore, they monitor elderly people. Community care conferences are considered important as a means of building the community-based integrated care system and were established as a law through revision of the long-term care insurance system in 2015. The members of community care conferences comprise local medical institutions, long-term care providers, welfare organizations, resident associations, and others who are led by the governments and comprehensive community support centers. These add to the importance of forming networks of actors in spheres of daily life. In addition, local issues that have been identified in community care conferences must be incorporated into local governments' policies. For this reason, the Ministry of Health, Labor, and Welfare requires the community care conferences to cover the entire municipalities that manage meetings held based on the spheres of daily life.

Such a way of building community-based integrated care systems has created marked disparities among communities (Niki 2017). The differences are largely attributable to the requirement of local governments to develop the system voluntarily according to regional characteristics. Regional variety in the community-based integrated care systems, therefore, is a natural result. It is more important to identify specific factors contributing to regional differences, and additionally, to link the characteristics and issues of respective differences from regional characteristics. Through these activities, it is also important to ascertain the mechanisms of community-based integrated care that are appropriate for the respective communities having different conditions.

The objective of Second Part, therefore, is to identify the conditions and factors of regional differences in the community-based integrated care system and to elucidate the characteristics and issues associated with the respective differences. In doing so, we specifically examine the perspective of local governance. The reason is that, as described earlier, the networking of various actors is regarded as important in community-based integrated care systems (Inoue 2011; Uenoya and Saito eds. 2015: 18). Actually, local governance is defined by Sato and Maeda eds. (2017: 14), from the standpoint of geography, as a form of governance of decision-making and operation of local autonomy through correlations such as negotiations and consensus-building, in which pluralistic actors, including governments, participate, in a geographical area smaller than a nation. Sato and Maeda eds. (2017) also pointed out that regional disparities are found in the participation of local governance actors in the networks and levels of resource reserves. The perspective of local governance is effective for analysis of regional variation in the community-based integrated care system, which emphasizes the networks of such various actors.

This chapter explains factors contributing to the regional variation in the community-based integrated care system, which is the first milestone in the achievement of the objective of Second Part. To this end, we classify the types of community-based integrated care system of local governments based on the placement of comprehensive community support centers and community care conferences, which are important elements in the development of the community-based integrated care

[1] Abuse, neglect, social withdrawal, and other issues that are difficult to resolve.

system, and identify the characteristics of the respective types. We will additionally examine the relations between the characteristics of community-based integrated care systems and regional attributes. The data used are the results of the authors' questionnaire survey administered to local governments across Japan (designated as the "Survey" in the following Second Part)[2] and public statistics and information provided by local governments. The conditions of local governance analyzed based on the results of fieldwork in the example are as described in subsequent chapters.

2 Bases and Conferences in a Community-Based Integrated Care System

2.1 Regional Disparities Found in Comprehensive Community Support Centers

There were 5,142 comprehensive community support centers[3] as of July 2018. The long-term care insurance act requires that each comprehensive community support center places at least one person each of public health nurse, social worker,[4] and senior care manager (officially designated a long-term care support specialist)[5] on duty for every 3,000–6,000 First Category persons insured (those aged 65 years and older) under long-term care insurance. Comprehensive community support centers are operated by municipal governments or local private-sector operators under contract. Because many of the operations of comprehensive community support centers are

[2]In November 2015, the authors administered a questionnaire Survey to Local Governments' Development of the community-based integrated care system. The Survey was completed by municipal governments across Japan, excluding those within the evacuation zone of the Fukushima Daiichi nuclear disaster. The questionnaires were distributed by mail. The responses were collected by mail and e-mail. The number and percentage of valid responses were 616 and 35.5%, respectively. By population size, however, approximately half of the municipalities having a population of 50,000 and above returned valid responses, which is considered sufficient to ensure the accuracy of the analysis.

[3]Information was collected from local government websites.

[4]Social workers are government-certified social welfare specialists. Their responsibilities include providing consultations to people with physical, mental, or economic handicaps and helping them carry out activities smoothly in their daily life and solve the difficulties they must confront (see also Chapter "Securing of Health, Medical, and Welfare Personnel and the Geographical Distribution in Japan").

[5]Certification to be a senior care manager is given to care managers providing consultations to people in need of care and helping them receive optimal care. The managers must have practical experience of more than five years and must have completed a specific specialized training program. They act as a liaison and coordinate the providers of long-term care insurance services and other health care and medical services and provide advice and guidance to other care managers (see also Chapter "Securing of Health, Medical, and Welfare Personnel and the Geographical Distribution in Japan").

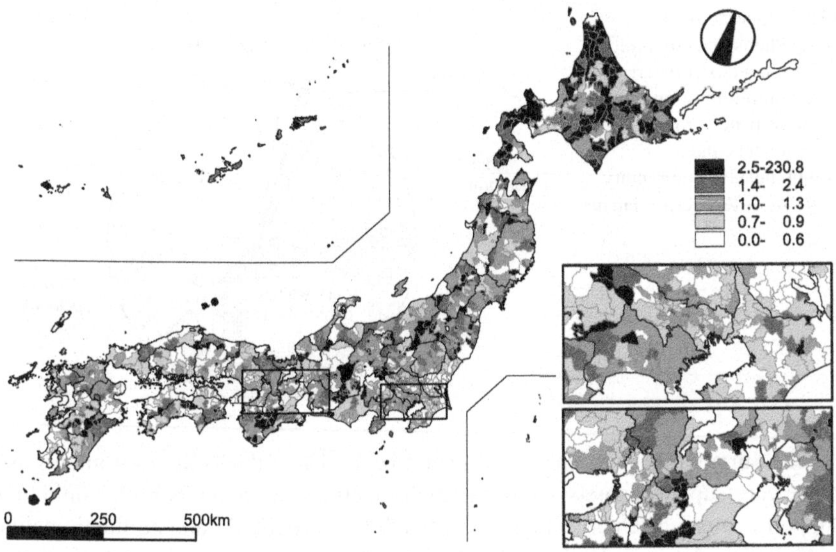

Fig. 1 Number of comprehensive community support centers per 6,000 individuals aged 65 years and older (by long-term care insurer). As of July 2018 (Prepared based on information related to the websites of local governments across Japan and population statistics in Basic Resident Register in Japan)

entrusted by municipal governments, the structure of local community-based integrated care systems varies depending on the relation between the comprehensive community support centers and their local governments.

Japan has 1,578 long-term care insurers (municipal governments, cross-regional federations,[6] etc.) that operate municipal special accounting for long-term care insurance. Of those long-term care insurers, 66.6% operate only one comprehensive community support center, although the rest manage multiple centers. To ascertain the sufficiency of comprehensive community support centers, we calculated the number of centers per 6,000 individuals aged 65 years and older for each insurer. The result indicated high sufficiency in underpopulated areas in Hokkaido and eastern Japan (Fig. 1). However, sufficiency was low in central metropolitan areas and underpopulated areas in western Japan. The study therefore found regional disparities in the sufficiency of comprehensive community support centers. Furthermore, sufficiency tends to be high in areas in which cross-regional insurers have been developed across multiple municipalities based on economies of scale.

[6] As a wide-area regional government formed by multiple municipal governments, a cross-regional federation develops wide-area plans, establishes communication and coordination as necessary, and comprehensively and systematically implements wide-area administration for affairs considered more appropriate if they are operated by multiple municipal governments to cover a broader area.

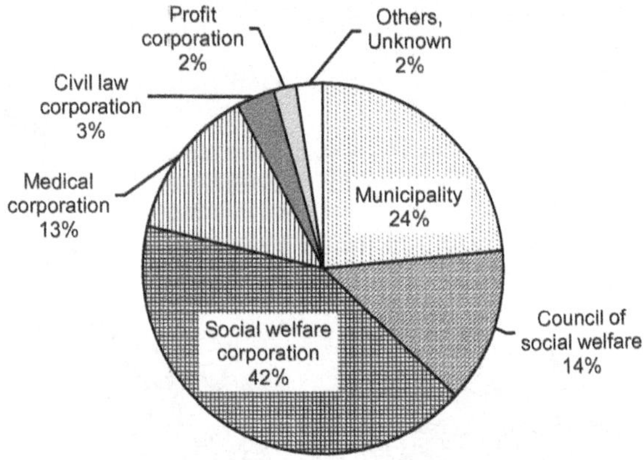

Fig. 2 Operators of comprehensive community support centers (Prepared based on information related to the websites of local governments and comprehensive community support centers across Japan)

The Survey results revealed, however, that 25.5% of the local governments used home care support centers[7] as sub-branch functions of comprehensive community support centers. The percentage exceeded 30%, particularly among municipalities having a population of 10,000 to 100,000. In other words, low sufficiency of comprehensive community support centers in these municipalities does not necessarily mean that a community-based integrated care system is not functioning. This point demands particular attention in the analysis.

The operation of comprehensive community support centers is performed directly by local governments in 23.5% of the municipalities. The operation of the centers in other municipalities is entrusted to social welfare corporations, councils of social welfare, medical corporations, and other organizations (Fig. 2). The case of entrustment to a social welfare corporation is particularly common. We used the modified Weaver's method to identify the dominant operators of comprehensive community support centers for each long-term care insurer (Fig. 3). This figure reveals that social welfare corporations and medical corporations are often the dominant operators in municipalities in the three largest metropolitan areas and other large cities. In non-metropolitan areas, however, municipalities in which the centers are operated directly by local governments or predominantly by councils of social welfare are more commonplace.

This distribution of councils and local governments is regarded as resulting of the following circumstances. Most insurers in non-metropolitan areas place only one comprehensive community support center in their municipalities because of small population size. In many such cases, the local government operates the center or a council of social welfare directly with a strong public nature. It is contracted to operate

[7] A home care support center is an institution that acts as a liaison between senior citizens staying home at all times, who require or might in the near future require protection, or their families and municipal governments, service institutions, care management centers, and other organizations to help such senior citizens receive health care and welfare services based on their need for long-term care or other services in response to consultations with them. A number of local governments abolished home care support centers when they established comprehensive community support centers.

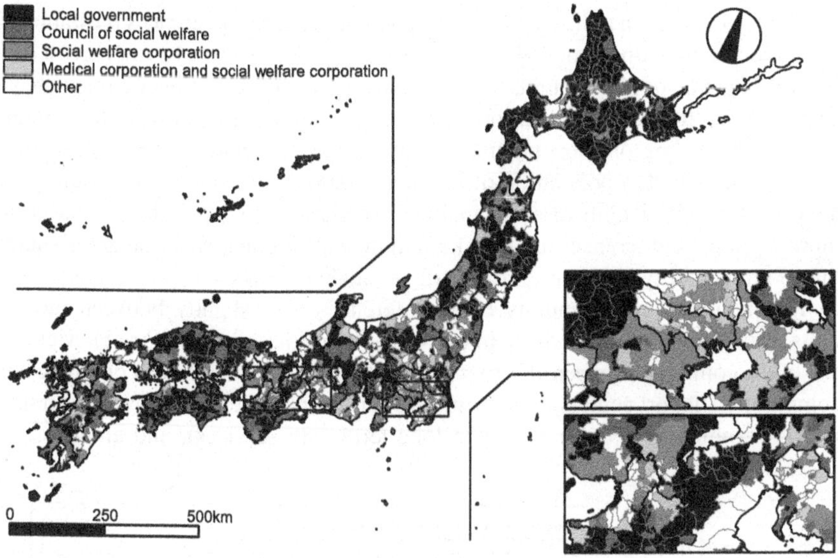

Fig. 3 Types of dominant operators of comprehensive community support centers (by long-term care insurer). As of July 2018 (Prepared based on information related to the websites of local governments across Japan and population statistics in Basic Resident Register in Japan)

the center. Insurers in metropolitan areas more often place multiple comprehensive community support centers in their municipalities because of their large populations. They commonly entrust social welfare corporations and medical corporations that have conventionally been providing long-term care insurance services to operate the centers.

2.2 *Regional Disparities in Community Care Conferences*

The Survey indicates that 84.1% of the municipalities have established community care conferences. Whether a community care conference has been established varies among the municipalities depending on their population size. According to the Survey, although all the local governments of the special wards of Tokyo and large cities having a population of more than 700,000 have established community care conferences, only 77.0% of small municipalities with a population of less than 10,000 have established their community care conferences (Hatakeyama and Miyazawa 2016). The percentage of municipalities which responded "having a plan to establish a community care conference in near future" was 15.3%. However, we expect to establish community care conferences in most municipalities by some future date.

The most common unit size in which a community care conferences is established is a municipality with the rate of 76.0%. This is followed in order by a sphere of daily life with the rate of 37.7%; units smaller than a sphere of daily life comprise 26.3%. The unit sizes for establishing a community care conference also vary among municipalities depending on their population size. Figure 4 demonstrates that, in small municipalities with a population of less than 10,000, community care conferences are operated only in units of municipalities. The larger the population, however, the more often the conferences are held both in units of municipality and areas smaller than a municipality (a sphere of daily life or smaller areas).

The members of community care conferences vary slightly between those in units of municipalities and those in units of spheres of daily life (Fig. 5). Members are often common between the two types; they are employees of comprehensive community support centers, long-term care service providers, employees of municipal governments (those engaging in long-term care services), and employees of

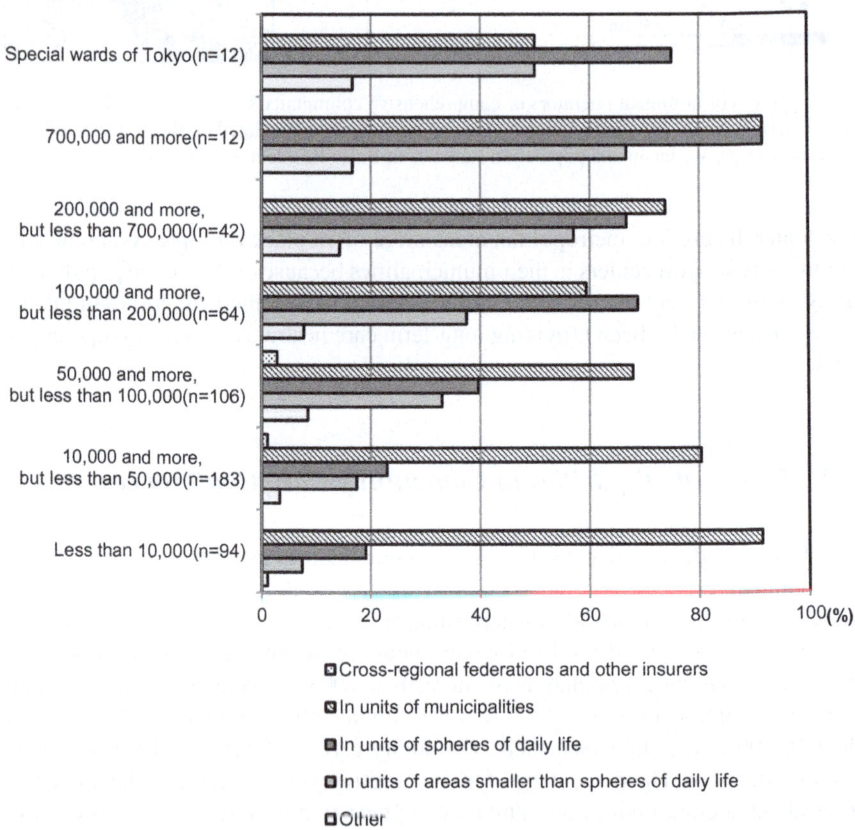

Fig. 4 Units of establishing community care conferences by population size. The total of responses exceeds 100% because of multiple choice questions (Prepared based on Survey responses)

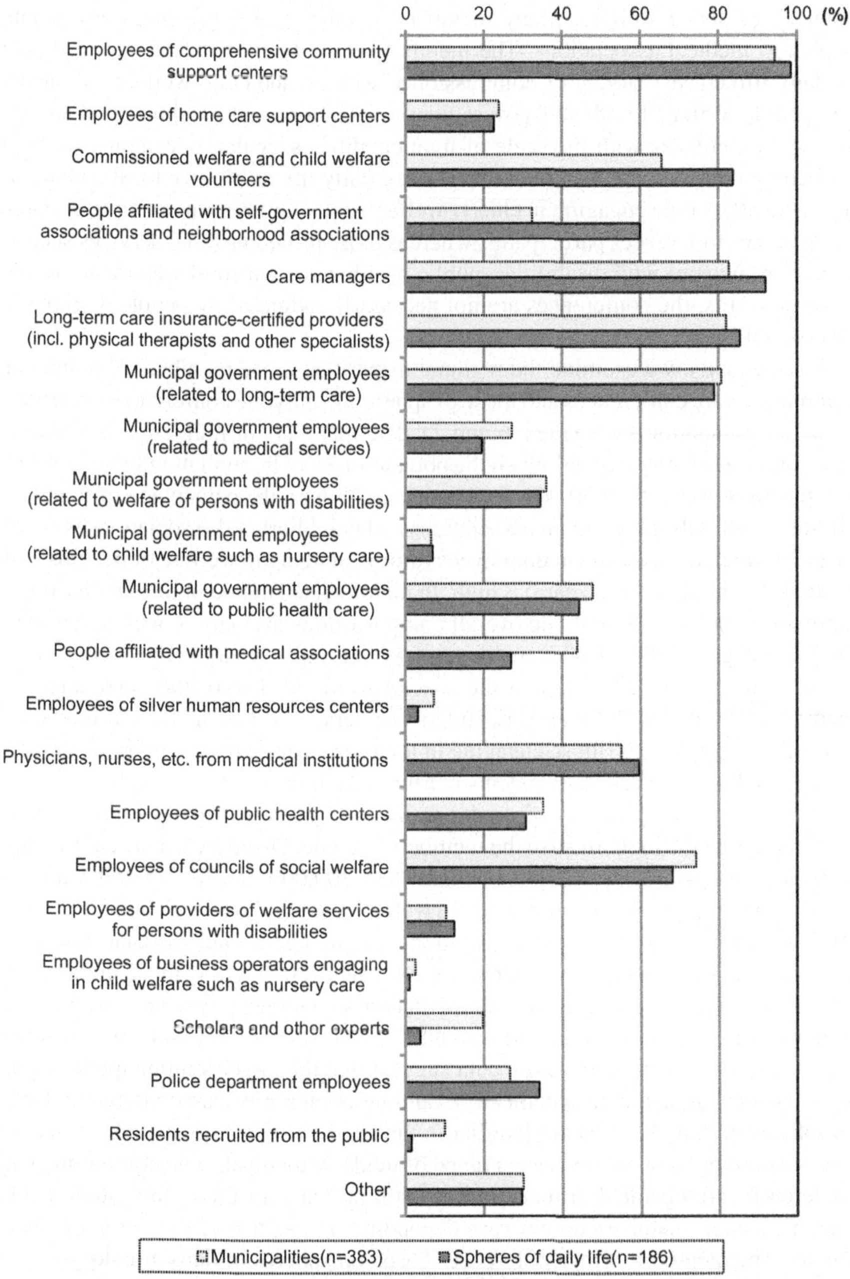

Fig. 5 Members of community care conferences (Prepared based on Survey responses)

councils of social welfare. Many members of municipal conferences are people related to medical associations. The members of conferences based on the sphere of daily life consist largely of commissioned welfare and child welfare volunteers and people affiliated with self-government associations and neighborhood associations. Conferences with the scale of municipalities specifically examine medical services and those with the scale of spheres of daily life emphasize local communities. Overall, people engaging in elderly welfare and long-term care comprise a large share of the conferences participants, whereas those providing other services such as welfare of persons with disabilities, public health care, and child welfare are fewer. In other words, the conferences are not necessarily attended by people working in diverse fields.

Subsequently, we examine the regional disparities in the members of municipal community care conferences and those of sphere-of-daily life conferences separately based on population sizes (Figs. 6 and 7). The members of municipal community care conferences vary depending on the population size. In small municipalities with a population of less than 50,000, for instance, although the percentage of members affiliated with self-government associations and neighborhood associations is small, that of municipal government employees (those engaging in welfare of persons with disabilities or public health care) is high. In middle-sized municipalities with a population of 50,000 to 200,000, the overall composition is average, in which, however, the percentage of public health center employees is high. In large cities with a population of more than 200,000 and the special wards of Tokyo, the percentages of members affiliated with home care support centers, care managers, and municipal government employees (those engaging in long-term care) are low, whereas those of experts such as scholars and residents recruited from the public are high.

As for community care conferences in units of spheres of daily life, the larger the population, the more diverse the members become. Diversity is particularly high in large cities with a population of more than 200,000 and the special wards of Tokyo. The proportions of commissioned welfare and child welfare volunteers, those affiliated with self-government associations and neighborhood associations, long-term care service providers, municipal government employees (those engaging in long-term care), those affiliated with medical associations, physicians, nurses, etc. from medical institutions, employees of public health centers, and scholars and other experts are especially high. These data suggest that the larger a municipality is, the more the local government specifically examines community care conferences based on spheres of daily life that are familiar to the residents and aims to address various issues by taking in members from a range of fields. Municipal scale community care conferences are operated primarily to develop policies for the entire municipality. Therefore, their members do not vary depending on the size of the municipality as much as the members of the conferences based on spheres of daily life do.

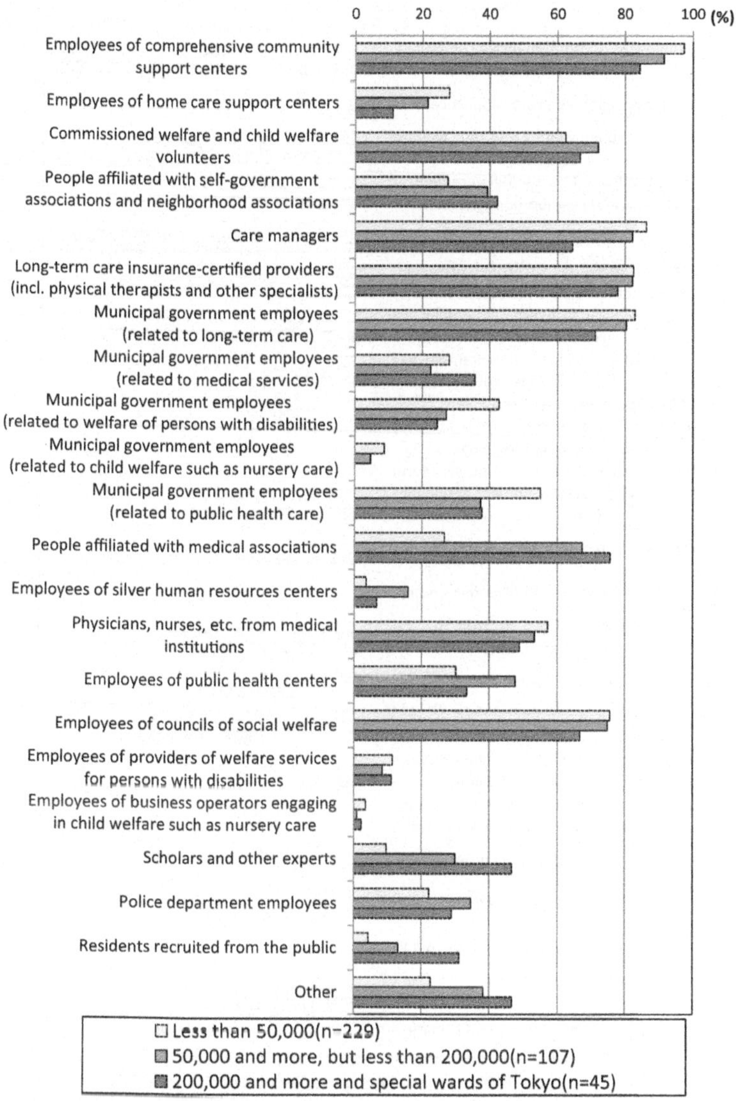

Fig. 6 Members of community care conferences in units of municipalities by municipal population size (Prepared based on Survey responses)

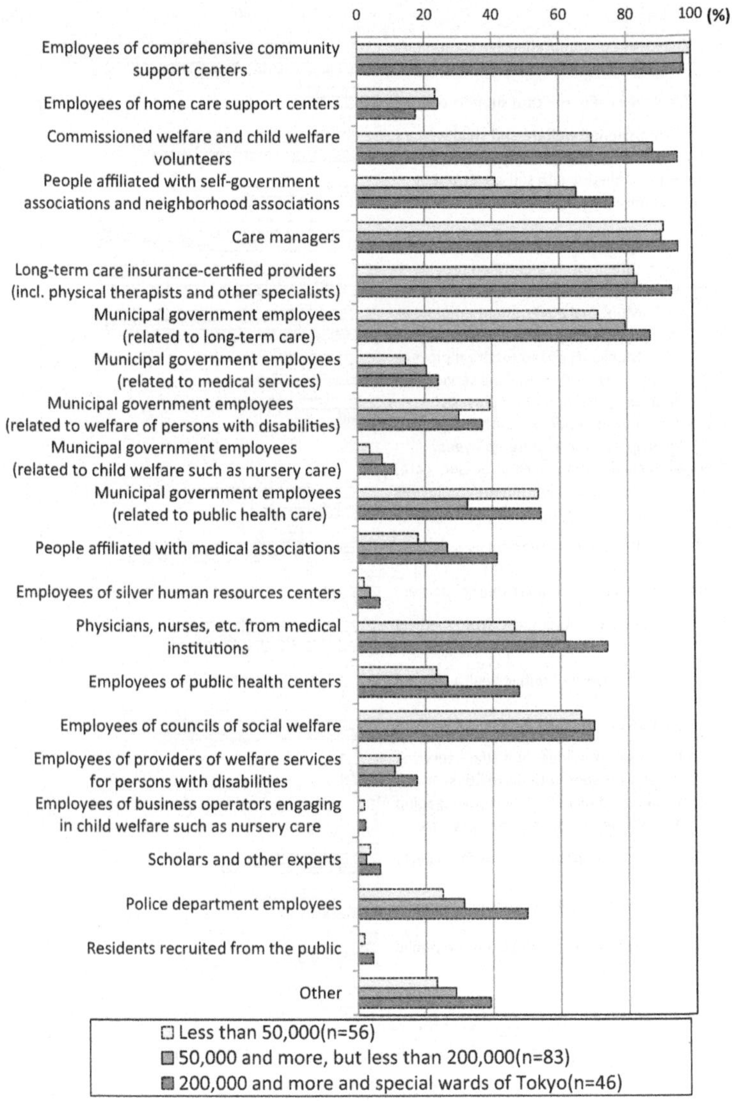

Fig. 7 Members of community care conferences in units of spheres of daily life by municipal population size (Prepared based on Survey responses)

3 Types of Community-Based Integrated Care System

3.1 Method of Classification

The classification of the types of community-based integrated care system in this chapter uses the levels of the establishment of comprehensive community support centers and the units of establishing community care conferences as indicators. The reason is that comprehensive community support centers are base facilities for the community-based integrated care system and community care conferences are important meeting bodies using the centers as secretariats, which help solve difficulties faced by elderly people, monitor them, and provide other assistance in their respective communities. Regional disparities are apparent in the levels of the establishment of comprehensive community support centers and the units of establishing community care conferences as discussed earlier, which are appropriate as indicators to identify factors contributing to regional variation in the community-based integrated care system. This classification is based on work reported by Hatakeyama et al. (2018).

The placement of comprehensive community support centers is divided into cases in which a single center serves the entire municipality and cases in which multiple centers are operated in their respective spheres of daily life. In the latter case, the operation of the centers is commonly entrusted to local corporations, which must be managed by the government. This creates a considerable difference in local governance from the placement of one center per municipality. An approach that is intermediate of the above two is the placement of a sub-branch function such as a home care support center in each sphere of daily life in addition to a comprehensive community support center. Community care conferences are placed either in each sphere of daily life (or an area equivalent to a sphere of daily life) only or in each municipality only. Furthermore, the conferences are held, in some cases, on a multilayered scale involving both of these types.

Based on the points raised above, we have classified the placement of comprehensive community support centers into three types, including Type 1, for which one center is placed in one area, Type 2, for which one center is placed in one area and sub-centers (branches) are also placed, and Type 3, for which the centers are placed in two or more areas. We have classified the placement of community care conferences into three area sizes, including Type A, a municipality, Type B, an area smaller than a municipality, and Type C, an area coming a municipality and a district smaller than a municipality. In this chapter, we describe our analysis and examinations based on nine types[8] created by crossing over these groups. Figure 8 is a diagram of these types. It illustrates the variation based on the spatial structures of the community-based integrated care system.

[8]Data used for the classification were Survey results (the above footnote 2).

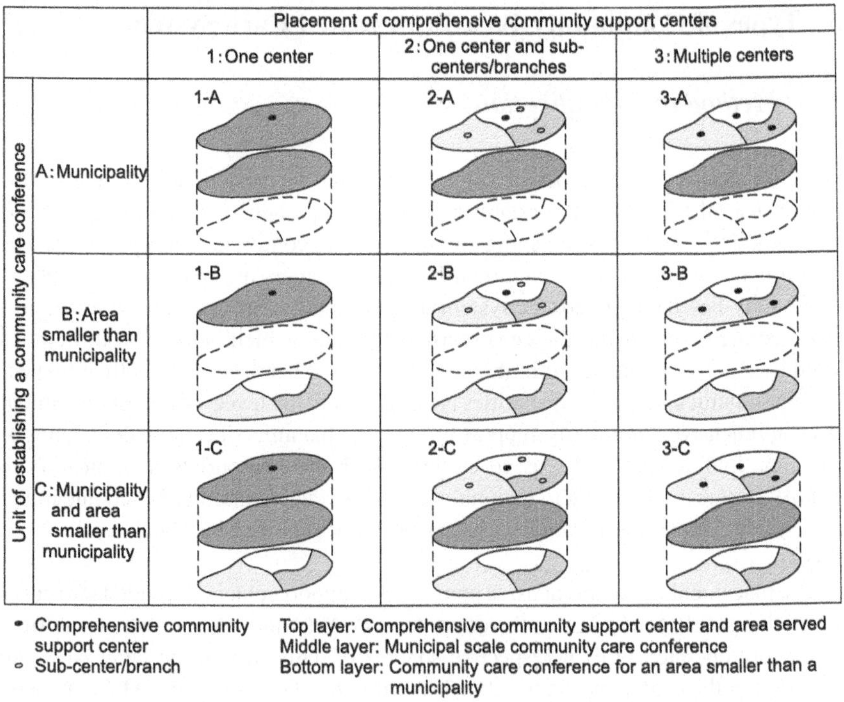

Fig. 8 Types of community-based integrated care system based on area structures (Reprinted from Hatakeyama et al. 2018 with permission of the Association of Japanese Geographers)

3.2 Number of Municipalities by Type

The municipalities corresponding to the types described in the earlier section were 507 out of the 616 municipalities which responded to the Survey. Many of the 109 municipalities excluded from the classification were not operating any community care conference at the time of the Survey.

Table 1 exhibits the numbers of municipalities based on the community-based integrated care system types. The types of placement of comprehensive community support centers reveal that the municipalities corresponding to Type 1 are the most numerous, 220, but followed, in order, by Type 3 and Type 2. The area sizes for placing community care conferences indicate that the municipalities corresponding to Type A are the most numerous with 231, followed by Type C and Type B.

In view of the nine types created by crossing these groups, the municipalities corresponding to both Types 1 and A (Type 1-A) are the most numerous with 148, comprising 29.2% of all municipalities. This is followed, in order, by Type 3-C (101 municipalities), Type 3-B (71 municipalities), and Type 2-A (60 municipalities). Many of the municipalities having only one comprehensive community support center operate community care conferences only in units of municipalities.

Table 1 Number of municipalities by Type

Unit of establishing a community care conference	Placement of comprehensive community support centers	1: One center	2: One center and sub-centers/Branches	3: Multiple centers	Total
	A: Municipality	148	60	23	231
	B: Area smaller than municipality	31	13	71	115
	C: Municipality and area smaller than municipality	41	19	101	161
	Total	220	92	195	507

Prepared based on survey responses

Those municipalities having multiple comprehensive community support centers, however, more commonly operate community care conferences in areas smaller than municipalities.

Figure 9 portrays distributions of municipalities by type. These images show that many municipalities of Type 1 and Type 2 (particularly Type A) are located in non-metropolitan areas, although those of Type 3 (particularly Type C) are generally located in the three largest metropolitan areas of Japan. This finding suggests that respective types have their regional characteristics.

3.3 *Characteristics of Respective Types*

We analyze the relevance of each type with the indicators of regional characteristics and attributes of community-based integrated care system to identify the characteristics of the types of community-based integrated care system. The indicators used are the following: The indicators of regional characteristics include municipal populations (total populations and populations of residents aged 65 years and older), areas of municipalities, financial capability index,[9] and municipal consolidations since April 1999 (the Heisei Municipal Mergers[10]). The indicator of the characteristics of

[9]The financial capability index is an indicator of the financial strength of a municipal government. It is calculated by dividing basic fiscal revenue (standard local tax revenue) by basic fiscal demand (expenditure required for administrative affairs). When the index value is below 1, a local allocation tax grant is provided by the national government.

[10]Mergers of municipalities have advanced since 1999 because of strengthening of administrative and financial bases as municipalities in Japan. This transformation has been called the "Great merger of Heisei". The number of municipalities was 3,232 on March 31, 1999, but 1,718 on October 10, 2016.

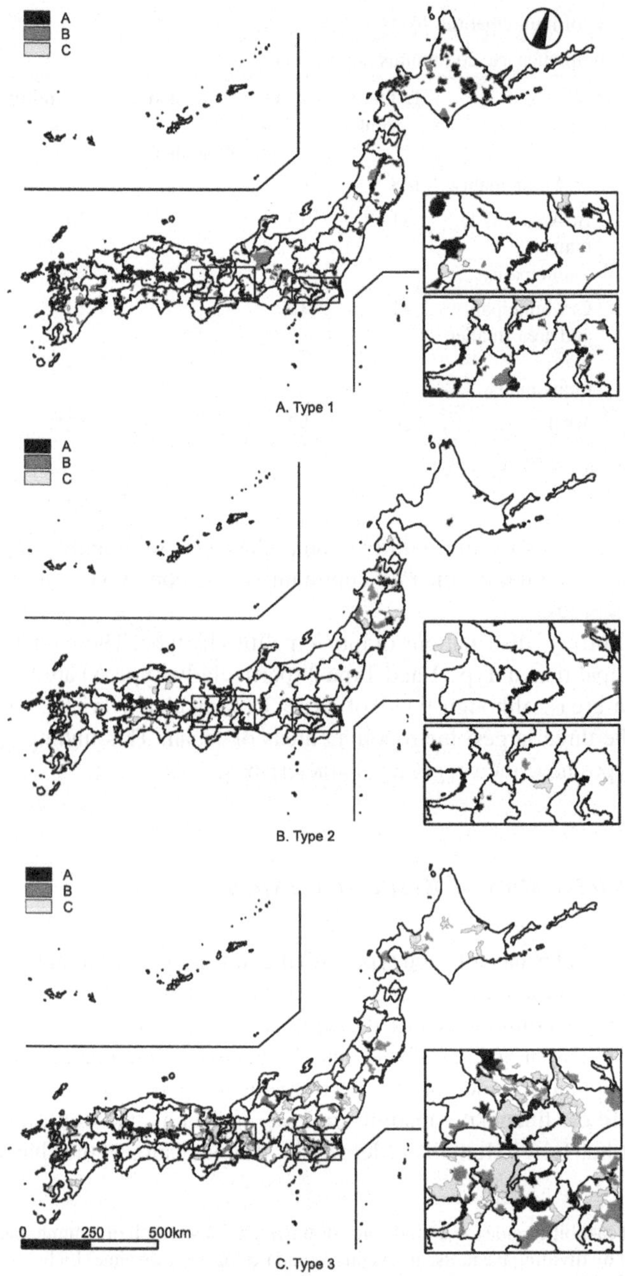

Fig. 9 Distribution of municipalities by Type (Prepared based on Survey responses)

community-based integrated care system is an operator of comprehensive community support centers (percentage of those operated directly by local governments), and the indicator of multi-field collaboration is the composition of participants in community care conferences (percentage of municipalities holding the conferences participated in by members from two or more sectors in addition to the public welfare sector).[11]

Tables 2 and 3 present the relations between the types and each of the indicators. The population of Type 1-A is the smallest of the nine types (Table 2a). This type also has the smallest values in the area size (Table 2b), the averages of the financial capability index (Table 3c), and percentages of merged municipalities (Table 2d). Type 1-A, therefore, evidently consists of small and financially less strong municipalities that did not undergo mergers during the Heisei Municipal Mergers.

Municipal populations are larger for Type 2 and Type 3 of the placement of comprehensive community support centers and Type B and Type C of area sizes for placing community care conferences. The percentages of merged municipalities differ between Type 1 and Types 2 and 3 of the placement of comprehensive community support centers. Within Type 1, the percentages of merged municipalities vary between Type A and Types B and C of area sizes for placing community care conferences.

Those local governments directly operating comprehensive community support centers tend to belong to Type 1 or Type 2 having one center placed in one area (Table 3a). However, the number of local governments within Type 3 that are directly operating all centers is considerably small. The local governments partially or entirely outsourcing the operation are dominant in this type. The local governments outsourcing all operations are particularly numerous in Type 3-B and Type 3-C.

In addition, the numbers of municipalities in which community care conferences have members participating from two or more sectors in addition to the public welfare sector are large for Type A and Type C, but the number is small in Type B (Table 3b). As explained earlier, the municipal scale conferences have participation by a large portion of members from the medical service sector. Local governments of Type B have constraints in their collaboration with people from the medical field. The reason is that community care conferences based on areas smaller than municipalities are attended by few members from the medical and other sectors that are unrelated to welfare services in their respective areas.

---

[11]Because community care conferences in all municipalities are participated by organizations from the long-term care sector (comprehensive community support centers, long-term care business operators, councils of social welfare, etc.), we specifically examined the participation of members from other sectors in this case. Each sector is defined as follows: resident service sector (commissioned welfare and child welfare volunteers, parties affiliated with self-government associations and neighborhood associations and residents recruited from the public), medical service sector (parties affiliated with medical associations, physicians, nurses, etc. from medical institutions, and employees of public health centers), sector for welfare of persons with disabilities (employees of providers of welfare services for persons with disabilities), and child welfare sector (employees of business operators engaging in child welfare such as nursery care).

Table 2 Characteristics of respective Types (Indicators of regional characteristics)

(a) Total population by municipality and the number of residents aged 65 years and older (average) by municipality

Unit of establishing a community care conference		Placement of comprehensive community support center			Total
		Type 1	Type 2	Type 3	
	Type A	18.9, 5.6	35.2, 10.9	102.8, 27.8	31.5, 9.2
	Type B	31.4, 9.2	56.8, 17.3	167.5, 42.1	118.3, 30.4
	excl. ordinance-designated	-	–	159.3, 40.2	112.8, 29.1
	Type C	28.0, 8.8	59.7, 15.9	357.6, 89.7	238.5, 60.4
	excl. ordinance-designated	-	-	214.5, 54.8	143.9, 37.4
Total		22.4, 6.7	43.3, 12.8	258.3, 65.0	116.9, 30.3

(b) Area (average)

Unit of establishing a community care conference		Placement of comprehensive community support centers			Total
		Type 1	Type 2	Type 3	
	Type A	207.6	218.9	214.0	211.2
	Type B	311.9	367.5	251.8	281.1
	excl. ordinance-designated	-	-	249.8	280.1
	Type C	225.5	311.8	374.4	329.1
Total		225.6	259.1	310.8	264.5

(c) Financial capability index

Unit of establishing a community care conference		Unit of establishing a community care conference			Total
		Type 1	Type 2	Type 3	
	Type A	0.39	0.44	0.71	0.44
	Type B	0.46	0.51	0.68	0.60
	Type C	0.41	0.60	0.68	0.60
Total		0.41	0.48	0.68	0.53

(continued)

Table 2 (continued)

(d) Percentage of merged municipalities

Unit of establishing a community care conference		Unit of establishing a community care conference			Total
		Type 1	Type 2	Type 3	
	Type A	22.3	48.3	52.2	32.0
	Type B	38.7	76.9	50.0	49.6
	Type C	43.9	52.6	55.4	52.2
Total		28.6	53.3	52.8	42.4

In the first table, (a), the values on the left-hand side of each column represent population by municipality and the values on the right-hand-side of each column represent the number of residents aged 65 years and older. Tables (a) and (b) also present values after excluding the impact of ordinance-designated cities, which is large (Prepared based on the Survey, the 2015 Population Census of Japan, Major Financial Indices of Local Public Bodies (FY 2008), and "Collection of Information Related to Municipal Mergers" of the Ministry of Internal Affairs and Communications.Reprinted from Hatakeyama et al. 2018 with permission of the Association of Japanese Geographers)

Table 3 Characteristics of respective Types (Indicators related to the community-based integrated care system)

(a) Percentages of municipalities directly operating a comprehensive community support center (%)

Unit of establishing a community care conference		Unit of establishing a community care conference			Total
		Type 1	Type 2	Type 3	
	Type A	75.7	86.7	0.0, 52.2	71.0, 5.2
	Type B	64.5	76.9	2.8, 26.8	27.8, 16.5
	Type C	85.4	78.9	5.0, 28.7	34.2, 18.0
Total		75.9	83.7	3.6, 30.8	49.5, 11.8

(b) Percentages of municipalities holding community care conferences with participation by members from two or more sectors in addition to the public welfare sector (%)

Unit of establishing a community care conference		Unit of establishing a community care conference			Total
		Type 1	Type 2	Type 3	
	Type A	39.2	41.7	56.5	41.6
	Type B	9.7	23.1	35.2	27.0
	Type C	61.0	57.9	67.3	64.6
Total		39.9	42.4	54.4	45.6

In the first table, (a), all values on the left-hand side of the Type-3 column represent the percentages of municipalities directly operating their comprehensive community support centers. Values on the right-hand-side are the percentages of municipalities that directly operate some of their comprehensive community support centers (Prepared based on survey responses. Reprinted from Hatakeyama et al. 2018 with permission of the Association of Japanese Geographers)

4 Impact of Municipal Population Sizes and Internal Diversity

The results above suggest that the population sizes of municipalities strongly affect the variation in the community-based integrated care system. This inference is related to the standards for the assignment of specialists in comprehensive community support centers and population sizes that become the criteria for determining spheres of daily life.

Each comprehensive community support center must have one person each of a public health nurse, social worker, and senior care manager on duty for every 3,000–6,000 residents aged 65 years and older (see the Sect. 1). A sphere of daily life is a regional unit in the community-based integrated care system and purportedly has a general population of 6,000 aged 65 years and older.[12] The average population of people aged 65 years and older in Type 1-A is similar to these standards. The Survey results also revealed that municipalities having only one sphere of daily life accounted for 44.9% of all respondents. Strong correlation (0.86) was found between the number of spheres and population of residents aged 65 years and older.

These findings point to the tendency of small municipalities to place one comprehensive community support center and community care conference in each municipality that constitutes one sphere of daily life. In many cases, small municipalities directly operate their respective comprehensive community support centers. This organization might be a result of their geographical condition, which indicates that a private-sector operator capable of undertaking the center operation is absent in the municipalities (Miyazawa 2017: 104).

The Survey results also revealed that those small municipalities having a large population or experienced mergers tended to place each community care conference in a smaller area. A reason for this might be the attempt by local governments and comprehensive community support centers to meet the needs of local communities gradually under the Type 1 condition, in which there is only one comprehensive community support center in each municipality.

Municipalities having a large population must establish multiple comprehensive community support centers and spheres of daily life in light of the standards described previously. Local governments are unable to operate their comprehensive community support centers independently when they have established multiple centers in their municipalities. Such local governments therefore often outsource some or all operations of their comprehensive community support centers. Another finding of the Survey was the many cases in which municipalities having a population of more than 100,000 placed a community care conference in each sphere of daily life or equivalent area. Merged municipalities, in some cases, determine spheres of daily

[12] The results of a survey conducted by the Ministry of Health, Labor, and Welfare in 2012 indicated that 35.5% of the spheres of daily life had 3,000 or more and fewer than 6,000 residents aged 65 years and older and 33.4% of the spheres of daily life had fewer than 3,000 residents aged 65 years and older. In other words, approximately 70% of the spheres of daily life had fewer than 6,000 residents aged 65 years and older (Ministry of Health, Labour and Welfare 2012).

life based on municipal borders before their mergers. This is an effort to maintain the characteristics of each municipality that existed before the mergers. As a result, the larger the population of a municipality, the more it tends to have multiple comprehensive community support centers that are operated by different parties within the municipality. Such municipalities also tend to use spheres of daily life as units of placing community care conferences.

Municipalities of Type 2 positioned between Type 1 and Type 3 in the placement of comprehensive community support centers commonly have a medium-sized population. In such municipalities, it is difficult for a single center to meet the needs of the entire municipality. Many of them are also merged municipalities that include various communities. These municipalities, however, might have markedly limited capacity to secure specialists or financial resources to establish multiple comprehensive community support centers within a single municipality. This is considered to be an important reason why medium-sized municipalities select the form of Type 2.

5 Conclusions

The study has uncovered regional variation in the development of the community-based integrated care system, which is considerably affected by population sizes of municipalities. In addition, the factors causing the regional variation, including the condition of establishing comprehensive community support centers, area sizes for placing community care conferences, and determining spheres of daily life, are affected considerably by historical, economic, and social conditions such as financial capacity, municipal mergers, and distribution of operators in the regions.

Local governments must build community-based integrated care systems according to their respective regional characteristics. Issues that arise in different regions are also expected to vary depending on the conditions of building community-based integrated care systems.

Accordingly, Chapter "Community-Based Integrated Care Systems in Municipalities Having One Comprehensive Community Support Center" through Chapter "Community-Based Integrated Care Systems in Municipalities Having Multiple Comprehensive Community Support Centers" specify regions as examples of the nine types identified in this chapter and perform analysis based on the fact-finding Survey of the community-based integrated care system conducted in these regions. The analyses will be based primarily on the condition of placing comprehensive community support centers and operating community care conferences. The results will be examined from the perspective of local governance. In this process, we also devote attention to the relations of respective municipalities with its regional characteristics. The activities described above will enable us to identify the conditions and issues associated with regional differences in the community-based integrated care system.

Chapter "Community-Based Integrated Care Systems in Municipalities Having One Comprehensive Community Support Center" presents examination of the community-based integrated care system in the town of Yusuhara (Type 1-A) in Kochi Prefecture, the town of Nanporo (Type 1-A) in Hokkaido, and the town of Shinkamigoto (Type 1-B and Type 1-C) in Nagasaki Prefecture as examples of having "only one comprehensive community support center in each municipality." Chapter "Community-Based Integrated Care Systems in Municipalities Having Sub-branchs of Comprehensive Community Support Centers" presents observations of the community-based integrated care system in the city of Goto (Type 2-A) in Nagasaki Prefecture, the city of Tsugaru (Type 2-B) in Aomori Prefecture, and the city of Fukutsu (Type 2-C) in Fukuoka Prefecture as examples of "municipalities that combine a comprehensive community support center and sub-centers or branches". Finally, Chapter "Community-Based Integrated Care Systems in Municipalities Having Multiple Comprehensive Community Support Centers" investigates the community-based integrated care system in the cities of Asago (Type 3-A), Hyogo Prefecture, Fujisawa (Type 3-B), Kanagawa Prefecture, and Naruto (Type 3-C), Tokushima Prefecture as examples of having "multiple comprehensive community support centers in one municipality".

References

Hatakeyama T, Miyazawa H (2016) Conditions of the development of community-based integrated care system: A geographical study of the results of a questionnaire survey taken by municipal governments. Commun Caring 18(14):65–68 (in Japanese)

Hatakeyama T, Nakamura T, Miyazawa H (2018) Community-based integrated care systems in Japan: focusing on spatial structures and local governance. E-Journal GEO 13:486–510 (in Japanese with English abstract)

Inoue N (2011) The functions of the integrated community care system and the role of community general support center. Stud Commun Welfare 39:12–23 (in Japanese)

Ministry of Health, Labour and Welfare (2012) The fifth questionnaire survey on the process of long-term care insurance planning, etc. in municipalities. http://www.mhlw.go.jp/stf/houdou/2r9852000002hvi8.html. Accessed 5 June 2018 (in Japanese)

Miyazawa H (ed) (2017) Mapping health, medical care, and welfare in Japan. Akashi Shoten, Tokyo (in Japanese)

Niki R (2017) Community-based integrated care and welfare reforms. Keisou Syobo, Tokyo (in Japanese)

Sato M, Maeda Y (eds) (2017) Local governance and region. Nakanishiya Shuppan, Kyoto (in Japanese)

Uenoya K, Saito Y (eds) (2015) Welfare governance and social work: an international comparison through vignette surveys. Minerva Shobo, Kyoto (in Japanese)

Community-Based Integrated Care Systems in Municipalities Having One Comprehensive Community Support Center

Tsutomu Nakamura and Teruo Hatakeyama

Abstract Based on the categorization presented in Chapter "Regional Variation in the Community-Based Integrated Care Systems in Japan", this chapter investigates the local governance of community-based integrated care systems in Type 1 areas, in which one comprehensive community support center is placed in each municipality. Because Type 1 areas consist largely of small municipalities, we have selected three small municipalities for this study: Yusuhara Town of Kochi Prefecture, Nanporo Town of Hokkaido, and Shinkamigoto Town of Nagasaki Prefecture. These three municipalities have established centralized local governance led by their respective comprehensive community support centers in the development of the community-based integrated care system. The centralization of local governance has, in some instances, failed to take in some local residents. The availability of medical institutions capable of providing medical services, particularly advanced medical services, are often limited in small municipalities because of their size. This inadequacy of services has made the comprehensive development of the community-based integrated care system difficult. Such a case requires wide-area cooperation with neighboring municipalities that have the necessary resources. Municipalities in Type 1 areas are therefore characterized by the development of single-layered, centralized local governance in connection with the community-based integrated care system. This characterization suggests that the future development of multilayered local governance is likely to be necessary if more localized needs of residents are to be served.

Keywords Community-based integrated care system · Comprehensive community support center · Community care conference · Sphere of daily life · Small municipalities

T. Nakamura (✉)
Ryutsu Keizai University, Tokyo, Japan
e-mail: nakamura-tsutomu@rku.ac.jp

T. Hatakeyama
Naruto University of Education, Naruto, Japan

H. Miyazawa and T. Hatakeyama (eds.), *Community-Based Integrated Care and the Inclusive Society*, International Perspectives in Geography 12,
https://doi.org/10.1007/978-981-33-4473-0_7

1 Introduction

This chapter presents the examination of the characteristics of local governance in building the community-based integrated care system in Type 1 municipalities introduced in Chapter "Regional Variation in the Community-Based Integrated Care Systems in Japan", i.e., municipalities having one comprehensive community support center. The chapter observes three small municipalities as examples of Type 1 municipalities. The reason is that Type 1 municipalities generally comprise small municipalities. When combined with the categories of area sizes for placing community care conferences, Yusuhara, Kochi Prefecture, in which the community care conference serves the entire town, corresponds to Type 1-A. Because many other municipalities correspond to Type 1-A, we selected another example (Table 1 in Chapter "Regional Variation in the Community-Based Integrated Care Systems in Japan"), Nanporo, Hokkaido, while being the same Type 1-A, has established community care conferences of two types, including Community Care Individual Conference serving individuals and the Community Care Promotion Conference serving the entire town. We examine two cases of Type 1 having community care conferences of different kinds. Type 1-B is exemplified by Shinkamigoto, Nagasaki Prefecture, in which community care conferences had been operated in units of spheres of daily life until 2015. In 2016, however, Shinkamigoto established a Community Care Promotion Conference that would serve the entire town in addition to the community care conferences based on spheres of daily life. The town therefore has been Type 1-C since 2016. It was selected to represent Type 1-B and Type 1-C.

The locations of these three cases are exhibited in Fig. 1. Descriptions of the community-based integrated care system of Shinkamigoto using Hatakeyama et al. (2018) were a framework.

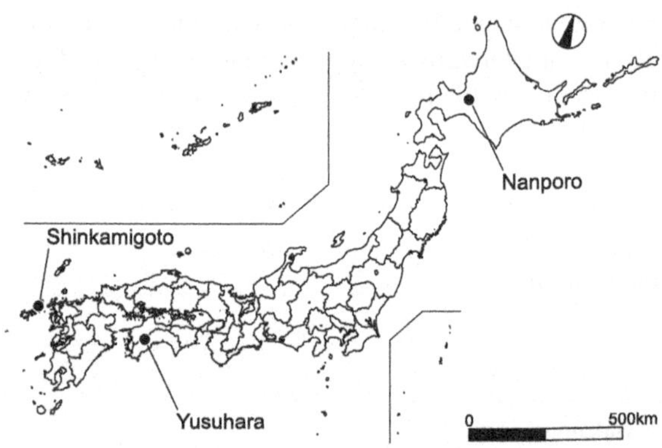

Fig. 1 Locations of towns of Type-1 cases

2 Community-Based Integrated Care System in Yusuhara, Kochi Prefecture: A Case of Type 1-A

2.1 *Reasons for Selecting Yusuhara and Overview of the Area*

Yusuhara is a municipality in which forest comprises 91% of the total area (236.5 km^2) (Fig. 2). The town of Yusuhara was established as the village of Nishitsuno in 1889 after a merger of six villages when the organization of municipalities was implemented. Subsequently, Nishitsuno became the village of Yusuhara (Yusuhara Mura) in 1912 and the town of Yusuhara (Yusuhara Cho) in 1960. Its population according to the 2015 Population Census of Japan was 3,608 with a total of 1,560 households, both of which had decreased approximately 10% from five years prior. The ratio of the population of residents aged 65 years and older as of 2014 was 43.2%, which tends to increase every year. The number of residents certified for the care-required benefits under long-term care insurance was 247 in 2017. Of those residents, those at the level of care required of 3 or above comprised 49.8% (123 residents). That fraction has been increasing gradually. The financial capability index of the town was 0.11 in 2016, indicating its low financial capability with an extremely low percentage of independent (self-generated) sources of income.

The entire town of Yusuhara is specified as a single sphere of daily life, considering that its total population and population of residents aged 65 years and older are below the levels assumed by the national government as the size of a sphere of daily life, which are respectively 20,000 to 30,000 and 3,000 to 6,000. In 2006, a comprehensive community support center operated directly by the local government was

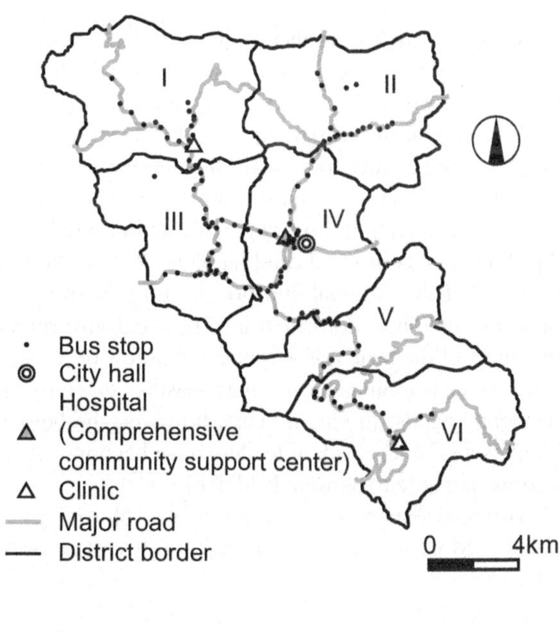

Fig. 2 Distribution of comprehensive community support center and medical facilities, and spheres of daily life in Yusuhara (as of August 2018)

established as a central institution to provide comprehensive assistance and support necessary for helping local senior citizens maintain their physical and mental health, improving public health care, welfare, and medical services, and stabilizing the residents' daily life (Yusuhara Health and Welfare Support Center 2018). The comprehensive community support center provides preventive care management services, comprehensive consultation services, and comprehensive and continuous management support for all town residents (Yusuhara Health and Welfare Support Center 2015). A community care conference for the entire town is held once a month as an opportunity to share information among people engaging in different operations.

Based on the explanation above, Yusuhara falls under the category of Type 1-A among the nine types specified in this study. The area of Yusuhara is almost identical to the average area of Type 1-A (208.5 km^2) (Table 2 in Chapter "Regional Variation in the Community-Based Integrated Care Systems in Japan"). Yusuhara experienced no mergers during the Heisei Municipal Mergers. Its low percentage of merged municipalities (22.3% of all Type 1-A municipalities) is also consistent with the characteristics of Type 1-A (Table 2 in Chapter "Regional Variation in the Community-Based Integrated Care Systems in Japan"). Indicators such as population (18,900) and financial capability index (0.39) are markedly below the averages of Type 1-A. Because, however, the characteristics of Type 1-A include the lowest average population and financial capacity among the nine types, we selected Yusuhara as an example of Type 1-A.

2.2 *Overview of Yusuhara's Community-Based Integrated Care System*

Yusuhara's local governance is characterized by the complementary role of its residents' autonomy system in the government, which is led by the heads of wards elected by the residents in elections (Fig. 3) (Sato 2014). The town has established a council system with strong leadership. The heads were originally nearly equivalent to the mayors of the former six villages that merged into the town. The heads of wards, who are able to deliver the requests and messages of the residents directly to the government, have played an important role as intermediaries among them. Every year in April, the heads of wards and hamlets, representatives of various groups, and local agencies of the national and prefectural governments hold a meeting to exchange opinions and share information. The local government explains its policies at the meetings of the heads of wards and hamlets of Yusuhara. Consequently, bottom-up governance is maintained so that measures taken by the government reflect the local residents' needs. In recent years, however, the burdens on the heads of wards and hamlets and other resident leaders have been increasing because of a lack of people willing and able to assume leadership.

Yusuhara experienced an absence of physicians in 1971 and strove to acquire doctors. Moreover, a mass outbreak of dysentery occurred around 1960, which had

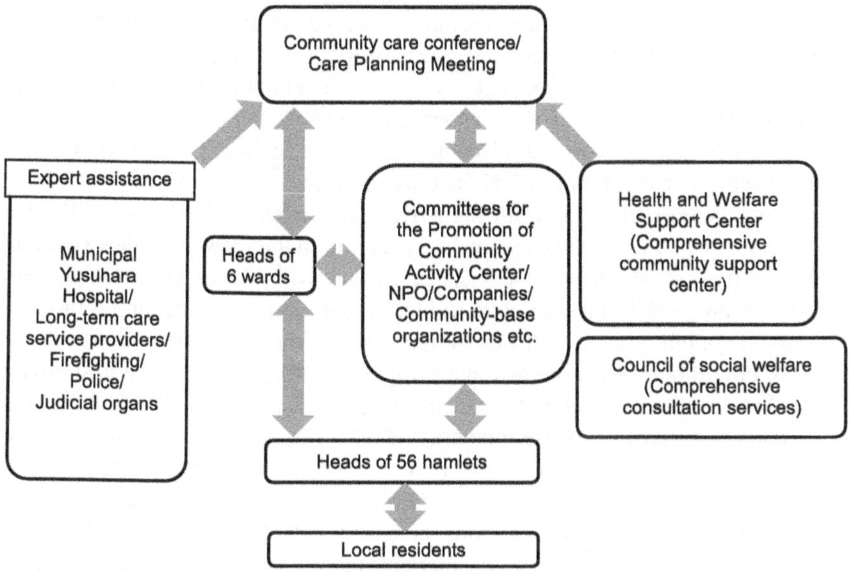

Fig. 3 Community-based integrated care system of Yusuhara (as of June 2018)

prompted the local government and residents to work in cooperation to promote local public health care and welfare for the disease prevention and health maintenance in addition to retaining doctors securely. The local autonomous system of Yusuhara described above has been the basis of its community-based integrated care system.

In terms of infrastructure, facilities that assemble municipal Yusuhara Hospital, Health and Welfare Support Center (health insurance, long-term care insurance, welfare, and health promotion sections, comprehensive community support center, care management center, and comprehensive support centers for childrearing generation), the welfare center for elderly people, and a day-care center run by the council of social welfare were founded in IV ward. Yusuhara, in 1996. This system facilitates efficient information sharing among various sections, services to satisfy needs for health, medical, welfare, and nursing care without physical and time-consuming movement, and various administrative procedures and consultations provided at a single place. The organization is led by the head of Yusuhara Hospital, who oversees the entire operation as the general manager of the Health and Welfare Support Center, to maintain the organizational management in a unified manner.

In April 2018, a new welfare compound was built in IV ward to allow people who were self-reliant yet needing monitoring and people not severely disabled yet needing nursing care to continue living with peace of mind in areas of their long-term residency. A day-care center, a fitness center, and a room for Yusuhara residents' communication were placed on the first floor. On the second floor, nine rooms of the care house (nursing care for tenants of specified community-based facility) for people requiring Levels 1 and 2 of care are placed. The third floor has 18 rooms for the welfare center for elderly people available for the temporary stay of people who are not certified for long-term care insurance care-required benefits or long-term care insurance assistance-required benefits, yet not confident of living at their homes. The council of social welfare acts as a designated manager[1] and oversees the operation of the entire institution (Yusuhara Health and Welfare Support Center 2018).

In Yusuhara, a federation of hygiene associations was established in 1958 to promote the health of the local residents actively. This federation comprises all households in the town. Each of the residents' associations of the six wards elects a hygiene leader. For the federation, the residents, after discussions, recommend one health and hygiene promoter (the current Kenko Bunka No Sato Zukuri Suishinin (promoter of developing communities with health culture)) with a three-year term of office from every 20 households based on the town's original health promoter system started in 1977. The selected residents participate in lectures to acquire knowledge of diseases and to carry out activities such as encouraging other residents to take annual physical examinations and cancer screening as part of their role as intermediaries between medical service providers and residents. Consequently, the percentage of residents who take annual physical examinations increased to 80.4% in 2015, which was the highest in Kochi Prefecture and considerably higher than the national average of 36.3%. Because of the ongoing decline in population, however, the number of residents capable of acting as a promoter is limited in some communities, leading to regional disparities in the burdens of promoters.

In terms of people's activities, monthly community care conferences and weekly care planning meetings have been held as opportunities for information sharing among various functions. The former meetings have been held regularly once a month since 1996 to coordinate services for senior citizens, which were renamed community care conference in April 2000 because of the inclusion of people with disabilities as a subject of discussions (Japan Research Institute 2014). The conferences have the participation of physicians and nurses of Yusuhara Hospital, physical therapists, representatives of comprehensive community support center and care management center, and commissioned welfare and child welfare volunteers to discuss the delivery of welfare services, issues held by local communities, and other relevant topics. In the past, the members have particularly discussed decisions related to admission to adult foster care facilities and welfare centers for elderly people and applications for the financial support provided to households giving home care, among other issues. At present, the meetings are joined also by members of the Council of Social Welfare and the welfare compound established in 2018. The community care conference, however, does not play the role of examining local issues.

Care planning meetings have been held since 1996 when an issue requiring discussions arose. Because a delay in information sharing caused problems in support for hospital discharge, however, care planning meetings have been held once a week since 2008 as regular meetings of health care, medical service, long-term care, and welfare service providers (Japan Research Institute 2014). In addition to physicians,

[1] A system that allows local governments to entrust the management and operation of public facilities under their jurisdiction to private-sector companies, NPOs, and other organizations.

nurses, physical therapists, and dietitians at Yusuhara Hospital, the care management center, comprehensive community support center, and public health nurses for health promotion have been promoting collaboration projects for home care medicine and for long-term care. They primarily discuss care plans for patients admitted to Yusuhara Hospital after home care and the support system for patients after their discharge from the hospital. In addition, they share information related to cases requiring home care assistance. According to them, care managers explain issues related to medical treatment and propose post-discharge service plans to facilitate patients' home care after discharge. Physical therapists join the team to help develop rehabilitation plans.

2.3 *Activities for Yusuhara's Community-Based Integrated Care System*

The council of social welfare was dissolved in 2004 because of the transfer of its conventional day care operations to another operator. Inadequate social support of families and communities for aged persons taking care of other aged persons, however, has persisted as a difficulty. The council of social welfare, then, was incorporated again in 2014 to play the role of a community social service provider capable of identifying issues varying among communities and finding solutions to them. As of August 2018, six members of the council of social welfare were working to monitor all households in each ward to improve community welfare. In Yusuhara, six public health nurses are promoting medical checkups and providing health guidance in the wards of which they are in charge. In response to issues such as poverty and money management that are not covered by insurance services, the members of the local council of social welfare recommend the use of the food bank of Kochi Prefecture council of social welfare or consider the benefits and shortcomings of applying for public assistance.

The personal histories of nine members of the council of social welfare as of August 2016 indicate that five of them (more than half) are from Yusuhara. The staff members, however, include three from outside Yusuhara, Kochi Prefecture, and one from outside Kochi Prefecture, partly because of recruitment inside and outside the town (Table 1). Members of the council of social welfare carrying out activities to monitor all households in each district comprise half from Yusuhara and half from outside the town. Members of the council of social welfare have stated that particularly those from outside the town, in some cases, find it difficult to build trust with residents, which is necessary for understanding their privacy-related needs in disability support services.

In 2017, the council of social welfare moved its corporate office to a new building adjacent to the Health and Welfare Support Center. Currently, the members of the council of social welfare visit one-person households of people aged 80 and older

Table 1 Backgrounds of Staff Members of the council of social welfare (as of September 2016)

Year joined	Position/responsible area	Hometown	Former employer	Reason for joining the council
2014	Director of Administration	Yusuhara	Council of social welfare	Kagura Preservation Society
	III ward	Yusuhara	Auto industry	Referral from acquaintance
	I ward	Yusuhara	Information industry	
	Accounting	Yusuhara	Technical college student	
	Retired	Outside Yusuhara	Welfare industry	
	VI ward	Outside Yusuhara	College student	College seminar
2015	IV ward	Outside Yusuhara	Council of Long-term Care and Social Welfare	Tour
	II ward	Yusuhara	College student	
2016	V ward	Hidaka, Kochi Pref.	College student	

The above does not include the backgrounds of general affairs staff members (Prepared based on interviews)

to understand their needs in disability support services. Many of them reportedly demand easier trash disposal because of limited access to garbage collection points.

In 2018, a conference that would oversee the entire town was established as part of a comprehensive support project. Additionally, the council of social welfare assigned one each of its members to the wards, who would work concurrently as disability support service coordinators. The coordinators attend round-‹ discussions and other meetings in their areas as the member of the conference. The council of social welfare and comprehensive community support center have been holding discussions once a month since 2017 on activities to implement. Moreover, the head of Health and Welfare Support Center and other relevant parties join and hold an annual First-layer conference. The development of assistance programs for independent living, however, has not changed markedly from the past.

3 Community-Based Integrated Care System in Nanporo, Hokkaido: A Case of Type 1-A that Holds Multiple Community Care Conferences

3.1 *Reasons for Selecting Nanporo and Overview of the Area*

The town of Nanporo is located on the Ishikari Plain in central Hokkaido (Fig. 1) according to the Population Census of Japan, its population in 2015 was 7,927. The residents actively engage in agricultural production of rice, cabbage, and other crops. The percentage of workers in primary industries is high: 18.9%. It is approximately 30 min by car to Sapporo. Its proximity to the city helped Nanporo develop as a bedroom community. Particularly, the town structure is simple: the resident population is concentrated in the area around the Nanporo municipal office; other areas are predominantly farmland. Although the population had increased to slightly less than 10,000 by the beginning of the 2000s, it has been decreasing since then. The ratio of elderly people is 29.8%, exceeding the national average, but which is yet low in Hokkaido. The financial capability index was 0.27 in 2016, suggesting low financial capacity. Despite such a financial condition, the town did not experience the Heisei Municipal Mergers. Its area is small at 81.5 km^2.

Because Nanporo is a small municipality, the entire town is specified as a single sphere of daily life in which one comprehensive community upport center operated directly by the local government was established to build its community- based integrated care system (Fig. 4). The town has community care conferences of two

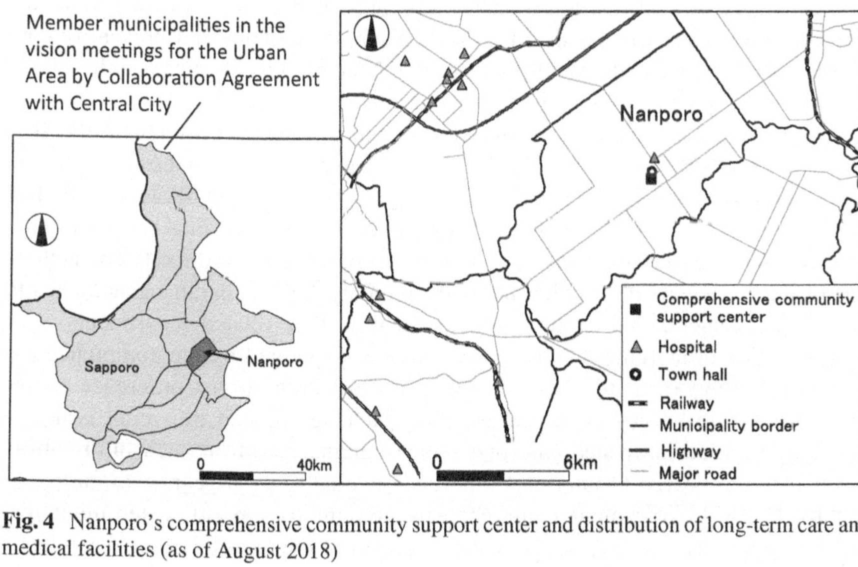

Fig. 4 Nanporo's comprehensive community support center and distribution of long-term care and medical facilities (as of August 2018)

types, including Community Care Promotion Conference serving the entire town and Community Care Individual Conference serving individuals.

Based on the above, Nanporo falls under the category of Type 1-A among the nine types specified in this study. The Nanporo population is close to the average population of Type 1-A communities (18,900). Its financial strength index approximately coincides with the overall average of Type 1-A communities (0.39). Its low percentage of merged municipalities (22.3% overall for Type 1-A), high percentage of direct operation of part of the comprehensive community support center (75.7% overall for Type 1-A), and a number of other characteristics are consistent with the characteristics of Type 1-A municipalities (Tables 2 and 3 in Chapter "Regional Variation in the Community-Based Integrated Care Systems in Japan"). The fact that Nanporo holds the community care conferences of two types, however, differs from Yusuhara's community-based integrated care system that corresponds to the same Type 1-A. Therefore, we selected Nanporo as the second example of Type 1-A.

3.2 Overview of Nanporo's Community-Based Integrated Care System

Nanporo had many residential long-term care facilities such as welfare facilities and health facilities for elderly people when the long-term care insurance system was introduced, which was causing long-term care insurance premiums to increase. The local government, then, shifted the focus of its long-term care policy from facilities to home care and has been placing importance on preventive care to help senior citizens prevent themselves from requiring care. Although the prevention of long-term care has been emphasized in Japan since 2006, Nanporo implemented preventive care ahead of the rest of Japan, which suggests that Nanporo had already established the foundation for the idea of community-based integrated care.

The town placed one comprehensive community support center in 2006. It is operated directly by the municipal government. The reason was that the infrastructure was inadequate because of a lack of private-sector operators and because of the fact that the council of social welfare had not initially been providing long-term care services. While the comprehensive community support center is directly operated by the local government, it is not located in the municipal office, but in a general health and welfare center called "Aikuru" together with welfare-related departments.

Both types of community care conferences are operated by the comprehensive community support center. The Community Care Promotion Conference serves the entire town of Nanporo for information exchange related to overall issues in community-based integrated care and policy-making based on such information. The Community Care Promotion Conference is held twice a year, with participation by the local government, comprehensive community support center, municipal hospital, long-term care service providers, council of social welfare, local residents, and community groups (Fig. 5).

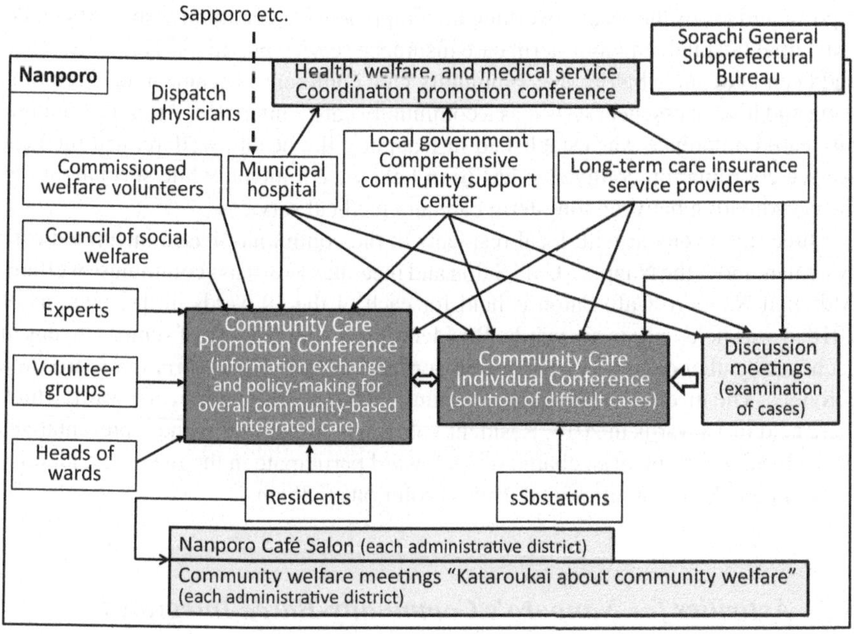

Fig. 5 Community-based integrated care system centering on community care conference in Nanporo (as of August 2018)

The Community Care Individual Conference covers the entire town of Nanporo and works to resolve difficulties faced by the residents. In other words, although the subject of Community Care Promotion Conference is the town, the subject of Community Care Individual Conference is the individuals. The Community Care Individual Conference is held once or twice a year. Although the participants vary depending on the case to be addressed, the meetings are generally joined by representatives of the local government, comprehensive community support center, municipal hospital, long-term care insurance providers, Sorachi General Subprefectural Bureau, and relevant substations. The results of the Community Care Individual Conference are reported to the Community Care Promotion Conference, which discusses solutions to issues and policy-making and which gives feedback to the Community Care Individual Conference.

Cases not developing into difficult issues are discussed within the comprehensive community support center rather than at the Community Care Individual Conference. This meeting has participation by representatives of the local government, comprehensive community support center, municipal hospital, and long-term care service providers, among others.

Additionally, a health, welfare, and medical service coordination conference is held for supporting partnerships among medical and long-term care providers to achieve comprehensive and continuous care at facilities and homes in the region. This conference is held once a month for the entire town of Nanporo. Participants include representatives of the local government, comprehensive community support center, municipal hospital, and long-term care insurance service providers. This conference had been held since before the community care conference became a legal requirement and had, in essence, served as a community care conference. After the community care conference was established in Nanporo, the health, welfare, and medical service coordination conference has been held once a month while specializing in the coordination between long-term care and medical services.

Direct involvement with local residents in the community-based integrated care system includes the Nanporo Café Salon and meetings to discuss community welfare. Although Nanporo Café Salon is held for each of the 19 wards in the town, as of 2018, it was held only in six wards. Residents consisting largely of senior citizens in each ward gather once every two weeks for the Café Salon and carry out spare-time activities. The meetings to discuss community welfare are held for each ward, which were held in 13 wards in 2018. Resident volunteers in each ward and representatives of the local government in charge of each ward participate in the meetings, identify issues in local communities, and discuss solutions to them.

3.3 Activities for Nanporo's Community-Based Integrated Care System

The following examines the two community care conferences, residents' participation, and coordination between medical and long-term care services to identify the activity status of Nanporo's community-based integrated care system.

In 2017, the Community Care Promotion Conference observed information related to local population aging, operations of the comprehensive community support center, and the delivery of community-based services. A meeting to address the prevention of abuse of elderly people and a meeting of the intensive support team for patients at an early stage of cognitive impairment were held simultaneously to increase coordination. This meeting allowed numerous local residents to be involved and had a system for receiving their opinions.

The Community Care Individual Conference is held once a year. In 2017, the members discussed difficulties of solitary senior citizens involving their neighbors. The comprehensive community support center leads meetings to discuss such cases that do not develop into difficult issues. Many recent issues concern the long-term care of solitary senior citizens and cognitively impaired senior citizens.

Nanporo Café Salon and meetings to discuss community welfare held in each ward have not been held in all wards. An interview with the comprehensive community support center has revealed that the wards in which such meetings are held have persons taking leadership in holding the meetings. Holding the meetings in the wards in which such leaders are absent is a challenge to be addressed. Even those wards in which the meetings have been held are facing issues. The primary purpose of the Café Salon is to facilitate the gathering and conversations of local residents, which rarely develop into specific activities. Meetings to discuss community welfare address local issues and seek solutions, which, however, consist largely of requests to the local government and rarely of cases of continuous activities conducted by the residents. Furthermore, these two types of meetings have not established collaboration with other conferences in the community-based integrated care system of Nanporo, as shown in Fig. 5. While the residents to participate in the Community Care Promotion Conference are collected from the general public, the participating residents are repeaters. As this suggests, residents' participation is a significant issue in the community-based integrated care system.

Subsequently, the following examines coordination between medical and long-term care services. Nanporo has Nanporo Municipal Hospital with a long history since its foundation[2] in 1947. It plays an important role in Nanporo's community-based integrated care system. This hospital, however, is small, with only 80 beds (26 general sickbeds and 54 beds for long-term care). Many of its physicians are dispatched from outside of Nanporo town (Fig. 5). It is also not capable of performing difficult surgeries.

For this reason, severely ill or injured Nanporo residents use large hospitals in other areas to undergo surgery or other treatment. Many Nanporo residents request post-surgery treatment and spend daily life in their home town, which requires cooperation among hospitals outside the town, Nanporo Municipal Hospital, health facilities for elderly people in Nanporo, and other institutions. Nanporo Municipal Hospital established the Community Medical Cooperation Section in 2010 and has been operating a partnership and health facilities for elderly people. Other long-term care facilities, too, have been taking steps individually for such a partnership. Such cooperation with hospitals outside the town has yet to be established with the town of Nanporo as a whole. At present, the institutions are individually taking measures. Nanporo must incorporate such partnerships with institutions outside the town into its community-based integrated care system.

In addition, Nanporo plans to join the Urban Area by Collaboration Agreement with Central City ("Urban Area")[3] led by Sapporo and attends meetings to discuss the vision of such collaboration as preparatory steps (Fig. 4). Wide-area cooperation in medical and long-term care services is also discussed as part of the Urban Area by Collaboration Agreement with Central City, which is likely to affect Nanporo's community-based integrated care system. Nanporo residents' current connection with the municipalities comprising the Urban Area is strong in their daily life, and particularly medical services. The town's participation in the Urban Area might therefore contribute to more effective cooperation in medical and long-term care services.

---

[2]It was a clinic when it was established; it became a hospital in 1953.

[3]A policy aimed at facilitating "leadership in economic growth," "integration and strengthening of high-level urban functions," and "improvement of services for life-related functions" of central cities acting as a core of an area that includes neighboring municipalities through compact-sizing and networking to revitalize communities, make the economy sustainable, and help residents maintain secure and comfortable life even in a society facing declining population and birth rate and aging population.

Nanporo, however, belongs to Sorachi General Subprefecture of Hokkaido. The Subprefectural system is Hokkaido's unique scheme unseen in other parts of Japan. Subprefectural Bureaus are agencies of the Hokkaido government. Many of its interactions with Nanporo's local government are conducted through the local Subprefectural Bureau. In fact, Nanporo's Community Care Individual Conference also has participation by the Sorachi General Subprefectural Bureau. Many of the municipalities comprising the Urban Area which Nanporo plans to join, however, belong to Ishikari Subprefecture, which has administrative procedures that differ from those of Nanporo. At present, Nanporo has strong ties with municipalities belonging to Sorachi General Subprefecture. Its ties with those belonging to Ishikari Subprefecture are weak. Therefore, a gap exists in the spatial frameworks of the current administrative procedures and medical services and other parts of the residents' daily life, which must be reduced from now on.

4 Community-Based Integrated Care System in Shinkamigoto, Nagasaki Prefecture: A Case of Type 1-B and Type 1-C

4.1 Reasons for Selecting Shinkamigoto and Overview of the Area

Shinkamigoto is a municipality with area of 214.0 km^2, as established by merging five towns in August 2004. It is located at the western tip of Nagasaki Prefecture and on the islands. It consists of seven inhabited islands and 60 uninhabited islands (Fig. 6). Much of the area is covered with large undulations with small areas of level ground along the shores. The Shinkamigoto population has been declining according to the Population Census of Japan: it was 38,140 in 1980, but fell by nearly half to 19,718 by 2015. Its financial capability index was 0.24 in 2016, suggesting the towns' low financial strength because of a decline in its basic industries and population. The ratio of the population of residents aged 65 years and older as of 2015 was 37.7%, indicating severe population aging. The southern part of the island has a higher ratio.

Shinkamigoto has set up spheres of daily life based on five towns that existed before the merger (Fig. 6). One comprehensive community support center was placed in 2006 following the merger as an institution directly operated by the local government. A meeting to discuss the operation of the center is held once a year: the members report the results and issues related to operations. In December 2017, the center was separated from the municipal office and was relocated to a building adjacent to Nagasaki Kamigoto Hospital to improve its functions and to increase convenience for the residents (Shinkamigoto, Nagasaki Prefecture, 2018). For the consultation service, the center assigns staff members to each of the spheres of daily life to improve its collaboration with relevant institutions and to organize local issues (Shinkamigoto, Nagasaki Prefecture 2018).

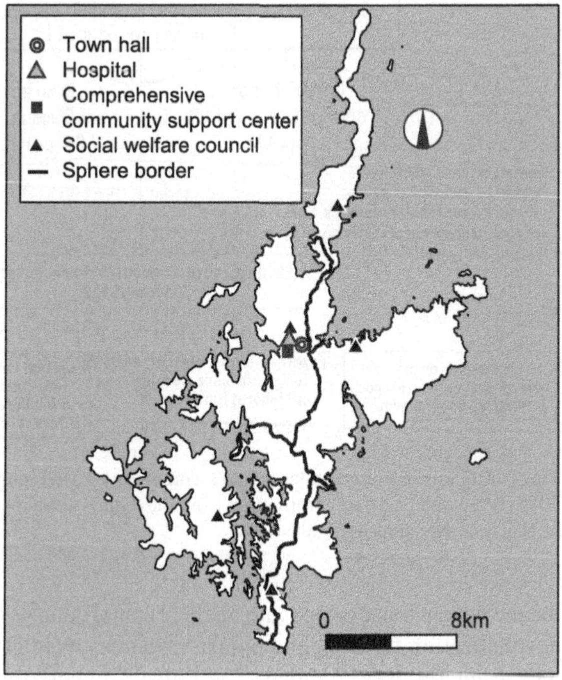

Fig. 6 Distribution of comprehensive community support center and medical facilities, and spheres of daily life in Shinkamigoto (as of July 2018)

The community care conference is held for each of the spheres of daily life, with discussion of solutions to issues such as difficult cases to support. In 2016, however, a new Community Care Promotion Conference was established to cover the entire town. The conventional community care conference became individual care conferences.

Based on the circumstances presented above, Shinkamigoto falls under a case that has changed from Type 1-B to Type 1-C. The population, area, financial capability index, and other attributes of Shinkamigoto are closer to the averages of Type 1-C (Tables 2 and 3 in Chapter "Regional Variation in the Community-Based Integrated Care Systems in Japan").

4.2 *Overview of Shinkamigoto's Community-Based Integrated Care System*

Figure 7 exhibits the organizational chart of Shinkamigoto's community-based integrated care system as of October 2015. Because Shinkamigoto lacks resources for medical and long-term care services and human resources, Kamigoto Hospital has been taking the lead in concentrating medical functions and developing long-term care resources (Nakamura 2014). At the time of the merger in 2004, two hospitals in

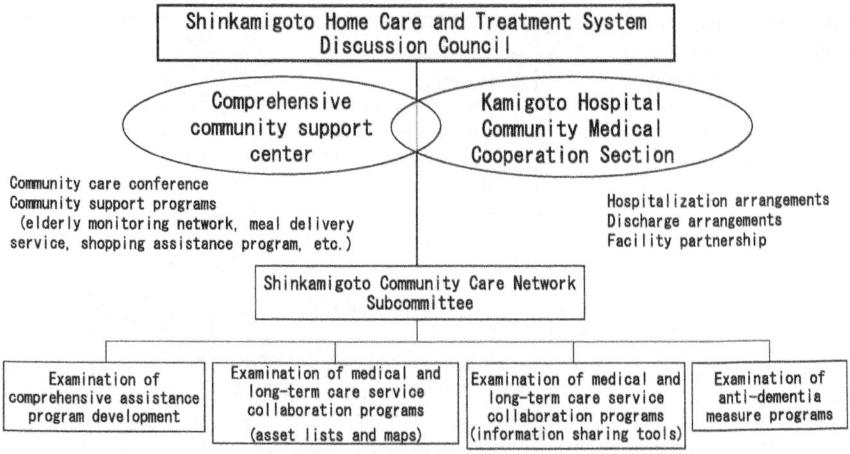

Fig. 7 Organization of Shinkamigoto's community-based integrated care system (as of October 2015) (Prepared based on information from the 37th Nagasaki Community Medicine Study Group of Nagasaki Hospital Agency)

the town were transformed into medical centers that would treat only outpatients. The hospitalization function and human resources were gathered at Kamigoto Hospital. As a result, services provided to cognitively impaired elderly people having no family and solitary aged persons requiring monitoring became important issues to continued home care and treatment, including preventive care and disability support services, particularly in areas near the town border. Given that background, the activities of Kamigoto Hospital have come to affect the development of the community-based integrated care system strongly in Shinkamigoto.

The process of developing the community-based integrated care system in Shinkamigoto is the following: First in 2013, a preparatory committee for community-based care network system was established based on Kamigoto Hospital's proposal to the municipal government. In 2014, the Home Care and Treatment System Discussion Council was established by developing this preparatory committee. This council confirmed its plan to improve cooperation with the comprehensive community support center and the Community Medical Cooperation Section of Kamigoto Hospital for the collaboration in the administrative Department of the Community-based Integrated Care System. Additionally, Community Care Network Subcommittee was founded as a group that would carry out the actual activities and information gathering. Organization of teams for working-level discussions was proposed in 2015 (the four teams in the lower section of Fig. 7). Subsequently, in December 2017, this council was replaced by Community Care Promotion Council, which had been established as a new council that would serve the entire town. The reason was an increase in persons certified for long-term care insurance care-required benefits, who needed disability support services rather than physical nursing. New workers, including senior citizens, who would provide disability support services, were needed because of a lack of human resources amid growing demand for an

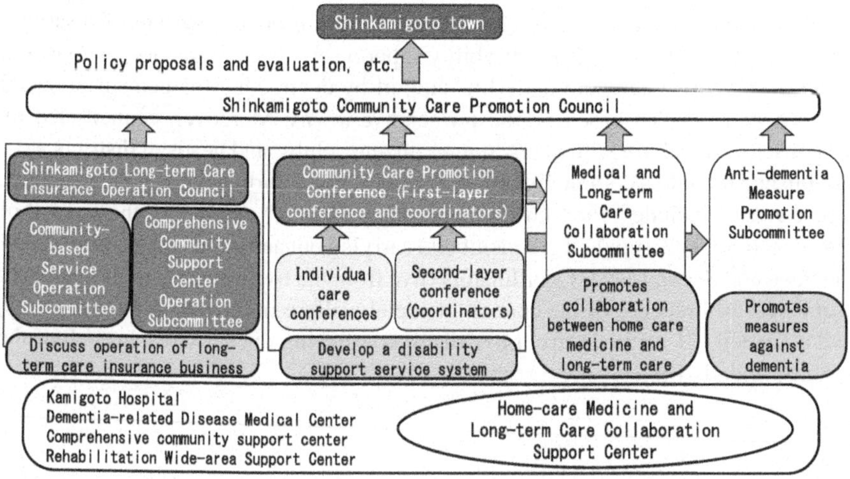

Fig. 8 Structure of Shinkamigoto's Councils Related to Promotion of community-based integrated care system (July 2018) (Prepared based on information from Shinkamigoto Community Care Promotion Council)

increase in domestic helpers for persons in need of specialized physical nursing at their homes.

Because of such issues, a shift was underway to a new system that would specifically examine the collaboration between home care medicine and long-term care and development of infrastructure for disability support and preventive care services (Fig. 8).[4]

The support center for the collaboration of home care medicine and long-term care was placed in the Community Medical Cooperation Section of Kamigoto Hospital in April 2016. The system of providing medical and long-term care services in a unified manner was established by entrusting the operation with the hospital. The hospital's nurses work concurrently at this support center to respond to inquiries from people engaging in medical and long-term care services. The center also examined methods of sharing information and promoting the use of such methods, developing partnerships with nursing care providers, and holding periodical discussions with the welfare and longevity section of the municipal government and comprehensive community support center. The comprehensive community support center was relocated to a building adjacent to Kamigoto Hospital partly for improving cooperation, which has caused an increase in the number of consultations.

For the development of infrastructure for disability support and preventive care services, the municipal government has been working to develop human resources

[4]This council consists of 17 members from the town, including healthcare practitioners (physicians, pharmacists, and physical therapists), long-term care providers (heads of long-term care facilities and care managers), welfare service providers (Council of Social Welfare, disability support service coordinators, and commissioned welfare volunteers), government-related parties, and parties specially approved by the mayor (disability support service providers and taxi business operators).

to support the independence of elderly people and to build a system of disability support services. In April 2016, disability support service coordinators,[5] who would provide preventive care and daily life support by contracting the council of social welfare, were placed based on the former five towns. Moreover, councils led by commissioned welfare volunteers, senior citizens' clubs, and heads of districts were established to build a community for wide-area support participated by the residents. These groups include a First-layer Coordinator (one staff member of the council of social welfare) and a council (18 members) carrying out activities in the entire area of the town and Second-layer Coordinators (five from the five districts, all of whom are employees of branches of the council of social welfare in their respective districts) and a council (61 members from five districts) carrying out activities in each of the spheres of daily life.

4.3 Activities for Shinkamigoto's Community-Based Integrated Care System

It has not been long since the Community Care Promotion Council was established. This section therefore examines the activities of the Community Care Network Subcommittee of the Home Care and Treatment System Discussion Council, which is the predecessor of the current Community Care Promotion Council, and individual Community Care Conferences held in each of the spheres of daily life during the period between 2015 and 2017.

4.3.1 Community Care Network Subcommittee

The team examining the development of comprehensive assistance programs of the Community Care Network Subcommittee held a meeting in 2015, studied the initiatives taken by forerunner communities, and examined local resources of Shinkamigoto. Because of the commissioning of the duties of disability support service coordinators to the council of social welfare in 2016, however, the role of the team was transferred to the council and activities as the team were ended. The disability support service coordinators investigate the status of disability support and preventive care services, identify the need for assistance, develop service providers, and match patients and service providers. Although the team did not have track records, the coordinators pointed out the following two issues at district round-table discussions and other meetings as their track records. One is that a system for linking community issues to specific assistance activities has not been established; the other

[5]Disability support service coordinators aim to promote the development of systems for disability support and preventive care services provided to elderly people and to coordinate the development of systems for providing disability support and preventive care services in communities.

is that the program of maintaining information required in an emergency for aged residents' households must be improved.

The team for examining the collaboration program of medical and long term care services is divided into a resource list and map section and an information-sharing tool section. The former held four meetings in 2015 and discussed details of the resource lists. The latter held one meeting in 2015 and three meetings in 2016. They discussed the implementation and result of a questionnaire survey of pharmacists primarily in charge of medical management. The necessity of regular meetings of both parties related to long-term care and medical services to understand mutual operations better, the inadequacy of appointment books at the municipal clinics, and the need for information maintained at the home of representatives of elderly people were pointed out among other issues.

The team for examining the program of promoting measures against dementia held one meeting in 2015 and three in 2016. The members exchanged opinions about the development of measures and system for dementia patients such as building a scheme for the early detection and treatment of dementia while sharing information particularly related to the condition of dementia patients in the town and issues related to them. They pointed out a lack of institutions that would undertake early treatment, the necessity of easily accessible liaisons for consultations, the need for improvement of collaboration between medical and long-term care providers and patients' families, attending physicians, medical specialists, etc., delay in explanations given to the residents, development of dementia supporters, and creation of opportunities for their activities.

4.3.2 Community Care Conference

Since 2015, the community care conference has been held once or twice a month on a regular basis in each sphere of daily life. Participants include staff members of the comprehensive community support center in charge of each sphere of daily life, members of home care support center, care managers, long-term care insurance-certified providers, municipal government employees, employees of public health centers, pharmacists, nurses, members of the council of social welfare, and residents related to patients. They are working to find solutions to difficult cases to support, build networks, and organize community issues. Issues that have been pointed out include a monitoring system, protection of rights, support for travel, and a system of assisting light housework (Shinkamigoto, Nagasaki Prefecture 2018).

Themes discussed at individual care conferences in 2017 included 1, disability support services for persons having delusions of theft, 2, support for elderly couples including their sons in need of employment support, 3, support for the cognitively impaired elderly living alone, 4, support for short stays of dementia patients having a strong desire to go home, and 5, support for elderly men having a strong desire to be independent, who are injured repeatedly from falling. Reportedly, all of these issues were addressed by defining policies for measures or developing assistance plans through information sharing among the relevant parties.

Community issues identified at individual care conferences included difficulty for elderly people whose physical abilities were declining, in going out and supporting their outing because of the land covered with large undulations and a small area of level ground. Weakening of family relationships and inadequate partnership for providing a combination of disability welfare and long-term care insurance services also contribute to the difficulty in assisting elderly people's outings.

5 Characteristics and Issues in Local Governance of Type-1 Municipalities

The Type 1-A town of Yusuhara has long been building local governance to facilitate an increase in the percentage of residents who take annual physical examinations and preventive care through activities to promote health and hygiene and mutual aid involving the local residents. Hospitals and welfare and health care facilities concentrate in the central area, allowing service providers to let all functions of the community-based integrated care for immediate mutual collaboration. Residents having access to the central area of the town benefit from being able to process applications and procedures for the use of services at one place.

However, a lack of human resources for health promotion and residents' associations caused by depopulation and population aging has been discovered. At present, when the mutual aid system in peripheral areas itself is becoming increasingly difficult to function, the council of social welfare, as a general advice desk, is expected to grasp the need for disability support services in the peripheral areas and find solutions. Although staff members of the council of social welfare concurrently work as disability support service coordinators in the project to build a system of disability support services as part of the comprehensive support project, they have yet to understand local issues specifically in each district to find solutions. It is fundamentally important that the council of social welfare gain thorough knowledge of local situations and build trust with residents to play the role of coordinators who support the residents' independence.

Nanporo, also a Type 1-A town, has built its community-based integrated care system by establishing the Community Care Promotion Conference covering the entire town and the Community Care Individual Conference serving individuals and holding discussion meetings based on one comprehensive community support center. The town is also developing collaboration between medical and long-term care services while working in cooperation with the municipal hospital in the central area and long-term care insurance service providers. Such practices of the community-based integrated care system are the result of making effective use of small town characteristics. Partly because of inactive residents' activities in communities, however, the town has yet to incorporate residents' activities successfully in the community-based integrated care system.

In the medical field, difficulty in responding to severe cases because of the small size of Nanporo Municipal Hospital and the recruitment of physicians from outside the town are issues that must be addressed. For this reason, Nanporo residents often use medical institutions outside the town. In many cases, however, the patients desire to return to Nanporo after undergoing surgery outside the town. An important difficulty is that medical institutions outside the town and medical and long-term care facilities in Nanporo must cooperate. That need for cooperation has not been incorporated into Nanporo's community-based integrated care system.

To respond to such issues, Nanporo plans to join the Urban Area through a Collaboration Agreement with Central City led by Sapporo. Sapporo and other cities have joined this scheme. Wide-area cooperation in medical and long-term care fields is considered, which might help solve the issues faced by Nanporo. Nanporo, however, has a route of administrative procedures different from that of the Urban Area because of its affiliation with Sorachi General Subprefectural Bureau. Necessary arrangements must be made.

Therefore, although Nanporo has built a single-layered, centralized local governance in its community-based integrated care system, it must build multilayered local governance comprising the Urban Area and Nanporo town by developing wide-area cooperation with other municipalities. This cooperation might make it difficult for Nanporo to take initiatives in the operation of its own community-based integrated care system. Building a new type of local governance will be another challenge.

Shinkamigoto, which has transformed itself from Type 1-B to Type 1-C, has formed centralized governance led by Kamigoto Hospital in medical services. This transformation, however, affected the town in terms of support systems for the continuation of home care and treatment including preventive care and disability support services, which attempted to build decentralized local governance led by disability support service coordinators from the council of social welfare and local residents providing informal support. The First-layer conference carrying out activities in the entire area of the town is one of four sub-themes under the Community Care Promotion Council, which is position as the core in the development of a system of disability support services and the town-level community care promotion conference. Under this, there are the Second-layer conference and individual care conferences carrying out activities in each of the spheres of daily life. In other words, a triple-layered structure placing the Community Care Promotion Council at the top is currently formed in the development of a system of disability support services, which is emphasized in Shinkamigoto's community-based integrated care (Fig. 8). At the base of the structure, the support center for the collaboration of home care medicine and long-term care is placed along with Kamigoto Hospital and comprehensive community support center, which is given the role of intermediating multiple institutions for collaboration.

Because such local governance functions effectively in Shinkamigoto, the role of disability support service coordinators who build a collaboration system that can respond flexibly to needs that vary among the districts is likely to become increasingly important. At present, however, the Second-layer conference is reportedly not functioning well. In many cases, individual care conferences are held based on communities. Yet, even the individual care conferences, in some cases, are purportedly limited to sharing information related to local issues. Such a situation is likely to result from the local governance with a strong nature of structure centralized at Kamigoto Hospital and the comprehensive community support center as the basis of the town as a whole, even while aiming for decentralized local governance in the field of disability support services. The understanding of volunteer activities conducted by various residents, including senior citizens, and development of human resources will be necessary to improve the function of the Second-layer conference. Additionally, physicians at hospitals and primary care doctors at clinics and organizations such as long-term care insurance providers, the municipal government, the council of social welfare, and the comprehensive community support center must make use of their specialized knowledge, efficiently share information related to a daily basis, and further develop a cooperative system that will prevent any person from being neglected in the assistance.

As described above, all three towns presented as examples of Type 1 have built centralized local governance led by one comprehensive community support center. They are small municipalities and are taking advantage of being able to organize cooperation among various institutions on a small scale. Particularly, Yusuhara and Nanporo have only one sphere of daily life each. They are therefore more centralized than Shinkamigoto.

They, however, are also facing challenges. While centralized governance facilitates collaboration among different fields, it might lead to an obstacle in carrying out activities corresponding to the characteristics of each community. Yusuhara lacks the access of residents in remote areas to the central area, and Nanporo has difficulty in taking its residents in the community-based integrated care system. These are characteristics of Type 1-A towns, in which the municipality is the only unit of placing a community care conference.

In Shinkamigoto, the comprehensive community support center has assigned personnel to each of the spheres of daily life based on the former towns and villages and held community care conferences in each of the areas. The support center also assigned disability support service coordinators to respond to needs in each area. While building such a centralized system with one comprehensive community support center, the town aims to develop a decentralized system to respond to more specific needs of local residents. This is a typical characteristic of Type 1-B communities. Because the decentralized system lacking specialized human resources makes it difficult to respond to the residents' needs, the town established the Community Care Promotion Conference based on the municipality using economies of scale. The purpose was to make policies from needs in each of the spheres of daily life on the scale of the entire town while using the decentralized system. In other words, Shinkamigoto transformed itself to Type 1-C, which can use economies of scale, to solve difficulties related to scattered human resources in Type 1-B. Shinkamigoto, however, faces difficulties in building collaboration between the first layer (the area of the town) and the second layer (spheres of daily life) and developing multilayered local governance.

6 Conclusions

The result of examining the local governance of Type-1 municipalities suggests that the local governance of Type 1-C is ideal for building a community-based integrated care system corresponding to the residents' needs, considering the concentration of local issues relevant to the residents and applying them to policy-making. Many Type-1 municipalities, however, are small towns and villages, which are likely to face financial and human resource difficulties. Therefore, it is more practical, like Type 1-A, to link the comprehensive community support center and relevant institutions, build centralized local governance on a small-scale, and maintain close cooperation with parties familiar with the residents such as residents' and neighborhood associations and commissioned welfare volunteers to make the community-based integrated care system function.

In this situation, difficulties in developing full-scale community-based integrated care system are expected to arise because of the small size of the municipalities. In this case, development of multilayered local governance, including wide-area cooperation of multiple municipal governments, must be addressed in future plans because wide-area cooperation often becomes necessary, as observed in Nanporo.

References

Hatakeyama T, Nakamura T, Miyazawa H (2018) Community-based integrated care systems in Japan: focusing on spatial structures and local governance. E-Journal GEO 13:486–510 (in Japanese with English abstract)

Japan Research Institute (2014) Subsidies for health promotion services for the elderly fiscal year 2013: Report of survey on development of community-based integrated care system by municipal governments "case of developing a community-based integrated care system". https://www.kaigokensaku.mhlw.go.jp/chiiki-houkatsu/files/mhlw_care_system_2014.pdf. Accessed 14 Nov 2018 (in Japanese)

Nakamura T (2014) Mechanism of health care restructuring in Kamigoto area, Nagasaki Prefecture. Jpn J Hum Geogr 66:405–422 (in Japanese with English abstract)

Sato M (2014) A study of contemporary meaning of the traditional self-government system: Kucho-sei system and the challenge to community revitalization of Yusuhara-cho, Takaoka-gun, Kochi. Shikoku Gakuin University Treatises 143:29–46 (in Japanese)

Shinkamigoto, Nagasaki Prefecture (2018) Shinkamigoto elderly welfare plan and the seventh phase long-term care insurance business plan (in Japanese)

Yusuhara Health and Welfare Support Center (2015) The sixth phase Yusuhara elderly welfare plan: 2015–2017 long-term care insurance business plan (in Japanese)

Yusuhara Health and Welfare Support Center (2018) The seventh phase Yusuhara elderly welfare plan: 2018–2020 long-term care insurance Business plan (in Japanese)

Community-Based Integrated Care Systems in Municipalities Having Sub-branches of Comprehensive Community Support Centers

Teruo Hatakeyama, Shin'ichiro Sugiura, and Hitoshi Miyazawa

Abstract This chapter identifies the characteristics of local governance in building the community-based integrated care system of municipalities corresponding to Type 2 among the types defined in Chapter "Regional Variation in the Community-Based Integrated Care Systems in Japan", which have established sub-centers or branches in addition to their comprehensive community support centers. Because many Type 2 municipalities are small cities, we have selected three cities: Goto in Nagasaki Prefecture, Tsugaru in Aomori Prefecture, and Fukutsu in Fukuoka Prefecture. These three municipalities sought to develop decentralized local governance based on sub-centers and branches placed in each sphere of daily life with the comprehensive community support center at the top to draw the local residents into the community-based integrated care system. Because of financial, human resource, and other issues, however, local governance in units of spheres of daily life did not function properly. Consequently, the cities built centralized local governance led by their local governments and comprehensive community support centers. This centralization suggests that limitations exist in the development of decentralized local governance through sub-branches in the community-based integrated care system of Type 2 municipalities.

Keywords Community-based integrated care system · Community care conference · Comprehensive community support center · Small cities · Sphere of daily life · Sub-center/branch

T. Hatakeyama (✉)
Naruto University of Education, Naruto, Japan
e-mail: thatakeyama@naruto-u.ac.jp

S. Sugiura
Meijo University, Nagoya, Japan

H. Miyazawa
Ochanomizu University, Tokyo, Japan

H. Miyazawa and T. Hatakeyama (eds.), *Community-Based Integrated Care and the Inclusive Society*, International Perspectives in Geography 12,
https://doi.org/10.1007/978-981-33-4473-0_8

Fig. 1 Locations of cities of Type-2 cases

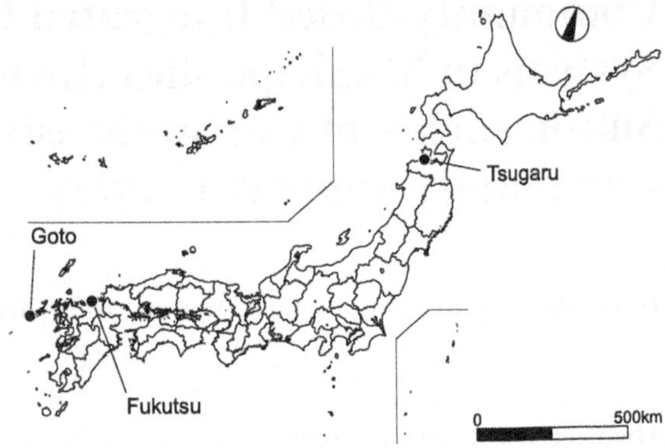

1 Introduction

This chapter presents examination of the characteristics of local governance in building the community-based integrated care system of Type 2 municipalities, which have placed sub-centers or branches of their comprehensive community support centers. The majority of Type 2 municipalities are small cities. Three small cities have therefore been selected as examples for this chapter. When combined with the types of community care conference, the city of Goto in Nagasaki Prefecture, which holds community care conferences for the entire city, corresponds to Type 2-A. An example of Type 2-B is the city of Tsugaru in Aomori Prefecture, in which community care conferences are held in each sphere of daily life. An example of Type 2-C is the city of Fukutsu in Fukuoka Prefecture, which holds community care conferences for both the entire city and each sphere of daily life. Figure 1 presents the locations of these three cities. Discussion of the community-based integrated care system of Tsugaru is based on a report of a study by Hatakeyama et al. (2018).

2 Community-Based Integrated Care System in Goto, Nagasaki Prefecture: A Case of Type 2-A

2.1 Reasons for Selecting Goto and Area Overview

Goto is located on islands separated by about 100 km distance to the west of the prefectural capital of Nagasaki Prefecture (Fig. 2). Its 11 inhabited islands exist in the city area. The City Hall is near the port on the eastern extent of the main island, which has area of 326.3 km^2. Goto's population according to the 2015 Population Census of Japan was 37,327. The city's population had decreased by 16.6% during the 10 years before 2015; the percentage of population aged 65 or older had increased

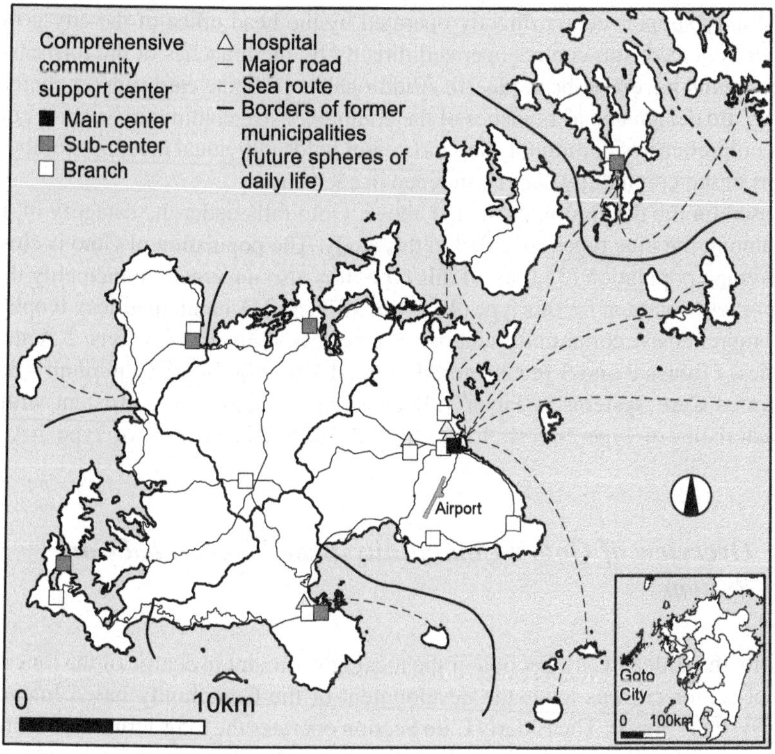

Fig. 2 Units of building community-based integrated care system and comprehensive community support center in Goto (as of September 2017)

from 33.4 to 36.7%. Goto's financial capability index is 0.24, indicating its low financial strength.

The percentage of residents certified for long-term care insurance care is more than 20%. The need for facility care services is particularly strong. This need is reportedly attributable to the low family care capacity of many residents caused by the outflow of their children's generations to outside the islands and a lack of rehabilitation services to reduce the need for care and assist patients' return to home care. Because numerous people use facility care services, the long-term care insurance premium paid by First-category insured persons is 6,233 yen, the highest in Nagasaki Prefecture. Strong demand for preventive care, home-care medicine, and long-term care is among Goto's challenges in its measures for senior citizens.

Goto city was established through the merger of one city and five towns in 2004. The regional divisions based on the former municipalities strongly remain even after the merger, which has affected the development of Goto's community-based integrated care system. Goto set the pre-merger municipalities as the basic unit in building its community-based integrated care system. Ten spheres of daily life that currently exist will be reorganized into these six areas. Whereas one comprehensive community support center directly operated by the head office of the city government was placed, sub-centers operated directly by the branches of the former local governments have also been placed. Additionally, 10 home care support centers in the city are designated as branches of the comprehensive community support center. The comprehensive community support center holds individual meetings for the city as part of the community care conference in each case.

Based on the discussion presented above, Goto falls under the category of Type 2-A among the nine types specified in this study. The population of Goto is close to the average population (35,000) of this type. It is also a merged municipality that is more or less common for this type (48.3% of all Type 2-A municipalities). It operates the comprehensive community support center directly (86.7% of all Type 2-A municipalities) (Tables 2 and 3 in Chapter "Regional Variation in the Community-Based Integrated Care Systems in Japan"). Because these factors are consistent with the characteristics of Type 2-A, we have selected Goto as an example of Type 2-A.

2.2 *Overview of Goto's Community-Based Integrated Care System*

In Goto, the Elderly Care Section of the local government in charge of the long-term care of senior citizens leads the development of the Community-based Integrated Care System (Fig. 3). The Elderly Care Section operates the long-term care insurance programs and the comprehensive disability prevention and assistance program for independent living and voluntary programs as part of the city's community support programs. The Comprehensive Community Support Center was placed in this Elderly Care Section in 2006; 16 staff members (including 10 part-time employees) of the Elderly Care Section concurrently operate the Support Center. The primary duties of the Comprehensive Community Support Center include preventive care management, comprehensive consultation services, advocate services, collaboration of home care medicine and long-term care, measures against dementia, and improvement of disability support services. Holding the Community Care Conference is part of the key operations of the Comprehensive Community Support Center, which has been holding individual meetings for each case since 2012.

As noted earlier, Goto has established five sub-centers and 10 branches of the comprehensive community support center. A public health nurse concurrently working at another section is assigned to the sub-centers. Operation of the home care support centers, which are branches of the comprehensive community support center, is commissioned to the council of social welfare, social welfare corporations, and medical corporations in the city. These sub-centers and branches provide comprehensive consultation services, which are expected to be accessible opportunities for consultations of the residents.

In March 2017, the liaison committee of the community-based integrated care system was established in Goto. The local government and a number of institutions

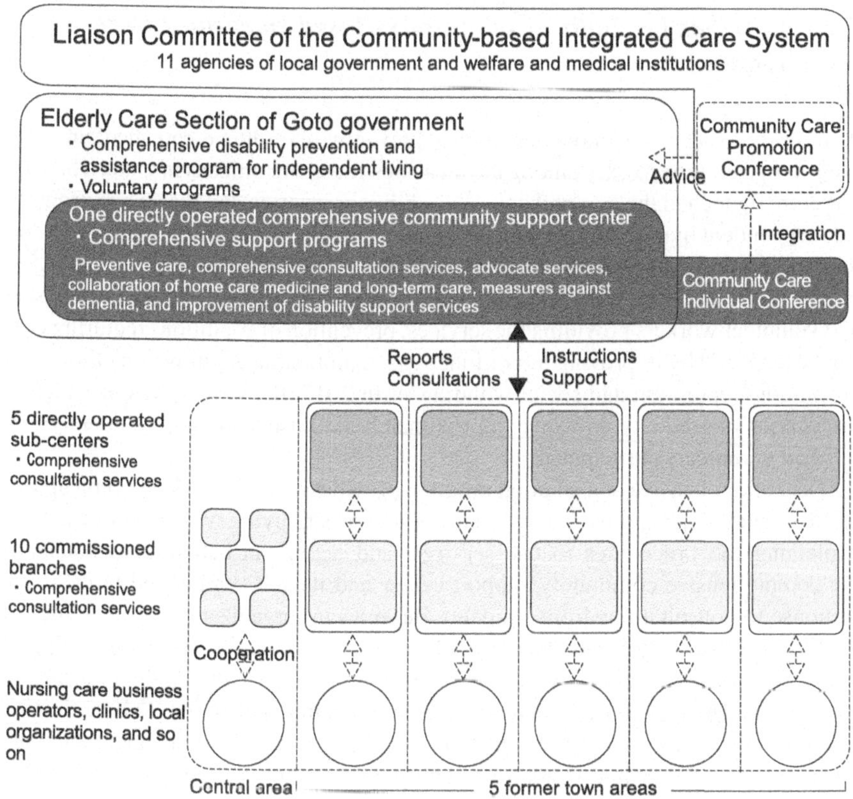

Fig. 3 Community-based integrated care system in Goto (as of September 2017) (Dotted line areas denote issues that should be solved or activities scheduled to be implemented. Prepared based on interviews with Goto government)

related to medical and long-term care services participated in the establishment of this committee to discuss and organize issues related to the development of Goto's community-based integrated care system. Multiple subcommittees are scheduled to be placed under this committee. One such subcommittee will be the Community Care Promotion Conference operated as the community care conference to address issues and make policy proposals for the entire city. This Community Care Promotion Conference established in addition to the individual meetings organized by the comprehensive community support center will make Goto's community care conference a multilayered organization.

2.3 Activities for Goto's Community-Based Integrated Care System

As described earlier, the implementation of preventive care, home-care medicine, and long-term care is necessary among the measures for senior citizens in Goto. The city, therefore, launched the comprehensive disability prevention and assistance program for independent living already in the first year of the program in 2015. In this program, the city has particularly emphasized "Community Mini Day Service," an operation of day services led by local residents with government support. For both the users and volunteer workers providing the services, prevention of conditions requiring care can be expected by improving motor functions, maintaining cognitive functions, and preventing depression and social withdrawal. In 2017, the service was provided at 36 locations in the city, in which approximately 620 users and approximately 240 resident volunteers participated.

The comprehensive consultation program provides support for developing appropriate services, programs, and organizations by supplying various information, explaining the procedures to use services, and acting on behalf of residents at the comprehensive community support center and its sub-centers and branches in response to consultations from residents and relevant organizations. Employees of the Elderly Care Section work concurrently at the comprehensive community support center. Public health nurses at the sub-centers work concurrently at the branches of the National Health Insurance and Health Policy Section and Social Welfare Section in addition to the Elderly Care Section. The council of social welfare and major care business operators such as residential facility operators providing multifaceted care services in their respective areas are commissioned to operate the branches. The city has therefore secured a basic system for providing various services and for developing institutions in response to consultations. Most actual responses to consultation and acceptance of applications, however, are reportedly provided at the comprehensive community support center at the city hall.

The community care conference holds several individual meetings each year through the comprehensive community support center. It held seven meetings in 2015 and nine in 2016. Each meeting addressed one case. All cases concerned difficult issues that had been selected by the comprehensive community support center or brought by care managers for consultations. Although the meeting participants are not fixed, but vary depending on the issues addressed in each case, employees of the comprehensive community support center organizing the meetings always attend them. The community care conference is generally held at the comprehensive community support center in the head office of the city government. When a meeting concerns a case in one of the former towns, the attendance of relevant parties from the area (relatives, commissioned welfare volunteers, representatives of a neighborhood association, etc.) is important. However, the meeting might be held at the sub-center closest to the area for the convenience of the participants. In this case, comprehensive community support center employees travel to the sub-center to host the meeting.

The Community Care Promotion Conference was at the stage of preparation and had not yet been held as of 2017.

Other activities include the improvement of measures against dementia and disability support services, which started in 2016, and the collaboration of home care medicine and long-term care, which was launched in 2017. As measures against dementia, one regional dementia support promoter was placed in the comprehensive community support center in 2016. The Initial-phase Intensive Support Team composed of medical specialists, public health nurses, and other nurses was established also in the comprehensive community support center in addition to the training of dementia supporters that had been conducted in 2017. To improve disability support services, a survey of needs in spheres of daily life was conducted in 2016. Based on the result, the city is currently preparing to establish the First-layer conference, which will address policy issues at a city-wide level, and the Second-layer conference, which will carry out activities to create disability support services in each of the former city and five towns that existed before the merger. The placement of disability support service coordinators has begun only recently, starting with the southwestern part of the city. Regarding the collaboration of home care medicine and long-term care, activities to promote the cooperation among parties such as medical institutions and long-term care insurance service providers began in 2017, considering the results of lectures held in 2016 and seminars with Nagasaki Prefecture.

Other noteworthy efforts include the development of an elderly monitoring network and community welfare activities through "Community Development Councils", community operation groups working in each district of the city. As for the former liaison committee of the local government, neighborhood associations, major care business operators, and companies in the city were established for the elderly monitoring network in 2013, which holds two meetings each year. Regarding the latter, Community Development Councils have been established in 13 districts based on the former municipalities which existed in the 1940s. Among them, two districts exhibiting sharp population declines and rapid population aging have welfare-related subcommittees, which are making active efforts for community welfare. The Community Development Councils in these districts are working in cooperation with influential social welfare corporations in their areas to carry out activities such as monitoring of elderly people, shopping assistance, salon activities, and creation of a welfare resource map.

3 Community-Based Integrated Care System in Tsugaru, Aomori Prefecture: A Case of Type 2-B

3.1 Reasons for Selecting Tsugaru and Area Overview

The small city of Tsugaru is located in western Aomori Prefecture (Fig. 4). Tsugaru's population according to the 2015 Population Census of Japan was 33,316. Because the entire city is located in the plain, agricultural production of rice, apples, and other products is robust, contributing to a high percentage, or 30.3%, of employment of

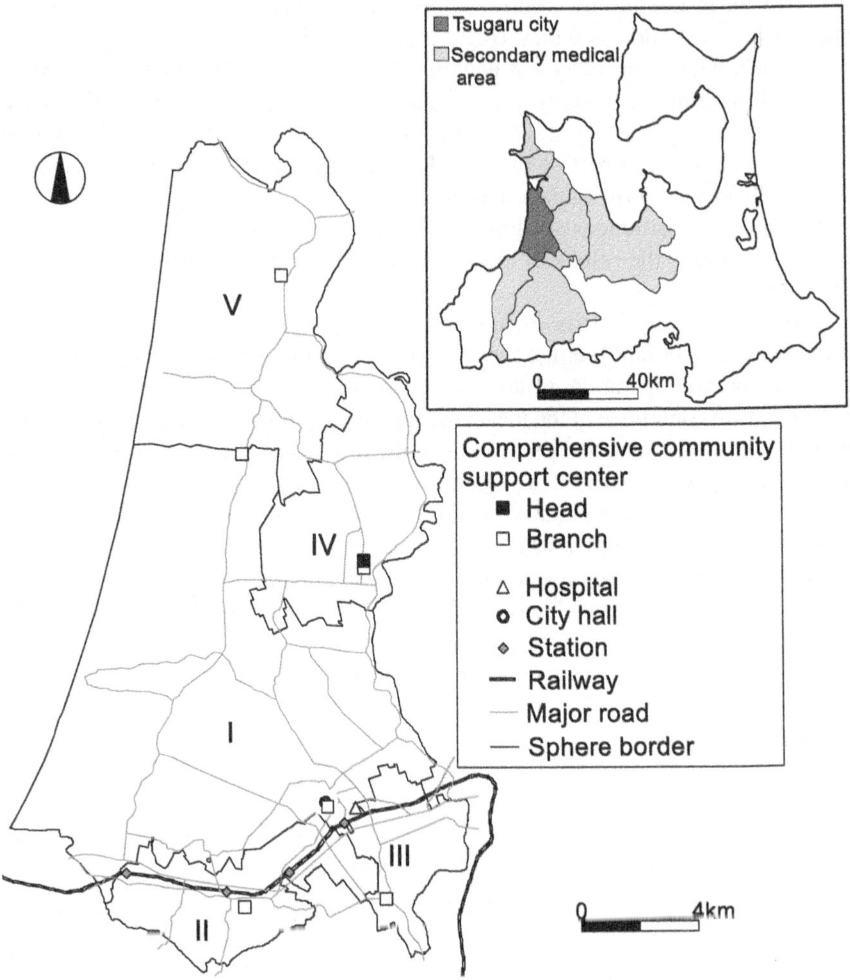

Fig. 4 Locations of comprehensive community support center and home care support centers and spheres of daily life in Tsugaru (as of September 2017)

residents in agriculture and primary industries. The ratio of the population of residents aged 65 years and above as of 2015 was 34.6%, indicating severe population aging. The financial capability index was 0.23 in 2016, suggesting low financial strength.

The city of Tsugaru was established through the merger of one town and four villages in 2005, which has a wide area of 253.9 km^2. The school districts, areas of neighborhood associations, and areas of community groups such as the council of commissioned welfare and child welfare volunteers generally follow the boundaries of the former municipalities, allowing the development of networks based on the former municipalities. Tsugaru, therefore, set up five districts based on the former municipalities to develop its community-based integrated care system. The city considered that one comprehensive community support center would be adequate based on its population applied to the standards of the Ministry of Health, Labour and Welfare. However, because the area expanded as a result of the merger, each of the former municipalities had mutually different regional characteristics. The city therefore considered placing offices in each sphere of daily life. Placing comprehensive community support centers in all areas, however, would have been difficult because of insufficient financial and personnel resources. For this reason, the city paced home care support centers with lower financial and personnel requirements as branches of the comprehensive community support center.

The community care conference is at two levels: one for the entire city and the other for the spheres of daily life. However, the city-wide conference (Community Care Promotion Conference) had not been established until the second half of 2015. Before this, the meetings had been held only for the spheres of daily life.

Base on the discussion presented above, Tsugaru fell under the category of Type 2-B among the nine types specified in this study, which held community care conferences for the spheres of daily life through the comprehensive community support center and its branches, until 2015 when we conducted the questionnaire survey. Since the latter half of 2015, the city has been holding the community care conference at two levels, one for the spheres of daily life and the other for the entire city, by adding the city-wide Community Care Promotion Conference to the conventional community care conference. This, therefore, is a case that has changed from Type 2-B to Type 2-C. Consequently, this section presents an analysis of the case as Type 2-B and also the conditions after the change to Type 2-C. We regard Tsugaru as an appropriate example of Type 2-B considering its population, which is close to the average population of Type 2-B municipalities (56,800), low financial capability (overall Type 2-B index is 0.51), and high percentage of merged municipalities (overall Type 2-B rate is 76.9%).

3.2 Overview of Tsugaru's Community-Based Integrated Care System

Tsugaru's government, comprehensive community support center, and council of social welfare established a preparation committee in 2015 to build the city's community-based integrated care system. The preparation committee divided itself into three groups, including home care medicine and long-term care, preventive care and living, and dementia, and built the foundation for the community-based integrated care system. After the current structure was established, the groups in this preparation committee have been maintained as a research subcommittee that meets five or six times a year and conducts activities to build a better community-based integrated care system.

Figure 5 presents the community-based integrated care system of Tsugaru as of 2017. Tsugaru has only one comprehensive community support center; it placed a home care support center in each sphere of daily life as a branch of the comprehensive community support center. The I district has a large population. However, one home care support center is placed in the eastern area and one in the western area to divide the operation. The comprehensive community support center was initially operated directly by the local government. Because of difficulty in securing public health

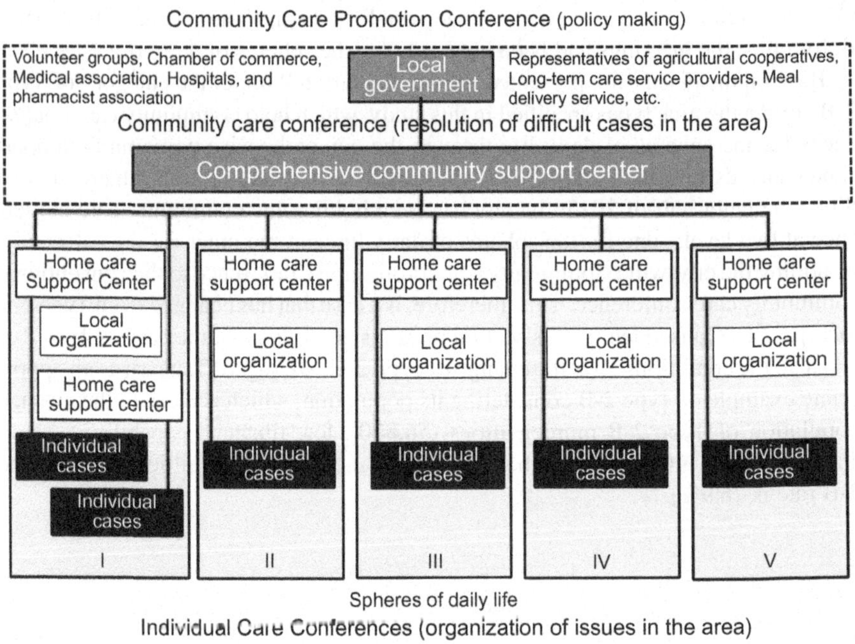

Fig. 5 Community-based integrated care system centering on community care conference in Tsugaru (as of September 2017) (prepared based on interviews with the long-term care section of Tsugaru government, comprehensive community support center, and home care support centers)

nurses in the direct operation, however, operation has been commissioned to the council of social welfare since 2008. Reasons for selecting the council of social welfare were that it was the only party that applied for the operation when applicants were sought from the public and that the operator would have to be a corporation capable of ensuring fairness and neutrality because the comprehensive community support center covered the entire city. Operation of the home care support centers is commissioned to the council of social welfare and social welfare corporations that operate welfare facilities for elderly people in their districts (Table 1). They are corporations that have long been rooted in local communities operating before the introduction of the long-term care insurance system. Such a triple-layer structure consisting of the government, comprehensive community support center, and home care support centers was built for the centers.

The home care support centers function primarily as a liaison for residents in their respective spheres of daily life. When inquiries are received from local residents, the home care support center, community organizations, families, etc. hold an individual care conference. If the case is difficult for the home care support center alone to solve, it is discussed at a community care conference held in each district. A community care conference comprises the home care support center in the district concerned and community organizations and relevant operators selected by the comprehensive community support center (Table 1). At community care conferences, the participants hold discussions to find solutions to difficult issues and gather information related to the monitoring of local residents and community issues.

Furthermore, the Community Care Promotion Conference was established to develop policies for solving community issues in each sphere of daily life and to carry out cooperative efforts among parties in different fields. This conference is led by the local government and the comprehensive community support center, with participation by volunteer groups, the chamber of commerce, agricultural cooperatives, and long-term care insurance service providers in the city and the medical association of the northwestern part of the Tsugaru region comprising the Tsugaru city and neighboring municipalities. Before the establishment of the Community Care Promotion Conference, discussions were held in each sphere of daily life, which, however, were not shared with the city.

3.3 Activities for Tsugaru's Community-Based Integrated Care System

Table 1 presents an overview of the Comprehensive Community Support Center, Home Care Support Centers, and meetings held for regional collaboration such as Community Care Conferences.

Table 1 Overview of comprehensive community support center and home care support centers of Tsugaru

Comprehensive community support center		Commissioned operator	No. employees	Conference aiming for regional cooperation	Participants
Head		CSW	Full time: 9	Community Care Promotion Conference (2 times)	Home care support center, C SW (city and districts), Long-term care provider, Meal delivery service (NPO), Home-delivery service, Senior citizens' employment guidance center, Pharmacy association, Honoboho Koryukai (social gathering), Chamber of commerce, Commissioned welfare volunteers, Hospitals, Medical association, Local government, etc.
Branch	Eastern I	CSW	Part time: 2	Home care support center (6 times)	Local government (head office and branches), Volunteer groups, etc.
	Western I	SWC	Full time: 1 Part time: 1	None	–
	II	CSW	Part time: 2	Information exchange of home care support center (2 times)	Self-government associations, Commissioned welfare volunteers, Volunteer liaison committee, Comprehensive community support center, Local government (head office and branches), C SW(districts), etc.
	III	SWC	Part time: 2	None	–
	IV	CSW	Part time: 1	Community care conference (2 times)	Senior citizens club, Commissioned welfare volunteers, Comprehensive community support center, Local government (head office and branches), Long-term care providers, Police substations, etc.

(continued)

Table 1 (continued)

Comprehensive community support center	Commissioned operator	No. employees	Conference aiming for regional cooperation	Participants
V	CSW	Part time: 2	Community care conference (6 times)	Comprehensive community support center, Medical institutions, Long-term care providers, Pharmacies, Local government (head office and branches), etc.

The commissioned operators and participants, CSW and SWC, respectively refer to the council of social welfare and social welfare corporation. Employees include full-time employees and part-time employees who concurrently engage in other duties. Information related to the conferences aiming for regional cooperation is from 2015 and other information is from 2016. II home care support center began holding community care conference in 2017 (Prepared based on interviews with long-term care section of Tsugaru government, comprehensive community support center, and home care support centers)

3.3.1 Community Care Promotion Conference

The Community Care Promotion Conference held for the entire city is led by the local government and comprehensive community support center. It initially engaged in the reporting of surveys addressing needs in spheres of daily life, identification of issues and examination of measures at the described previously research subcommittee, information exchange, and other activities. In 2016, the conference discussed the collaboration of home care medicine and long-term care in addition to conventional issues.

Medical institutions in Tsugaru include only one hospital in the I district; the rest are clinics. Because of this, the Tsugaru residents use hospitals in the neighboring municipalities. The low traffic accessibility from the V district in the northernmost part of Tsugaru makes it difficult for the residents to go to the hospital. Moreover, it snows in Tsugaru during winter; strong monsoons cause blizzards, making outings for any reason difficult. Such circumstances create major hindrances to the use of home care medicine and succession from medical to long-term care services.

The Community Care Promotion Conference benefits from participation by both representatives of the long-term care field, including the council of social welfare and long-term care insurance service providers, and the representatives of the medical service field, including the pharmacy association, hospital in the city, and medical association of the northwestern part of the Tsugaru region, which works to improve collaboration between medical and long-term care services. Challenges are arising, however, in the medical service field. First, whereas the conference is joined by the hospital in the city, many residents use hospitals outside the city, which causes the discussions to contradict with the reality. Secondly, because the medical association of the northwestern part of the Tsugaru region services the secondary medical area comprising the municipalities of 5 nearby municipalities, in addition to Tsugaru, the discussions do not specifically address Tsugaru even when they are held in Tsugaru. Because of such locational conditions of medical resources in the city and inconsistency in the scales of participating organizations, efficient collaboration has not been achieved.

3.3.2 Community Care Conference

The community care conference expected in Tsugaru will be led by the home care support centers in their respective spheres of daily life and aim to solve the residents' difficult issues.

The following discussion specifically pertains to the V district in the northernmost part of the city. In the V district, the community care conference is held primarily by the home care support center, which is commissioned to the council of social welfare. The community care conference meets six times a year to provide and share information related to local residents and to discuss difficult cases and other issues. It also prepares a senior citizen register with the cooperation of the V branch of the council of social welfare, commissioned welfare volunteers, senior citizens' clubs, and other parties to assess the real conditions faced by the local residents (Hatakeyama et al. 2018). Such a local network in the V district is built by the application of the network built before the time of the former V village. Its establishment is a result of consistency between Tsugaru's spheres of daily life and the area division of various local organizations. The area division of spheres of daily life and local organizations does not match in many other municipalities in Japan (Hatakeyama and Miyazawa 2016), which, in some cases, leads to difficulties in the development of local networks (Hatakeyama 2009). In this sense, Tsugaru is regarded as having an advantage in building a local network for having set up spheres of daily life based on the former municipalities that existed before the merger and having many local organizations that match the spheres.

The status of holding meetings such as the community care conference aiming for regional cooperation in Table 1, however, suggests that the condition varies considerably depending on the home care support center. The community care conferences in the IV and V districts meet on a regular basis. Meetings on regional cooperation similar to the community care conference are held also in the II and eastern I districts. Additionally, the II district established its community care conference in 2017. The western I and III districts, by contrast, have not been holding community care conferences or other meetings related to regional cooperation. One reason for their lack of meetings is the difference in the willingness or personnel placement of corporations commissioned to operate the come care support centers.

The commissioned operator of all home care support centers that hold meetings on regional cooperation is the council of social welfare. The merged municipalities had one each of the council of social welfare before the merger: they were combined into one because of the municipal merger. At present, a branch is placed in each of the former municipalities and takes charge of the operation. The post-merger council of social welfare exchange information and accumulates expertise internally and among the branches, The community care conference and other meetings on regional cooperation, therefore, also exchange information. Because of the characteristics of the council of social welfare responsible for social welfare of the local communities, it has been building local networks since the Ministry of Health, Labor and Welfare began proposing the community care conference. It was therefore unlikely that Tsugaru faced any major obstacle in holding the community care conference and similar meetings.

The districts in which meetings related to regional cooperation are not held, however, are those in which welfare facilities for elderly people have been founded. Social welfare corporations operating the facilities are commissioned to operate the local home care support centers. These corporations engage primarily in the operation of long-term care insurance services. They have postponed the development of local networks. In interviews, they acknowledge the needs and challenges in building the networks and report that they were yet to establish the community care conference.

Additionally, it is given to be concerned about securing human resources. As revealed in Table 1, each home care support center has only one or two employees. The majority of them concurrently engage in duties in the corporations other than the centers. One reason is the low commissions paid by the city. The amount of commissions paid in 2015 was 1.3 million yen per center. That amount is insufficient to pay an employee to work full time at the center. In Tsugaru, the expenses for the home care support centers operated as the branches of the comprehensive community support center are included in the commissions paid to the comprehensive community support center. The commissions are thereby treated as expenses for community support programs in the municipal special accounting for long-term care insurance, of which the upper limit is specified in the Long-Term Care Insurance Act (Hatakeyama 2009). Although the expenses might be covered by the general account budget, Tsugaru is not participating because of financial strain. Such a financial constraint is forcing the home care support center employees to hold multiple positions concurrently. Conditions under which most employees hold multiple positions lead to decisions of whether to hold the community care conference or meetings on regional cooperation to depend on the willingness of the commissioned corporations.

4 Community-Based Integrated Care System in Fukutsu, Fukuoka Prefecture: A Case of Type 2-C

4.1 Reasons for Selecting Fukutsu and Area Overview

Fukutsu, a small city located in the northern part of Fukuoka Prefecture, has a population of 58,781 (the 2015 Population Census of Japan) and area of 52.8 km^2 (Fig. 1). It sits between the two ordinance-designated cities, Fukuoka and Kitakyushu, of Fukuoka Prefecture, through which a railroad connecting these large cities runs. Fukutsu was created in January 2005 through a merger of two towns. The city has an urban area developed along the railroad line and an area having the characteristics of agricultural and fishing village. The southern part of the city along the railroad line is being developed into housing land. This development has contributed to growth in population, which was 64,282 (the Basic Resident Registration in September 2018). The ratio of the population of residents aged 65 years or older is 27.8%, which is close to the national average.

In 2015, Fukutsu had one comprehensive community support center and four liaison offices (Fig. 6). The four offices comprised two home care support centers operated as sub-centers and two Elderly Comprehensive Consultation Service Centers for elderly people operated as branches. The community care conference was held not only for the entire city, but also for smaller areas to address spheres of daily life. In 2016, however, the four sub-centers and branches were integrated into the comprehensive community support center, details of which are described later.

Based on the discussion presented above, Fukutsu corresponded to Type 2-C among the nine types until 2015, when we conducted questionnaire surveys, which had a comprehensive community support center, sub-centers, and branches and held community care conferences for the entire city and spheres of daily life. It is an

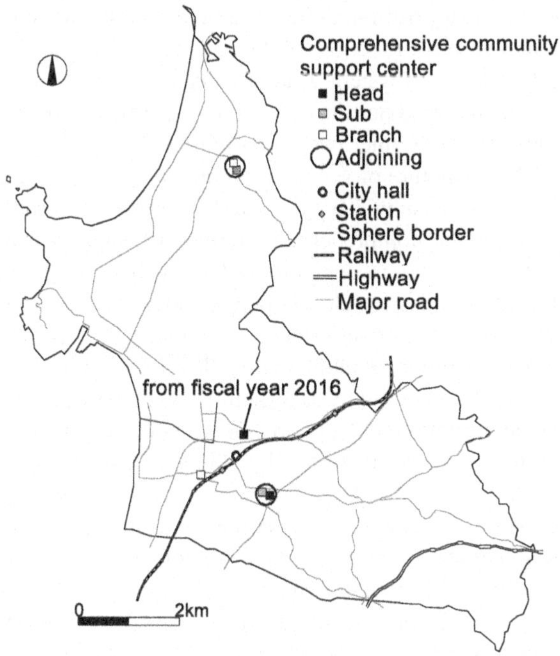

Fig. 6 Locations of comprehensive community support center and spheres of daily life in Fukutsu (as of March 2016)

example that changed to Type 1-C in 2016, in which one comprehensive community support center holds the above community care conferences. This section treats Fukutsu as a case of Type 2-C and also examines the conditions after the transformation to Type 1-C. The Fukutsu population is close to the average population (59,700) of Type 2-C municipalities. Its area, however, is smaller than the average area (311.8 km^2) of Type 2-C municipalities (Table 2 in Chapter "Regional Variation in the Community-Based Integrated Care Systems in Japan").

4.2 *Overview of Fukutsu's Community-Based Integrated Care System*

The government of Fukutsu has led the development of its community-based integrated care systems. In 2006, when the city founded its comprehensive community support center, it was operated directly by the Fukutsu government. The reasons are, first, that most residents were unfamiliar with the principles of the community-based integrated care system and the system of the comprehensive community support center in 2006. The local government decided that it should operate the center rather than entrusting it to other organizations. Secondly, the Fukutsu government considered that the community-based integrated care system must be operated by a party in a neutral position.

Fukutsu positioned the comprehensive community support center as part of its administrative functions and an agency playing a core role in its community-based integrated care system. Additionally, it positioned the two home care support centers in the city as advice desks. Each of these two home care support centers was located in one of the former two towns that existed before the merger.

The structure of Fukutsu's community-based integrated care, however, changed in two respects in 2010. One is that the city government terminated its direct operation of the comprehensive community support center and contracted a medical corporation for the operation. The medical corporation operates a few hospitals, health facilities for elderly people, adult day care centers, group homes for people with dementia, private elderly care homes, and assisted living residences in Fukutsu and a neighboring municipality. The reason for changing from direct operation to contracted operation was difficulties with the placement of personnel in the comprehensive community support center under the direct operation. The national government has established rules by which specialists of three types must be assigned to a comprehensive community support center according to the community population size. Because the local government carries out regular personnel change, however, maintaining such specialists at the comprehensive community support center under its direct operation is difficult. The commissioned medical corporation, by contrast, has 1,000 employees assigned to hospitals, welfare institutions for elderly people, and other facilities, which include those qualified to be specialists of three types required at the comprehensive community support center. The Fukutsu government therefore resolved the issue of personnel placement in the comprehensive community support center by entrusting it to this competent medical corporation.

The other change to the structure of Fukutsu's community-based integrated care made in 2010 was the establishment of two new branches called the Elderly Comprehensive Consultation Service Center with an aim to increase the liaison function of the comprehensive community support center. One service center was also placed in each of the two towns that had existed separately before the merger. The operation of one of these centers was commissioned to a social welfare corporation that operates welfare facilities for elderly people. The other was entrusted to an NPO corporation. Both branches were operated by the medical corporation described above, which was subcontracted by the two home care support centers.

The addition of the four advice desks to the comprehensive community support center in 2010 was intended to help refer various needs of the residents in their daily life to the comprehensive community support center to the greatest extent possible in an effort to address their issues. In Fukutsu, however, the four liaisons were abolished and integrated into the comprehensive community support center in 2016. The reasons for the integration included 1, liaisons having three names (comprehensive community support center, home care support center, and Elderly Comprehensive Consultation Service Center) coexisted, which confused the residents in determining who to consult, 2, the number of inquiries received by the four sub-centers and branches combined was low, and 3, issues that could not be resolved by a sub-center or branch with limited functions had to be brought to the comprehensive community support center after all. This reform emphasized the benefit of providing one-stop services in the consultation function by integrating the liaisons into the single comprehensive community support center. The comprehensive community support center is placed in a building complex located close to the center of the urban area. This building complex is positioned as a comprehensive center that becomes the basis of the function to promote the residents' health and welfare in Fukutsu.

The contract for the previously described medical corporation to operate the comprehensive community support center that began in 2010 was a private contract that was reviewed annually and either renewed or not. In 2016, however, that contract was replaced with a multi-year contract covering a three-year period. The city also decided to seek a contractor from the general public. The multi-year contract was adopted to facilitate the systematic personnel placement and human resource development of the comprehensive community support center in contrast to the single-year contracts. The contractor was selected from applications from the general public based primarily on its financial stability and ability to secure human resources necessary for operation of the comprehensive community support center. The contractor selected eventually was the same medical corporation group that had been commissioned until 2015. In interviews with the local government, the personnel in charge stated that this contractor had a reliable advantage in addition to its financial stability and personnel placement. The management of this corporation group would not possibly instruct or pressure Fukutsu's comprehensive community support center, the operation of which was entrusted to the group, to guide senior citizens' consulting with the center to use the services offered by various long-term-care-related businesses of the group. This decision is likely to be based on the exceptional presence of the corporate group operating large hospitals, other businesses in the area covering Fukutsu and neighboring municipalities, and the group's financial strength derived from that presence.

4.3 *Activities for Fukutsu's Community-Based Integrated Care System*

The following examines the activities of community care conferences of the five types exhibited in Table 2 and the status of collaboration with relevant institutions to identify the conditions of activities for Fukutsu's community-based integrated care system.

According to the 8th Fukutsu Elderly Welfare Plan and 7th Long-term Care Insurance Plan announced in March 2018, Fukutsu has established five community care conferences that serve in respectively differing roles. They include the Community Care Conference for Assistance for Independent Living, Community Care Individual Conference, Community-based Integrated Care System Council, Community Care Promotion Conference, and the Health and Welfare Council.

The Community Care Conference for Assistance for Independent Living is held jointly by the Fukutsu government and comprehensive community support center.

Table 2 Activities of community care conferences in Fukutsu (2017)

	No. times the conference was held	Major activities
Community Care Conference for Assistance for Independent Living	48	• Support care management that facilitates assistance for independent living
Community Care Individual Conference	28	• Provide consultation and advisory services to help solve issues in cases such as those which are difficult to support
Community-based Integrated Care System Council	12	• Share, define, and organize the community issues Community Care Promotion Conference
Community Care Promotion Conference	0	• Discuss how to apply community issues that have been identified to community development and the city's policies
Health and Welfare Council	8	• Develop resources needed for problem solving, make policies based on solutions, apply solutions to long-term care insurance planning, and carry out other activities for policy-making • Evaluate the comprehensive community support center's operation

Prepared based on information provided by the Senior Citizen Service Section of Fukutsu government

Its primary purpose is to support care management that facilitates assistance for independent living. The members comprise local government employees and those of the comprehensive community support center, occupational therapists, physical therapists, speech therapists, dietitians, dental hygienists, pharmacists, public health nurses, and care managers, and service providers. This conference meets 48 times a year (four times a month).

The Community Care Individual Conference is organized by the comprehensive community support center, with participation by the center staff members and employees of organizations relevant to individual cases. The conference provides consultation and advisory services to help solve issues in cases such as those which are difficult to support. This conference meets 28 times a year. The conference serves areas that are smaller than the entire city. The participants vary depending on the case, which might include individuals such as neighbors. For this reason, the meetings are held, in some cases, at community centers scattered throughout the city. The conference might invite members of specialized institutions to seek advice. Specialized institutions include public health centers and Fukuoka Prefectural Dementia Medical Centers located approximately in every secondary medical area.

The Community-based Integrated Care System Council is a place to share, define, and organize the community issues identified in the above two community care conferences. This conference has participation by the city government employees and comprehensive community support center and meets 12 times a year (once a month).

The Community Care Promotion Conference is a conference organized by the city government, which assumes its employees and representatives of relevant agencies as members. This conference is held to discuss how to apply community issues that have been identified to community development and the city's policies. In reality, however, this conference had not been held in Fukutsu as of 2017.

The Health and Welfare Council is a subsidiary agency of the city government, consisting of parties in specialized positions such as physicians, dentists, pharmacists, Fukuoka prefectural agencies, welfare organizations, and long-term care providers. The council held eight meetings during the fiscal year 2017. This council aims to develop resources needed for problem-solving, to make policies based on solutions, to apply solutions to long-term care insurance planning, and to carry out other activities for policy-making. It also plays the role of examining the evaluation of the comprehensive community support center's operation prepared by the city government.

The relations of these five community care conferences can be organized as follows (Fig. 7). Although the Community Care Conference for Assistance for Independent Living and Community Care Individual Conference are not mutually related, both report individual issues for discussion to the Community-based Integrated Care System Council. The Community-based Integrated Care System Council particularly provides the Community Care Conference for Assistance for Independent Living with feedback on information for supporting its care management operation. The Community-based Integrated Care System Council is intended to report the issues it organized to the Community Care Promotion Conference and Health and Welfare Council to contribute to Fukutsu's policy-making related to the community-based integrated care system.

Furthermore, Fukutsu works in cooperation with various organizations in and outside the city in the process of making the community-based integrated care system function. In the city, the Senior Citizen Service Section in charge of the community-based integrated care system holds monthly meetings with other sections in the city government to address elderly households facing problems such as disabilities, need for public assistance, and poverty. Particularly, aged persons living with a child having a disability, suffering from social withdrawal, or being unemployed are facing multiple complex problems. Because the Senior Citizen Service Section alone is not able to solve issues of such elderly households, mutual cooperation with other sections becomes necessary. The section also shares information with the city's council of social welfare.

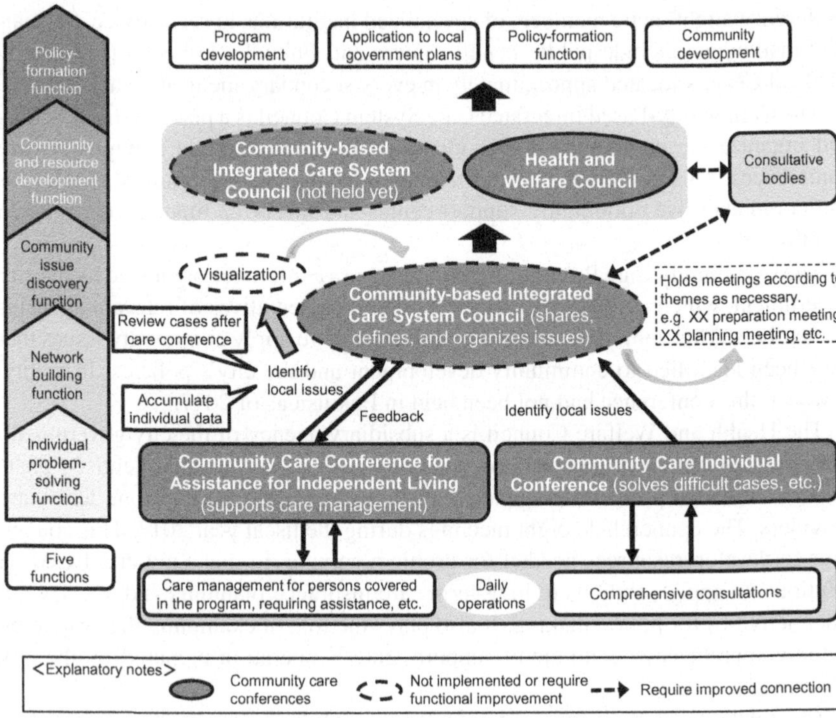

Fig. 7 Relations of community care conferences in Fukutsu (as of 2017) (prepared based on Fukutsu City 2018)

Cooperation with other organizations in and outside the city includes that with the police, public health centers, medical associations, medical institutions, and other municipal governments. The section shares information with the police about wandering elderly people and conflicts with neighbors, the police provide the comprehensive community support center with information when some difficulty arises. The public health center of Fukuoka prefectural government, which is located in a neighboring municipality, provides consultation services particularly for patients with mental disorders. The Senior Citizen Service Section holds six meetings a year with the medical association which covers the two-city area on the project to promote collaboration between home care medicine and long-term care services. The home-care medicine and long-term care collaboration project specifically examines the needs of elderly people for unified support through medical and long-term care and aims to develop a foundation for collaboration in various communities with an intention to build "face-to-face relationships" (Kuroda 2018). The Senior Citizen Service Section shares information with medical institutions, not only including the hospitals in Fukutsu, but those in two neighboring cities, for consultations with inpatients and status of medical examinations (medical histories, whether consultations have been suspended, etc.). Furthermore, the Senior Citizen Service Section employees observe community care conferences in other municipalities to acquire useful knowledge.

5 Characteristics and Issues in Local Governance of Type-2 Municipalities

The city of Goto, an example of Type 2-A municipalities, was established in 2004 through a merger of one city and five towns. Further back in the 1950s, the one city and six villages merged. The pre-merger regional division still persists strongly today. It is strengthened also by the fact that Goto is an island municipality. To build a community-based integrated care system in such a region, the spheres of daily life must be set up as small areas. Moreover, multiple comprehensive community support centers must be placed in each sphere. Additionally, it is desirable to adopt a decentralized system in which a community care conference is held in each sphere. Goto should place multiple comprehensive community support centers, considering also its senior citizen population. The actual centers, however, are still only one. The community care conference is held only for the city.

The reason for that limitation of size is a significant constraint on acquiring employees. Many municipalities placing multiple comprehensive community support centers are achieving it by commissioning private-sector operators. Securing specialists to place multiple comprehensive community support centers in Goto, however, is difficult for both the public and private sectors. The island region, with its small population, has no organization that develops specialists. Developing and retaining such personnel would therefore be a severe burden and hindrance. As explained earlier, the employees of the comprehensive community support center work concurrently at the Elderly Care Section. Although this system is intended to centralize information at the government, the constraint on securing employees is also a factor. Each sub-center has only one health nurse assigned as an employee, who also works concurrently at another section. The amount of commissions paid to the home care support centers is small, which prevents the commissioned corporation from placing full-time employees, forcing the employees to work at two places. Such constraints on personnel limit the sub-centers' operation to comprehensive consultation services. The corporation's jobs tend to be prioritized at the branches. The local governance related to the development of Goto's community-based integrated care system seems invariably centralized.

Given this situation, the inquiries and applications of the residents and relevant organizations have been submitted to the comprehensive community support center at the head office of Goto's government. In Goto, competent business operators provide multifaceted long-term care insurance services using residential long-term care facilities as a business base in each sphere. The city also has wide-ranging medical institutions, including acute care hospitals, chronic care hospitals, and private-sector clinics providing home care medicine, among the island regions. The current system is likely to be effective for the local government to unify its collaboration with these related institutions through the comprehensive community support center. When facing difficulties in improving the functions of the sub-centers and branches, however, it would be more important to establish a network of parties involved such as long-term care providers, medical institutions, and Community Development Councils and to develop a decentralized system through working closely with the comprehensive community support center to build a the community-based integrated care system in each of the former, pre-merger city and five towns. The distribution of welfare resources in Goto suggests that the roles played by competent long-term care insurance service providers in each area will be important for network development.

The city of Tsugaru, which transformed itself from Type 2-B to Type 2-C, placed one comprehensive community support center, and additionally, home care support centers as branches of the comprehensive community support center in each sphere of daily life, considering the characteristics of each of the former municipalities that existed before the municipal merger. Through this, the city aimed to build decentralized local governance based on the home care support center in each sphere of daily life. Although Tsugaru initially held community care conferences based only on the spheres of daily life, it established Community Care Promotion Conference for the entire city in addition to the conventional community care conferences for applying information gathered from each area to the city's policies. This multilayered local governance has enabled the city to centralize the activities done in each sphere of daily life and exchange and share information also with community associations and relevant organizations working for the entire city at the municipal head office. Although future developments must be monitored since it has not been long since the establishment of the Community Care Promotion Conference and because specific policies have yet to be made, a system to develop policies from local community issues has been set up. Because the medical association serves an area larger than Tsugaru, however, gaps in the scale of discussions constitute an issue that must be addressed.

As a result of placing the branches with consideration for the characteristics of each sphere of daily life in Tsugaru, the community-based integrated care system was built, in some cases, to exchange information at community care conferences and other meetings and solve difficult issues using a local network that is complete in a sphere of daily life. At the home care support centers operated as branches, however, operations expected by the city government could not be completed, in some cases, because of limited personnel placement for financial reasons. Unlike the case of comprehensive community support center, the law does not specify a clear standard for the placement of full-time personnel at a home care support center. Personnel placement is done largely at the discretion of the organization commissioned to operate the center. At present, home care support centers in Tsugaru are operated primarily based on the good intentions of the commissioned operators. Differences in their willingness strongly affect the development of the community-based integrated care system in each sphere of daily life. Particularly, disparities in the local governance among the spheres of daily life have persisted as a difficulty.

In Fukutsu, which changed from Type 2-C to Type 1-C, the city government is responsible for leading various collaborative efforts with organizations in and outside the city. Many of its community care conferences are held collectively on a city-wide scale. In addition, the sub-centers and branches that had been established were integrated into the comprehensive community support center in 2016. Such developments are seemingly indicative of the city's general intention to operate its current community-based integrated care system on the scale of the entire city. One factor contributing to the integration of the centers in 2016 was the small area of Fukutsu. Although two towns merged to become Fukutsu, it only takes about 30 min to drive from one end to the other.

It is inappropriate, however, to infer that Fukutsu's the community-based integrated care system is operated through centralized local governance based only on the scale of the entire city. One reason is that Community Care Individual Conferences, which play a fundamentally important role in the process from consultations to specific advice and policy-making for measures, are held for areas smaller than the entire city. Such a policy of not holding the conference only on the scale of the entire city is also clearly stated in the 7th Fukutsu Long-term Care Insurance Plan announced in March 2018 as part of the functional improvement and multilayer development of community care conferences (Fukutsu City 2018: 84).

Another reason is that local activities included in the community-based integrated care system are conducted by dividing Fukutsu into eight districts. In its First Comprehensive Plan (2007–2016), Fukutsu aimed to create communities based on eight districts established (by dividing the city based on seven elementary school districts) in a project titled "hometown development". Based on this Comprehensive Plan, the hometown development council in each district has organized various local activities with participation by the residents. The eight districts receive subsidies (a total of 130 million yen) from the city government. Two fundamentally important undertakings that demand equal and diligent effort are disaster prevention and community welfare. The home-care support project, for example, is implemented in each of the eight hometown development districts. In all, 100 self-government associations in the city operate under the eight hometown development districts. Among them, approximately 50–60 self-government associations engage in preventive care services and have established connections with the city government and the council of social welfare.

Challenges to be addressed in the development of the community-based integrated care system in Fukutsu include the following. One is an issue of organizational management. Of the five conferences positioned as community care conferences, the Community Care Promotion Conference had not been held as of 2017. This conference has an important role of connecting community issues and administrative policies.

Furthermore, challenges exist from a geographical perspective. The 2016 reform that integrated the sub-centers and branches, which had the advice desk function, into one comprehensive community support center was partly a result of the small area of Fukutsu. As described initially, however, Fukutsu has both an urbanized area and an area largely in the northern part of the city that has the characteristics of an agricultural and fishing village. With regard to community-based integrated care, the consultations are few in the agricultural and fishing village area (hometown development district). Issues are often resolved to some degree within the agricultural and fishing district. At the stage of consulting with the comprehensive community support center, however, the issues have become markedly severe in some cases. This severity is a characteristic of the agricultural and fishing village area. In the more urbanized area (hometown development district), however, issues are solvable to a certain degree in some cases; apart from this, residents in some areas tend more to rely on the city government. Some note must be taken of the fact that such diversity among the districts invalidates a uniform approach to the development of the community-based integrated care system taken in the entire city.

6 Conclusions

As described above, most Type 2 municipalities are small cities with populations that are often larger than Type 1 municipalities. Interviews conducted in the sample municipalities revealed personnel and financial obstacles in becoming Type 3, in which multiple comprehensive community support centers would be established. As an intermediate measure, they developed the system by adding sub-centers and branches to the comprehensive community support center placed at the top. Thereby, they aimed to build decentralized local governance based on sub-centers and branches located near the local residents. In Goto and Tsugaru, however, sub-centers and branches became dysfunctional because of inadequate specialist human resources and finance, causing their decentralized local governance to be unable to function in some cases. In Fukutsu, sub-centers and branches became unnecessary because of the small area of the municipality. As a result, Fukutsu integrated its sub-centers and branches into a comprehensive community support center and transformed itself to Type 1-C.

The observation described above indicates that although the municipalities in all three cases aimed to establish decentralized local governance, local governance centralized by comprehensive community support centers was ultimately built.

Community care conferences tend to be built in a multilayered system, comprising centers based on spheres of daily life and one covering the entire city, when multiple spheres of daily life have been established as in Type 1 municipalities. Tsugaru, which was Type 2-B, also changed to Type 2-C. Goto, too, while remaining as Type 2-A, has developed meeting bodies in two layers, including the Community Care Promotion Conference, which makes policy proposals and the Community Care Individual Conference that discusses individual cases, which still serve the entire city.

In Type 2 cases such as those described above, local governance in developing a community care system involves several challenges. These municipalities aimed to establish decentralized local governance, which, however, is inadequately functioning. In other words, limitations exist to a community-based integrated care system that combines sub-centers and branches which are unable to place sufficient personnel. Although Fukutsu changed to Type 1-C in 2016, a proposal to place another comprehensive community support center in 2022 to shift to Type 3-C has not been ruled out in the relevant department in the city's government. Similarly, Tsugaru maintains a proposal to shift to Type 3-C. As of 2018, however, these ideas have yet to become reality. The transformation to Type 3 would be desirable provided that adequate financial and human resources are secured; otherwise, it would be more realistic for municipalities having a small area to change to Type 1, rather than Type 2.

As in Type 1 cases, some Type 2 municipalities establish wide-area partnership with other municipalities for being unable to prepare community resources completely on their own. As in Nanporo of Type 1, such a partnership is observed primarily in the medical care field in Tsugaru and Fukutsu. In Goto, too, the liaison committee of the community-based integrated care system is joined by members of a university faculty of medicine from mainland Japan. Such wide-area cooperation requires the development of local governance on a scale that is larger than the municipality areas. How municipalities take the initiative in implementing their community-based integrated care system, therefore, represents an issue to be addressed.

References

Fukutsu City (2018) The seventh long-term care insurance plan (in Japanese)

Hatakeyama T (2009) The expansion of municipal authority and regional influence resulting from long-term care insurance system reform: the case of Fujisawa City, Kanagawa Prefecture, Japan. Jpn J Hum Geogr 61:409–426 (in Japanese with English abstract)

Hatakeyama T, Miyazawa H (2016) Conditions of the development of community-based integrated care system: A geographical study of the results of a questionnaire survey taken by municipal governments. Commun Caring 18(14):65–68 (in Japanese)

Hatakeyama T, Nakamura T, Miyazawa H (2018) Community-based integrated care systems in Japan: focusing on spatial structures and local governance. E-Journal GEO 13:486–510 (in Japanese with English abstract)

Kuroda K (2018) Integration of medical and long-term care. In: Sumida Y, Fujii H, Kuroda K (eds) Clear descriptions of community-based integrated care. Minerva Shobo, Kyoto (in Japanese)

Community-Based Integrated Care Systems in Municipalities Having Multiple Comprehensive Community Support Centers

Teruo Hatakeyama

Abstract This chapter identifies the characteristics of local governance of the community-based integrated care system of municipalities corresponding to Type-3 among the types defined in Chapter "Regional Variation in the Community-Based Integrated Care Systems in Japan", which have established multiple comprehensive community support centers. Many Type-3 municipalities are cities with populations that vary from small to large cities. The study therefore has selected three cities of different sizes, including Asago in Hyogo Prefecture, Fujisawa in Kanagawa Prefecture, and Naruto in Tokushima Prefecture. These three municipalities have populations larger than many of Type-1 and Type-2 municipalities. For this reason, they have built decentralized local governance based on comprehensive community support centers placed in each of their spheres of daily life. This governance provides the benefit of being able to establish a community-based integrated care system that incorporates consideration of the characteristics of each sphere of daily life. Activities in each area related to the community-based integrated care will rely on the willingness of the corporation entrusted to operate the comprehensive community support center. This reliance becomes a factor of instability in the development of local governance. Additionally, some Type-3 municipalities change from Type A or Type B to Type C or some Type-2 municipalities shift to Type 3-C in terms of the units of establishing community care conferences. This shift requires, on the whole, the development of multilayered and decentralized local governance in the future.

Keywords Community care conference · Comprehensive community support center · Community-based integrated care system · Sphere of daily life · Urban municipalities

---

This chapter is a revised version of Hatakeyama (2017) and Hatakeyama et al. (2018).

T. Hatakeyama (✉)
Naruto University of Education, Naruto, Japan
e-mail: thatakeyama@naruto-u.ac.jp

H. Miyazawa and T. Hatakeyama (eds.), *Community-Based Integrated Care and the Inclusive Society*, International Perspectives in Geography 12,
https://doi.org/10.1007/978-981-33-4473-0_9

Fig. 1 Locations of cities of Type-3 cases

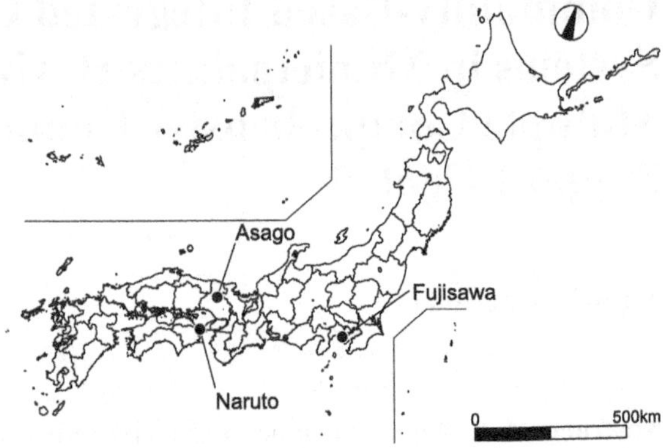

1 Introduction

This chapter examines characteristics of local governance in building the community-based integrated care system of Type-3 municipalities, which have placed multiple comprehensive community support centers. Type 3 is characterized by the inclusion of urban municipalities having various population sizes from small to large cities. The selected three cases are, thereby, cities with different population sizes. Asago, Hyogo Prefecture, was selected as an example of Type 3-A municipalities that establish multiple community care conferences for the entire city. Fujisawa, Kanagawa Prefecture, was selected to represent Type 3-B municipalities that have a community care conference in each sphere of daily life. Fujisawa, however, shifted to Type 3-C in 2016 when it placed a community care conference for the entire city in addition to those based on spheres of daily life. Naruto, Tokushima Prefecture, was selected as a case of Type 3-C municipalities established both for the city and for each sphere of daily life. Figure 1 presents the locations of these three cities. The discussion of the community-based integrated care system of Fujisawa is based on a report of Hatakeyama et al. (2018); that of Naruto is based on Hatakeyama (2017).

2 Community-Based Integrated Care System in Asago, Hyogo Prefecture: A Case of Type 3-A

2.1 *Reasons for Selecting Asago and Overview of the Area*

Asago is a small city located in the northern part of Hyogo Prefecture (Fig. 1). Its population according to the 2015 Population Census of Japan was 30,805. The entire city is in a mountainous region. Snow covers the area in winter. The ratio of the population of residents aged 65 years and older is 33.3%, indicating severe

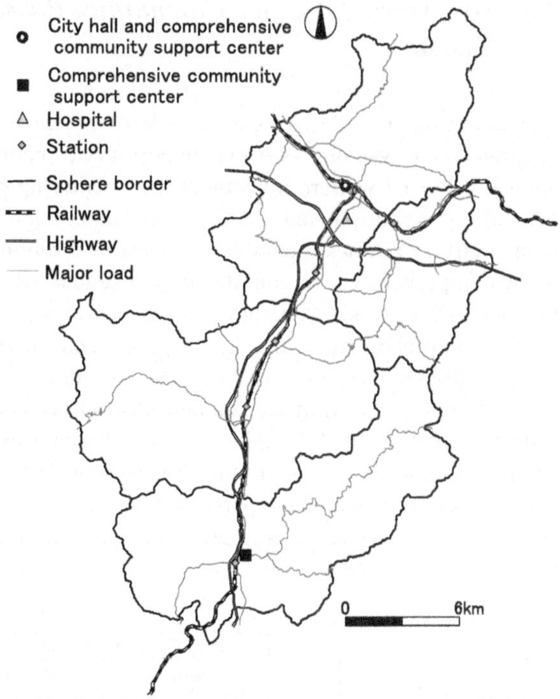

Fig. 2 Locations of comprehensive community support center and medical facilities and spheres of daily life in Asago (as of August 2018)

population aging. The population decreased by 16% during the 20 years from 1995, suggesting increasing depopulation. Asago's financial capability index is low at 0.41.

Asago is a city created through a merger of four towns in April 2005. Its area increased to 403.0 km^2 as a result of the merger (Fig. 2). Because of this, the former towns having varied characteristics were specified as the spheres of daily life in the community-based integrated care system. Comprehensive community support centers were established at two locations (one at the city hall and the other in the southern district) rather than in each of sphere of daily life, considering the balance between the total population and the number of facilities in Asago. Five conferences are positioned as the community care conferences covering the entire city.

Based on the circumstances presented above, Asago falls under the category of Type 3-A among the nine types specified in this study. Although Asago's population is smaller than the average population (102,800) of Type 3-A municipalities, many other characteristics, including the average area (214.0 km^2), being a merged municipality (52.2% of all Type 3-A municipalities), and direct operation of part of comprehensive community support centers (52.2% of all Type 3-A municipalities), are consistent with the characteristics of Type 3-A (Tables 2 and 3 in Chapter "Regional Variation in the Community-Based Integrated Care Systems in Japan"). We have therefore selected Asago as an example of Type 3-A.

2.2 *Overview of Asago's Community-Based Integrated Care System*

The government of Asago leads the development of the city's community-based integrated care system. Disability support service coordinators commissioned to the council of social welfare implement the monitoring program of local residents by networking commissioned welfare volunteers, neighboring residents, post offices, home-delivery business operators, financial institutions, etc. (Figure 3). They report to the comprehensive community support centers when anything unusual occurs in the monitoring. In addition, they discuss measures at a community care conference.

The comprehensive community support center in the southern district is responsible for the same district; the one at the city hall covers the remainder of the city. The center was placed in the southern district because of the great distance of the district from the city hall. Although a comprehensive community support center should preferably be located in each sphere of daily life, the city established only two because of financial and personnel constraints. However, the central role in the community-based integrated cCare system played, in effect, by the center at the city hall.

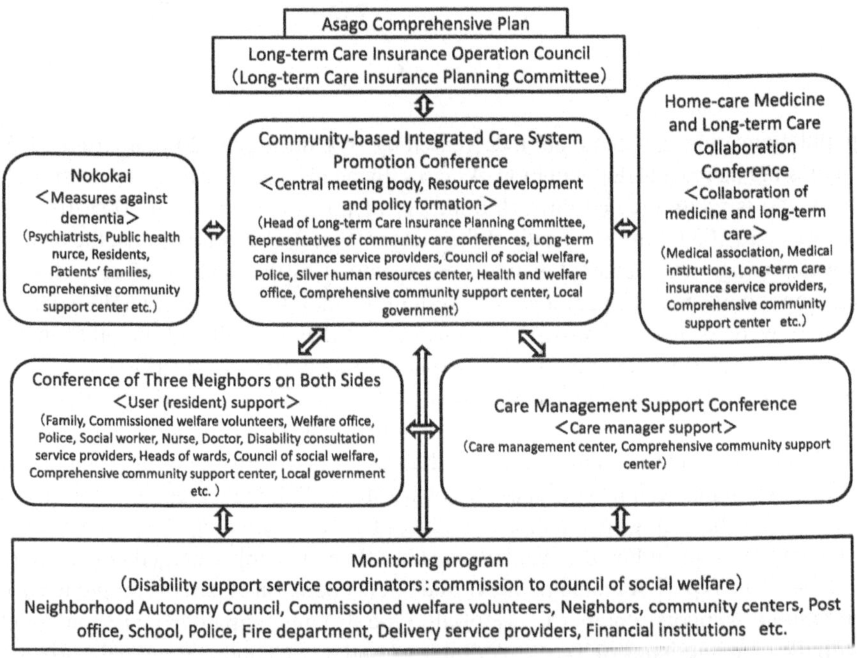

Fig. 3 Conceptual diagram of Asago's community-based integrated care system (as of June 2018) (Prepared based on information provided by Community-based Integrated Care System Promotion Conference of Asago)

Asago has established five community care conferences, including the Community-based Integrated Care System Promotion Conference as the core meeting body, the Conference of Three Neighbors on Both Sides, Care Management Support Conference, Nokokai, and Home-care Medicine and Long-term Care Collaboration Conference.

The Conference of Three Neighbors on Both Sides discuses solutions in response to consultations with local residents. It is organized by the comprehensive community support centers. It summons appropriate participants for the respective cases. This conference, which is based on the individual conferences in the cases at the time when the comprehensive community support centers were established in 2006, was named the Conference of Three Neighbors on Both Sides. It is positioned as a community care conference in 2013.

The Care Management Support Conference, which is held by care managers working at care management centers and comprehensive community support centers, is intended to improve the quality of care managers. Although this conference aims to help care managers solve issues, it originally succeeded the system of conference-style care plan checking for the rationalization program for long-term care insurance expenses, which has been conducted since 2007. It was renamed the Care Management Support Conference in 2013 and was positioned as a community care conference. This conference might discuss cases of the Conference of Three Neighbors on Both Sides. Both conferences are linked.

Nokokai discusses community issues related to dementia. This conference is organized by the comprehensive community support centers and participated by dementia patients, their families, and relevant specialists. This conference was originated in 2003 from daily meetings held by doctors at a psychiatric hospital and public health nurses in the former town of Asago that existed before the municipality merger. It was subsequently positioned as a community care conference in the preparation of the operation manual for the community care conferences.

The Home-Care Medicine and Long-term Care Collaboration Conference discusses community issues related to the collaboration between medical and long-term care services. This conference is organized by the comprehensive community support centers and participated by specialists from medical associations, medical institutions, long-term care insurance-certified providers, etc. After the comprehensive community support centers were established in Asago, residents often complained about the collaboration between medicine and long-term care. Initially, persons in charge responded individually to those complaints. However, they were unable to handle them because of time limitations. Therefore, the meeting body expected to handle them as issues of the entire community was founded in 2009.

The Community Care Promotion Conference discusses solutions to community issues identified in the above four community care conferences and the monitoring program. The discussions were aimed to make policy proposals that will be included in the long-term care insurance planning and Asago's comprehensive planning. This conference is organized by the comprehensive community support centers and consists of representatives of diverse fields, including the community care conferences, long-term care insurance-certified providers, police departments, council of social welfare, and silver human resources center. This conference was established in 2014 to connect the four existing community care conferences and to apply them to policy formulation.

Asago built its community-based integrated care system based on the five community care conferences described above, which have different purposes and participants. Many of them are organized by the comprehensive community support centers. All of the conferences cover the entire area of Asago. No discussions are held based on the spheres of daily life.

2.3 *Activities for Asago's Community-Based Integrated Care System*

Table 1 presents the status of holding the community care conferences in 2017 to help examine the activities conducted in the community-based integrated care system. All community care conferences are held multiple times each year, suggesting ongoing vigorous activities.

The Conference of Three Neighbors on Both Sides was held as needed at the request of local residents and specialists (care managers, medical institutions, etc.). This conference was held 66 times in 2017. Many discussions were related to

Table 1 Activities of community care conferences in Asago (2017)

	Number of times the conference was held	Major activities
Community-based Integrated Care System Council	5	• Report activities and planning of each community care conference • Exchange opinions and examine application to policies
Conference of Three Neighbors on Both Sides	66	• Discuss solutions based on consultations with local residents and specialists
Care Management Support Conference	10	• Provide care managers with training and seminars
Nokokai	9	• Consider solutions to the wandering of elderly people with dementia • Establish a working group for practicing friendly greetings
Home-care Medicine and Long-term Care Collaboration Conference	5 meetings 4 working group sessions	• Prepare individual disaster response manuals • Prepare medication information sheets

Prepared based on information from the 2018 Asago Community-based Integrated Care System Promotion Conference issues involving medical care and symptoms of dementia (wandering alone, pica disorder, delusion, etc.). Issues identified in cases discussed at the meetings include increased frequency of traffic accidents caused by elderly persons driving automobiles, increased frequency of accidents involving elderly persons wandering alone, and difficulties in decision-making for medical treatment of persons with no family. Therefore, because increased frequency of accidents caused by elderly drivers had been a particular difficulty up to 2017, a working team for voluntary relinquishment of driver's licenses was established in 2017 along with these meetings to encourage elderly people to surrender their driver's license voluntarily.

The Care Management Support Conference held 10 meetings. Its primary activity was to hold seminars for care managers. Issues identified in the meetings included inadequate assistance provided to families of elderly persons requiring support, numerous aged persons isolated from their community, and the need for high-quality care management because of an increased frequency of difficult cases.

Nokokai held nine meetings at which the members discussed measures for wandering by elderly people with dementia and established a working group for practicing friendly greetings. As part of the measures for elderly wandering, the members identified the number of aged persons who were able to go wandering from questionnaires taken by care managers.

The Home-care Medicine and Long-term Care Collaboration Conference met a total of nine times, including five meetings and four working group sessions. Its primary activities included the preparation of individual disaster response manuals and medication information sheets. The activities above were conducted in 2017, when the themes of the discussions were to ensure the evacuation of elderly people requiring support and appropriate medical and long-term care and to develop collaboration between pharmacists and care managers.

The activities implemented through the above four community care conferences were reported and discussed at the Community-based Integrated Care System Promotion Conference. At the first meeting in 2018, plans for the year were reported. Opinions about activities of the community care conferences were exchanged across different meeting bodies.

3 Community-Based Integrated Care System in Fujisawa, Kanagawa Prefecture: A Case of Type 3-B

3.1 Reasons for Selecting Fujisawa and Overview of the Area

Fujisawa is a medium-sized city of the Tokyo suburbs, located in the southern part of Kanagawa Prefecture (Fig. 1). Many of its residents commute to Tokyo, Yokohama, and other large cities. It therefore has the characteristics of a commuter town. The Population Census of Japan revealed that its population in 2015 was 423,894. The ratio of elderly residents is 23.4%, which is lower than the national average. Because many of the residents are baby boomers, however, the city is experiencing rapid population aging. Its financial capability index was 1.07 in 2016, indicating its high financial strength. For this reason, the city did not experience the Heisei Municipal Mergers. Its area is small: 69.6 km^2.

Fujisawa has districts with various characteristics including the central urban area, residential areas developed before World War II, new towns developed during and after the high economic growth period, and agricultural areas. The municipality therefore has decentralized itself based on the districts in the city since the 1980s and has been building a community in each district (Omizu 2005). Currently the city is divided into 13 districts circumjacent to the community centers and building communities under the name Hometown Development Promotion Council, with the participation of residents. This conference holds regular meetings of representatives of local organizations and residents gathered from the public. It carries out activities corresponding to issues and characteristics of each district.

Fujisawa has set up spheres of daily life based on these 13 districts and sought to build its community-based integrated care system based on community networks in each sphere. The reason for the distributed services is the area divisions of groups such as commissioned welfare volunteers, the council of social welfare, neighborhood associations, and the federation of senior citizens' clubs. These main actors become the key players in the community networks. In fact, the local government conducted planning consistent with the 13 districts, supporting the efficient development of community networks.

As of April 2018, Fujisawa had 15 comprehensive community support centers and one sub-center in addition to the core comprehensive community support center placed in the city hall (Fig. 4). Except for the core comprehensive community support

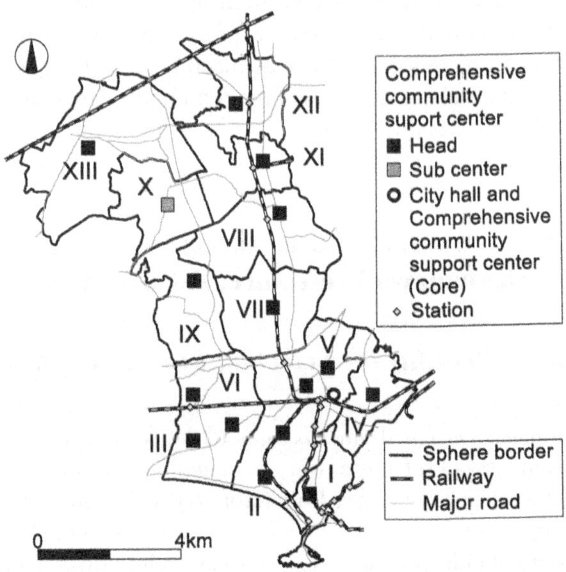

Fig. 4 Distribution of comprehensive community support centers and spheres of daily life in Fujisawa (as of August 2016)

center, the operation of all these centers is commissioned to social welfare corporations, medical corporations, and other organizations. Since 2011, the comprehensive community support Center in each sphere has been holding small meetings of community care conference based on the spheres of daily life. Until 2015, bloc meetings had also been held by dividing the city into four districts, including north, south, east, and west. In 2016, the bloc meetings were reorganized into theme-based community care conferences covering the entire city. Furthermore, in August 2015, the city founded the Fujisawa Community-based Integrated Care System Promotion Conference, a conference aimed at collaboration across different fields for the entire city.

Based on the above, Fujisawa fell under the category of Type 3-B among the nine types specified in this study until 2015, which held Sub-community Care Conferences for the spheres of daily life through multiple comprehensive community support centers and the bloc meetings for areas smaller than the entire city. In 2016, the city shifted to Type 3-C by holding the community care conferences at two levels, one for the spheres of daily life and the other for the entire city, after the reorganization of the bloc meetings and the full-fledged launch of Fujisawa Community-based Integrated Care System Promotion Conference. This section, thereby, analyzes the case as Type 3-B and also the condition after the transformation to Type 3-C. We considered Fujisawa appropriate as an example of Type 3-B municipalities, considering its high financial capability (overall Type 3-B index is 0.68) and low percentage of municipal bodies directly operating comprehensive community support centers (overall Type 3-B rate is 26.8%) despite its population, which is larger than the average population of Type 3-B municipalities (167,500).

3.2 *Overview of Fujisawa's Community-Based Integrated Care System*

When the comprehensive community support centers were founded in Fujisawa in 2006, there were only eight of them to cover 13 spheres of daily life because of constraints associated with the long-term care insurance special account (Hatakeyama 2009). Because of this, the community care conference held meetings by dividing the city into four blocks. The spheres of daily life and areas covered by the comprehensive community support centers did not match, which led to obstacles hindering the efficient implementation of community-based integrated care systems.

The city added comprehensive community support centers in 2011 when the constraints from the long-term care insurance special account were removed.[1] One center each was placed in the spheres. This center became the basis of operating the community-based integrated care system in each sphere of daily life (Fig. 4). The

[1] The general framework of the municipal special accounting for long-term care insurance was expanded because of increases in long-term care benefits and expenses caused by a growing number of users of long-term care services.

city, then, created the Sub-community Care Conference based on spheres of daily life while maintaining the Bloc-based Community Care Conference. The operation of the Sub-community Care Conference is led by the comprehensive community support centers. Partly because of the solitary death of elderly people that had become a social problem, the Sub-community Care Conferences specialized their discussions in the monitoring of senior citizens. Common issues raised in the Sub-community Care Conferences were taken to the bloc meetings. In this way, the Sub-community Care Conference and bloc meetings divided roles between them.

The bloc meetings' lack of expertise in the subject matters discussed, however, came to present a difficulty. Consequently in 2016, the bloc meetings were reorganized into Theme-based Community Care Conferences based on four themes, which would cover the entire city. The four themes included dementia-related issues, collaboration between medical and long-term care, disability support services, and individual cases. At the Theme-based Community Care Conferences, employees of the comprehensive community support centers and specialists such as care managers and public health nurses examine various cases and provide feedback to each sphere.

Furthermore, Fujisawa Community-based Integrated Care System Promotion Conference was established in 2015. This conference sets 2025, the year in which baby boomers will be 75 years old or older, as a medium-term target for its efforts to build Fujisawa Community-based Integrated Care System with an aim to developing a town in which "all residents will be able to remain in their familiar communities with peace of mind." Having learned from the excessive emphasis of past community care conferences on the welfare of elderly people, this conference cooperates with people in various fields such as medicine, childcare, welfare of persons with disabilities, schools, and resident associations, in addition to the welfare of elderly people. Through this cooperation, it aims to build a community-based integrated care system that can serve all generations. The full-scale launch of this conference, however, will take place after 2016.

Based on the circumstances presented above, Fujisawa has built a community-based integrated care system based on three types of community care conferences on two scales since 2016, including Fujisawa Community-based Integrated Care System Promotion Conference and Theme-based Community Care Conferences covering the entire city and Sub-community Care Conferences based on spheres of daily life (Fig. 5).

3.3 Activities for Fujisawa's Community-Based Integrated Care System

The following examines the activities of the community care conferences to identify the condition of activities for Fujisawa's community-based integrated care system.

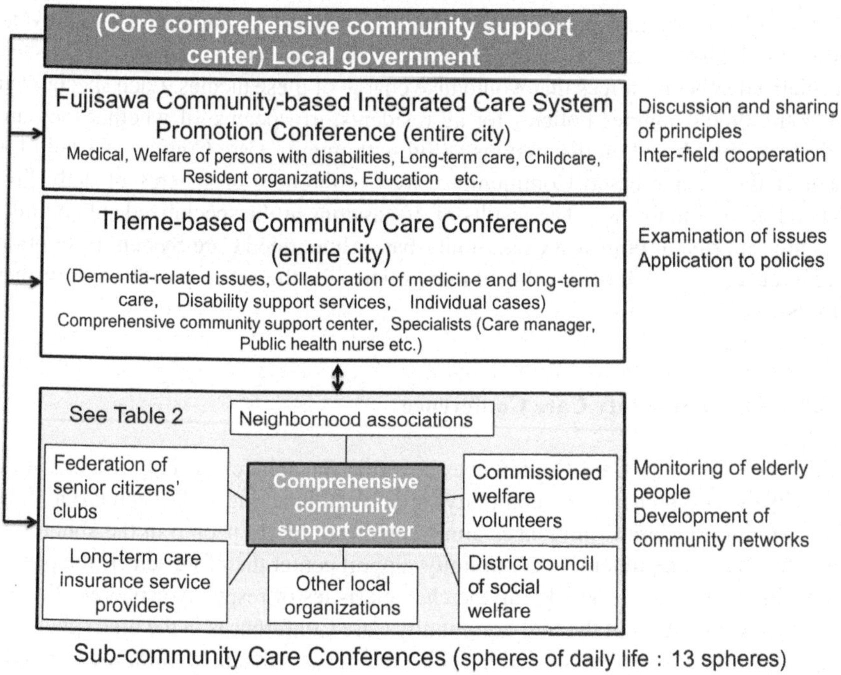

Fig. 5 Community-based integrated care system centering on community care conferences in Fujisawa (as of August 2016) (Prepared based on interviews with the senior citizen support section of Fujisawa government and comprehensive community support centers)

3.3.1 Community Care Conferences Covering the Entire City

The Theme-based Community Care Conferences divide issues into the previously described four themes and then examine inquiries and difficult cases brought to the comprehensive community support centers. These conferences, organized by the comprehensive community support centers, include the core comprehensive community support center covering the entire city. Common themes in recent years include long-term care and adult guardianship provided by aged or impoverished persons. These themes cannot be solved by a comprehensive community support center alone. They require cross-organizational responses in the local government.

The Fujisawa Community-based Integrated Care System Promotion Conference is held two to four times a year. This conference is placed in the local government and operated in cooperation with the internal examination committee for the promotion of Fujisawa Community-based Integrated Care System. The results of this conference are subject to reporting to the municipal assembly. The activities are conducted by the entire city.

This conference has set up six focal themes including the development of a local consultation support system, support for local activities and developing people to take charge, health care and spending rewarding life, support for home care life, prevention of social isolation, and development of environment. The conference established specialized subcommittees that would take charge of these themes. Each specialized subcommittee examines policies for all residents, irrespective of whether they are elderly people, based on discussions at Sub-community Care Conferences held for each of the Theme-based Community Care Conferences or spheres of daily life, depending on the themes. The results of discussions at the specialized subcommittees are reported at Fujisawa Community-based Integrated Care System Promotion Conference. They will be developed into policies for the entire city through further discussions.

3.3.2 Sub-community Care Conference

The Sub-community Care Conferences covering the spheres of daily life are organized by the comprehensive community support center acting as the secretariat.[2] The Sub-community Care Conferences aim to monitor elderly people in the spheres of daily life. The comprehensive community support center therefore selects the participants in the meetings considering the characteristics of respective spheres. Table 2 presents the members of the Sub-community Care Conferences in the nine spheres in which the information was available. This table reveals that the federation of neighborhood associations, council of commissioned welfare and child welfare volunteers, municipal councils of social welfare, district councils of social welfare, federation of senior citizens' clubs, long-term care insurance-certified providers, municipal government, and comprehensive community support centers provide members to all spheres. The welfare volunteers in the sphere of I, IV, VII, and VIII, the community exchange salons in sphere of VII and XIII, the commercial business operators in the sphere of VIII, and the welfare subcommittee of Hometown Development Promotion Council in the sphere of IX have selectively joined the members based on local characteristics. In all cases, parties from within their respective spheres comprise the members.

The following examines activities of the Sub-community Care Conference in the sphere of IX. The sphere of IX is located in the western part of Fujisawa, which largely comprises new towns developed in and after the 1970s. Many of the residents are baby boomers. Therefore, the average age has been rising rapidly in recent years. The ratio of elderly residents in this area, which was the lowest in the city in 2005, had risen to 31.2%, the highest in the city, by 2018.

Table 2 presents the members of the community care conference in sphere of IX. A unique characteristic of this district is participation by the representatives of the welfare subcommittee of Hometown Development Promotion Council and the municipal councils of social welfare as members of the conference. The Hometown Development Promotion Council includes various subcommittees and working

[2]The conference is held jointly by two centers in a sphere of daily life requiring multiple facilities.

Table 2 Members of Sub-community Care Conferences of Fujisawa (as of 2016)

Sphere of daily life	Federation of Neighborhood Associations	Council of commissioned welfare and child welfare volunteers	Council of social welfare		Federation of Senior Citizens' Clubs
			M	D	
I	○	○	○	○	○
II	○	○	○	○	○
III		○	○		
IV	○	○	○	○	○
VI	○	○	○		
VII	○	○	○	○	○
VIII	○	○	○	○	○
IX	○	○		○	○
XIII	○	○	○	○	○

Sphere of daily life	Long-term care insurance-certified provider	Administration	CCSC	Others
I	○	○	○	Welfare volunteers
II	○	○	○	
III		○	○	
IV	○	○	○	Welfare volunteers
VI		○	○	
VII	○	○	○	Community exchange salons, Welfare volunteers
VIII	○	○	○	Providers and commercial business operators in welfare of persons with disabilities, Welfare volunteers
IX	○	○	○	Welfare subcommittee of Hometown Development Promotion Council
XIII	○	○	○	Community exchange salons

The local government includes residents' centers and welfare centers for elderly people in addition to the senior citizen support section. Council of social welfare M and D respectively refer to the Municipalities and District. CCSC refer to the comprehensive community support center (Prepared based on interviews with the senior citizen support section of Fujisawa government and comprehensive community support centers)

groups corresponding to local characteristics. Those similar to the welfare subcommittee exist in all districts. In sphere of IX, the welfare subcommittee joined the members of the Sub-community Care Conference to improve its cooperation with the conference.

The meetings specifically address the monitoring and promotion of the monitoring of elderly residents in the area. More specifically, the participating groups report their monitoring and promotional activities, share information by identifying issues, and consider solutions. In addition to the above Sub-community Care Conference, the comprehensive community support centers receive consultations from local residents as necessary. Whereas some cases are solvable within the centers through collaboration with commissioned welfare volunteers depending on the issue, they work in cooperation with the local government to solve difficult cases such as abusive behavior.

Interviews with the comprehensive community support centers revealed that the centers had expanded after 2011, which caused a general match between the spheres of daily life and areas covered by the centers. This matching facilitates cooperation between the centers and participating groups in the areas. The centers' employees consider that the recognition of the centers has been increasing through the promotional activities. Such formation of local networks based on the spheres of daily life allows the development of a more functional community-based integrated care system. Efforts such as the Sub-community Care Conference responding to the characteristics of each community are observed also in other areas (Hatakeyama 2018).

There are, however, challenges that must be confronted. Mutual cooperation with the Hometown Development Promotion Council is limited to a few districts such as sphere of VI. The Hometown Development Promotion Council holds discussions equivalent to Sub-community Care Conference in other spheres through groups similar to the welfare subcommittee. The members often overlap with the members of the Sub-community Care Conference. Cooperation between the council and the conference, however, remains weak. This weakness derives from the shortcomings of the vertical structure of the local government organization, in which the personnel of the Sub-community Care Conference belong to the welfare section, although the personnel of the Hometown Development Promotion Council belong to the citizen autonomy section. One opinion holds in the community care conferences that they should be unified, which persists as a key issue that must be addressed.

4 Community-Based Integrated Care System in Naruto, Tokushima Prefecture: A Case of Type 3-C

4.1 Reasons for Selecting Naruto and Overview of the Area

Naruto is a regional city located in the northeastern part of Tokushima Prefecture. It is approximately 30 min away by car from the city of Tokushima, the prefectural capital, implicating Naruto as a Tokushima commuter town. The 2015 Population Census of Japan revealed its population as 59,101. The ratio of elderly residents is 27.6%, which is slightly higher than the national average. The city's financial capability index was 0.65 in 2016, a high value for a regional city. Naruto did not experience the Heisei Municipal Mergers. It has area of 135.7 km^2.

Naruto has districts with various characteristics, including the central urban area, agricultural and fishing village areas, industrial areas, and new residential areas. The city therefore placed five comprehensive community support centers and five spheres of daily life based on the centers (Fig. 6). The operation of these centers is commissioned to social welfare corporations and medical corporations in the city. The core comprehensive community support center was established to oversee the five comprehensive community support centers in September 2015 to be operated by the council of social welfare. A total of six comprehensive community support centers were in operation as of August 2018.

Community care conferences were placed in 2012 based on the spheres of daily life centering on the comprehensive community support centers. The Independent Living Support Care Conference was launched in November 2015 for standardization and skill improvement for the care management of long-term care insurance and on-the-job training (OJT) of care managers, long-term care insurance service providers, etc. This conference covers the entire city of Naruto.

As described above, Naruto operates its community care conferences at two levels, one for the spheres of daily life and the Independent Living Support Care Conference for the entire city, through multiple comprehensive community support centers. The city therefore corresponds to Type 3-C among the nine types specified in this study. We considered Naruto appropriate as an example of Type 3-C municipalities, considering its financial capability index, which is close to the average (overall Type 3-C index is 0.68) and low percentage of municipal bodies directly operating comprehensive community support centers (overall Type 3-C rate is 28.7%) despite its population, which is smaller than the average population of Type 3-C municipalities (214,500, excluding ordinance-designated cities).

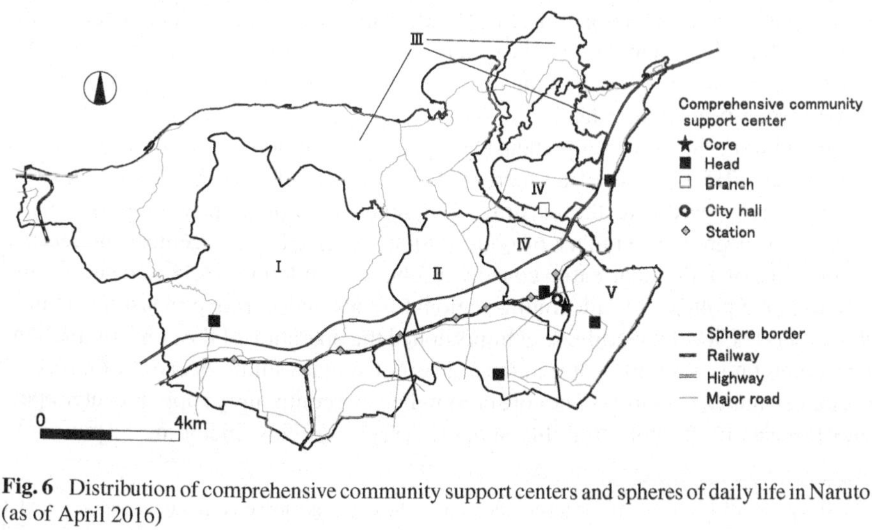

Fig. 6 Distribution of comprehensive community support centers and spheres of daily life in Naruto (as of April 2016)

4.2 Overview of Naruto's Community-Based Integrated Care System

Naruto adopted the viewpoint of the community-based integrated care system after third-phase long-term care insurance planning (2006–2008). It has been implementing various programs. The city has particularly emphasized support for cognitively impaired elderly residents and projects to prevent conditions that would require nursing care. To support cognitively impaired elderly residents, the city provides training courses for dementia supporters, prepares a community resource map, establishes regional dementia support promoters, presents an advice desk for adult guardianship, and supports other activities to improve its support system for elderly people with dementia and to help connect residents with the local government through information sharing and understanding.

In 2006, the city established five comprehensive community support centers as the bases for its community-based integrated care system and set up five spheres of daily life having each of the centers as their respective cores. The following is the reason for placing five comprehensive community support centers. The number of centers appropriate for the population of Naruto based on the government's criteria would be two to five. The city established comprehensive community support centers at five locations as the maximum number based on the criteria. Naruto considered that it would be efficient to abolish home care support centers, which had been functioning as local advice desks, and to transfer the function to the comprehensive community support centers. The city subsequently decided to entrust the operation of the comprehensive community support centers with the operator of the home care support centers, which had a long history of operating in the community[3] (Table 3). For the spheres of daily life, too, the city transferred the areas covered by the home care support centers to assign, in general, one center to one sphere. The sphere of IV, however, has a large population. The sphere is divided into an island and the mainland. Therefore, it placed a branch of the health facility for elderly people operated by the same corporation on the island.

In addition, the core comprehensive community support center was founded in September 2015 because of an increased need for an agency that would support, connect, and coordinate the existing comprehensive community support centers because of a growing number of consultations received at the centers, increasing complexity of individual cases consulted (abuse, wandering, etc.), and the necessity to make policies by integrating regional issues and improve measures against dementia. The local government commissioned the operation of the core comprehensive community support center to the council of social welfare because of a lack of specialist human resources. The core comprehensive community support center operates projects to develop disability support service systems, including responses to

[3]In the sphere of III, the commissioned operator of the comprehensive community support center was operating a home care support center in the northwestern part of the same sphere. It decided to open the comprehensive community support center in the eastern part of the sphere, in which its head office was located.

Table 3 Overview of comprehensive community support centers and community care conferences of Naruto (as of April 2016)

	Core	Head				
		I	II	III	IV	V
Year of establishment	2015	2006	2006	2006	2006	2006
Operator	CSW	SWC	MC	MC	SWC	SWC
Year of operator's establishment	1989	1994	1990	1977	1978	1985
Number of employees	5	3	3	3	5	3
Number of times Community Care Conference was held (2014)	–	Unknown	2	0	0	5

Commissioned operators CSW, SWC, and MC respectively refer to the council of social welfare, social welfare corporation, and medical corporation (Prepared based on information provided by elderly people and social section of Naruto government and interviews with comprehensive community support centers)

difficult cases (abuse, advocacy, general consultation support, etc.) primarily through community care conferences, coordination and backup support for partnership and improvement of the centers in providing skill training, etc., holding Independent Living Support Care Conference, promotion of measures against dementia by setting up initial-phase intensive support team, identification of local resources, and support for the launch of disability support services.

Community care conferences have been held generally at the comprehensive community support centers since FY2012. Significant differences in the frequency of holding community care conferences among the comprehensive community support centers, however, have come to present difficulties (Table 3). More specifically, the perception of community care conferences varied substantially among the comprehensive community support centers: (1) positioned the conference as a key measure for solving community issues; (2) emphasized the meetings of personnel holding three positions in the center and would not hold a conference because of time constraints, and (3) expressed concern that the personal information of the related residents would spread in the community through the conference and make them reluctant to hold it. Naruto, therefore, had not yet come to hold community care conferences based on a common understanding within the city.

Given those circumstances, the Independent Living Support Care Conference, which would be organized by the core comprehensive community support center, was established in November 2015. This conference, which covers the entire city of Naruto, examines details of care plans prepared by the preventive care support section of the comprehensive community support centers and considers the subsequent care services. Members of this conference are employees of the comprehensive community support centers who have prepared care plans related to each case and the core comprehensive community support centers. Dentists, physical therapists, nutritionists, pharmacists, occupational therapists, and other specialists participate in the conference as advisors.

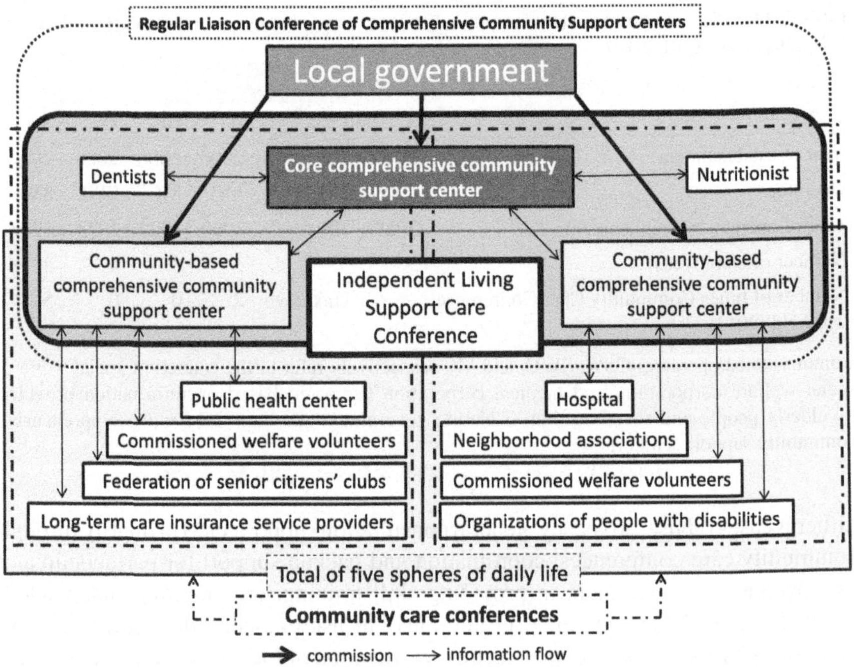

Fig. 7 Community-based integrated care system centering on community care conference in Naruto (as of April 2016)

The city aims to establish a Community Care Promotion Conference to help improve the collaboration of relevant parties at the level of spheres of daily life, share information related to the residents' needs and the status of care resources, and examine measures for the entire city during the seventh long-term care insurance planning (2018–2020).[4]

Naruto has therefore, through the comprehensive community support centers and community care conferences, built its community-based integrated care system at two levels, including the Independent Living Support Care Conference led by the core comprehensive community support center at the city level and community care conferences led by the comprehensive community support centers at the level of spheres of daily life (Fig. 7). Moreover, the city plans to strengthen this structure further by holding the Community Care Promotion Conference.

[4]Based on the 7th elderly health and welfare plan and long-term care insurance plan of Naruto.

4.3 Activities for Naruto's Community-Based Integrated Care System

4.3.1 Conferences Covering the Entire City

The Independent Living Support Care Conference has been held once a month since November 2015. Employees of comprehensive community support centers other than the one in the area relevant to a case attended the conference as observers in the early phase after the establishment. The attendance, however, decreased gradually. Consequently, since 2016, employees of comprehensive community support centers other than the one in charge have been participating in the conference as advisors to maintain collaboration and information sharing.

Furthermore, the Regular Liaison Conference of Comprehensive Community Support Centers is held as an opportunity for representatives of all the comprehensive community support centers in the city to meet together. This conference is held once a month, with participation by employees of the core and other comprehensive community support centers and the Elderly Care Section of the local government. The activities of the conference include description of the Independent Living Support Care Conference, other notifications, activity plans of regional dementia support promoters, and workshops to improve employees' skills. This conference, however, consists largely of reports from the local government rather than opinions of the comprehensive community support centers provided or exchanged. Since 2016, the liaison conference members have been exchanging opinions about cases discussed in the community care conferences to promote the active involvement of the comprehensive community support centers, developing networks among the comprehensive community support centers, and improving center management skills.

4.3.2 Community Care Conference

The community care conferences serving the spheres of daily life hold meetings to resolve issues and subsequently monitor the residents. The two comprehensive community support centers operating the community care conferences address issues that are not limited to elderly people and convenes heads of self-government associations, physicians, representatives of the police, public health centers, and welfare offices, supporters of persons with disabilities, long-term care insurance-certified providers, employees of the local government and comprehensive community support centers, and other relevant parties appropriate for each case (Hatakeyama 2017).

As noted earlier, however, only two comprehensive community support centers are holding community care conferences. Differences in the willingness of commissioned operators of community care conferences are causing disparities in the frequency of holding the conferences.

Since establishment of the core comprehensive community support center, it has gradually taken over the administration of the community care conferences because of insufficient personnel at the comprehensive community support centers. Many difficult cases handled by the core comprehensive community support center, however, are elevated from the comprehensive community support centers, public health centers, community support centers,[5] and other relevant agencies and counselors such as commissioned welfare volunteers. Issues of local residents handled by the community care conferences in their respective spheres of daily life have decreased. Furthermore, the core comprehensive community support center hopes to convene multiple comprehensive community support centers from other spheres of daily life to share information and to improve their skills to handle specific cases. The comprehensive community support centers, however, are reluctant to cooperate because, aside from their busy schedules, the commissioned operators of the centers share a competitive relationship for providing long-term care insurance services.

5 Characteristics and Issues in Local Governance of Type-3 Municipalities

Asago, a Type 3-A city, is building its community-based integrated care system through five community care conferences having different characteristics based on the comprehensive community support center placed in the city hall. Disability support service coordinators monitor elderly people in the local communities through cooperation with various organizations.

All of these community care conferences and monitoring activities are conducted for the entirety of Asago city. In other words, Asago has developed centralized local governance of the entire city with the Community-based Integrated Care System Promotion Conference placed at the top. When multiple spheres of daily life are established as in the case of Asago, community care conferences are generally held at comprehensive community support centers placed based on the spheres of daily life. In many cases, therefore, multilayered local governance of the entire municipality and spheres of daily life is built. In Asago, however, the comprehensive community support center placed in the city hall takes charge of all community care conferences. Because of this, local governance is more single-layered, being unable to make effective use of the spheres of daily life. Although local residents occasionally attend the Conference of Three Neighbors on Both Sides and Nokokai, these participants are mostly the subjects of the meetings or their families.

Despite its large area, Asago is building a centralized community-based integrated care system without using its spheres of daily life effectively. As a result,

[5] Welfare institutions support the daytime activities of persons with disabilities who have difficulties being employed by ordinary places of business because of their disabilities as specified in the Services and Supports for Persons with Disabilities Act.

self-government associations, neighborhood associations, Autonomy Council,[6] and other resident organizations have not been participants in local governance. The perspective of community development is important in the community-based integrated care system (Fujii 2018). Therefore, a system in which the residents can actively take part is desired. Asago is considering incorporation into the Neighborhood Autonomy Council, enthusiastically carrying out residents' autonomous activities, into the community-based integrated care system by making it a member of the Community-based Integrated Care System Promotion Conference. The city must start building a community-based integrated care system that incorporates consideration of the local characteristics through residents' participation without delay.

Fujisawa, which shifted from Type 3-B to Type 3-C, planned to make its decentralized community-based integrated care system functional by setting up several spheres of daily life and placing vomprehensive community support centers there, which were operated by commissioned operators in the city. Expansion of the comprehensive community support centers allowed the centers to function as the base in each sphere, which facilitated more effective use of local networks in each sphere. The community care conferences, which had been capable only of holding bloc meetings by dividing the city into four sections, became able to hold meetings in each sphere of daily life by virtue of the expansion of the comprehensive community support centers. The block meetings were subsequently reorganized into theme-based meetings, which would cover the entire city. Inadequate expertise in handling difficult cases became a difficulty at block meetings, which indicates that Fujisawa transformed itself from Type 3-B to Type 3-C to resolve issues it faced while being Type 3-B. The city also established the Fujisawa Community-based Integrated Care System Promotion Conference with an aim to developing its community-based integrated care system into a cross-field scheme, which would include not only elderly people, but also welfare and medical care, and to apply issues of the spheres of daily life to the policies of the entire city. Fujisawa's community care conference, thereby, developed into a multilayered organization. Through this, the city also established a system of sharing local community issues in the entire city and applying them to policy-making.

Fujisawa has therefore formed local governance on a multilayered scale of the entire city and spheres of daily life while restructuring the spatial framework of the community care conference and expanding the comprehensive community support centers in its development of a community-based integrated care system. Considering its large population and diverse regional characteristics, the city further placed a comprehensive community support center in each sphere of daily life by commissioning corporations in the city and held community care conferences. These have become the basis of decentralized local governance. Community networks can be

[6]This residents' autonomous organization was established to examine and solve local issues within the community and make the community sustainable in the future based on the principle of "building a community where we think and act for ourselves, help one another, and create together." Its groups are generally based on the city's elementary school districts.

developed based on spheres of daily life in Fujisawa's community-based integrated care, particularly because, as in the case of Tsugaru (Chapter "Community-Based Integrated Care Systems in Municipalities Having Sub-branches of Comprehensive Community Support Centers"), the spheres of daily life and the activity areas of local organizations are generally the same. This case underscores the importance of the areas of daily life and their spatial framework in setting up the spheres of daily life, which determine the area structure of the community-based integrated care system, and the policy for placing comprehensive community support centers.

In Fujisawa, discussions and overlapping of members, its Sub-community Care Conferences, and Hometown Development Promotion Council had come to pose some difficulties. This was a shortcoming of the vertical structure of the local government organization. To resolve this issue, Fujisawa Community-based Integrated Care System Promotion Conference was established as a cross-departmental group that is expected to help solve issues going forward.

Naruto, a Type 3-C city, arranged comprehensive community support centers at the center of local governance in the spheres of daily life and commissioned various long-term care insurance-certified providers in the city to operate the centers. Additionally, the city planned to absorb issues arising in each sphere of daily life and build its local governance on a higher level scale managed by the local government and operators of comprehensive community support centers. Thereby, Naruto aimed to develop multilayered and decentralized local governance on the two scales, i.e., the entire city and spheres of daily life.

As a result, the city has begun to carry out activities that effectively use local characteristics while using community networks through comprehensive community support centers operated by corporations rooted in the spheres of daily life. The local government that was responsible for coordinating the comprehensive community support centers, however, faced difficulties in building local governance based on the above system caused by insufficient personnel. The core comprehensive community support center was consequently placed by commissioning the council of social welfare to bridge between the local government and comprehensive cCommunity support centers. As a result, the core comprehensive community support center became the core of the community-based integrated care system, which would organize the Independent Living Support Care Conference, Comprehensive Community Support Center Liaison Council, and community care conferences based on the spheres of daily life while incorporating consideration of the city's policies. The community care conferences, which were organized by the comprehensive community support centers in their respective areas before the establishment of the core comprehensive community support center, are now operated jointly by the central and local comprehensive community support centers through the operators of the local centers, which entrust the core comprehensive community support center to operate the community care conferences. In other words, the independence of comprehensive community support centers in the spheres of daily life, the most local level, and those at the city level on a larger scale have gradually been lost. The community-based integrated care system at the level of the spheres of daily life and that at the city level have approximated each other.

Before placing the core comprehensive community support center, Naruto had aimed to build multilayered and decentralized local governance, which nevertheless became difficult because of disparities in the willingness of the commissioned operators of the comprehensive community support centers and inadequate human resources of the local government that should have coordinated the entire community-based integrated care system. The city then commissioned the council of social welfare to operate the core comprehensive community support center having a coordinating function with an aim to build a centralized system based on this center. In other words, Naruto strengthened its government function, although in the form of commission to the council of social welfare, to make the governance function.

6 Conclusions

As explained up to this point, many of the Type-3 municipalities place multiple comprehensive community support centers to cover their large populations. The community-based integrated care systems of the municipalities examined in this chapter suggest that Type-3 municipalities often develop decentralized local governance.

Fujisawa, which was building decentralized local governance, changed from Type 3-B to Type 3-C in an effort to overcome the difficulty of inadequate expertise of community care conferences that it found in the spheres of daily life alone. This change suggests that many of the medium-sized and large municipalities might eventually become Type 3-C. Such a form of decentralized and multilayered local governance is consistent with the condition of the community-based integrated care system sought by the Ministry of Health, Labor and Welfare. While the author also regards the development of local governance based on such a community-based integrated care system desirable, some issues remain to be addressed. Under decentralized local governance, activities in the spheres of daily life are affected by the willingness of corporations entrusted to operate the comprehensive community support centers. This point is exemplified by Tsugaru, a Type 2-B city. Among the Type-3 cities, Fujisawa was not experiencing important differences in the activity levels among the spheres of daily life thanks to the efforts of the corporations. Naruto, by contrast, was confronting disparities in community activities. Such influence of the motivation of corporations on the activities in the spheres of daily life poses the greatest uncertainty in building decentralized local governance. The role of a municipal government as meta-governance, which checks and coordinates local governance centering on the comprehensive community support center in each sphere of daily life while building decentralized local governance, considering the characteristics of each sphere of daily life is therefore important.

Furthermore, Type-3 municipalities, by virtue of their large populations, have a wealth of community resources for the community-based integrated care system. Wide-area cooperation with other municipalities observed among the Type-1 and

Type-2 municipalities is rare. Type 3 is therefore characterized by the completion of local governance within the municipalities.

Asago, having a small population among the Type-3 municipalities, was categorized into Type 3-A and building centralized local Governance based on multiple community care conferences covering the entire city through the comprehensive community support center placed in the city hall. The inability to incorporate local residents into Asago's system, however, came to present a difficulty. Although centralized local governance can be developed for a smaller population or area, as exemplified by Type 1-A, Asago's size requires that it shift to Type 3-C and build decentralized local governance in the future.

The result of analyzing the characteristics and issues of the nine types of community-based integrated care system from Chapter "Community-Based Integrated Care Systems in Municipalities Having One Comprehensive Community Support Center" through Chapter "Community-Based Integrated Care Systems in Municipalities Having Multiple Comprehensive Community Support Centers " suggests that Type-2 municipalities, including those which have shifted to another type because of the small number of available specialists, might be divided between the other two types, Type 1 and Type 3, depending on their population sizes in the future. Regarding categorization based on units of establishing community care conferences, Type-B municipalities might also change to Type A or Type C, depending on their population sizes, because of a lack of expertise in the spheres of daily life and challenges in applying local issues to policy-making of the municipalities. Because many Type-3 municipalities have a large population, however, it is more likely, in general, that they will shift to Type C. In other words, such municipalities are likely to be divided into Type 1-A, Type 1-C, and Type 3-C among the types defined for this study. As revealed in this analysis, each type has both benefits and shortcomings. It is therefore desirable that each municipality build local governance that considers its local characteristics and develops its own community-based integrated care system.

References

Fujii H (2018) Community development and community welfare in community-based integrated care. In: Sumida Y, Fujii H, Kuroda K (eds) Understand community-based integrated care. Minerva Shobo, Kyoto (in Japanese)

Hatakeyama T (2009) The expansion of municipal authority and regional influence resulting from long-term care insurance system reform: The case of Fujisawa City, Kanagawa Prefecture, Japan. Jpn J Hum Geogr 61:409–426 (in Japanese with English abstract)

Hatakeyama T (2017) Development of community-based integrated care system that considers local characteristics in regional cities and the roles of local governments. In: Sato M, Maeda Y (eds) Local governance and region. Nakanishiya Shuppan, Kyoto (in Japanese)

Hatakeyama T (2018) Welfare and long-term care systems in a super-aging society. In: Yagasaki N, Morishima W, Yokoyama S (eds) Series Topographic Topics 3: sustainability: challenges of the Earth and Human Race. Asakura Shoten, Tokyo (in Japanese)

Hatakeyama T, Nakamura T, Miyazawa H (2018) Community-based integrated care systems in Japan: Focusing on spatial structures and local governance. E-Journal GEO 13:486–510 (in Japanese with English abstract)

Omizu Y (2005) Daily life and community development conference: From participation to autonomy of residents. In: Aoki S (ed) Fujisawa municipal government researcher (eds) Fujisawa residents' participation and cooperation: Towards the Sea of Shonan. Gyosei, Tokyo (in Japanese)

Regeneration of Housing Estates by the Community-Based Integrated Care Systems

Yoshimichi Yui, Hitoshi Miyazawa, Yoshiki Wakabayashi, and Leng Leng Thang

Abstract A great number of housing estates created in the suburbs of urban cities in the 1960s are aging rapidly in recent years. This accelerating obsolescence is attributable to the fact that many monotonous houses without diversity or vitality in room arrangements were supplied at sales prices and rent values that did not vary considerably. Therefore, their residents have remained homogeneous with respect to life stage, age distribution, and social position. In many housing estates in the suburbs, 30 years after their move-in, the ratio of elderly people increased markedly together with the decline of young people caused by employment and marriage. At the beginning of the development, housing estates in the suburbs expected residents of younger generations such as child nurturing generations. Many facilities in the estate are for kindergarten children's playgrounds, educational facilities such as elementary schools, and commercial facilities for the purchase of food and daily necessaries. Hospitals are arranged by design, although it is characteristic that medical facilities and nursing care facilities for elderly people are few. Many middle-rise old apartments are not equipped with elevators. In fact, many steep sloping roads and steps cause hindrance today for the daily living of elderly people. Focusing on housing renewal projects incorporating welfare viewpoints, the aim of this study is to identify the meaning and tasks of regeneration of housing estates for activation of the community through integrated streamlining of welfare service facilities including elderly people welfare facilities. The study area is Nagayama estate in Tama new town and Kashiwa Toyoshikidai, which introduced the community-based integrated

This chapter is a translated version of Yui et al. (2018).

Y. Yui (✉)
Hiroshima University, Higashi-Hiroshima, Japan
e-mail: yyui@hiroshima-u.ac.jp

H. Miyazawa
Ochanomizu University, Tokyo, Japan

Y. Wakabayashi
Tokyo Metropolitan University, Tokyo, Japan

L. L. Thang
National University of Singapore, Singapore, Singapore

H. Miyazawa and T. Hatakeyama (eds.), *Community-Based Integrated Care and the Inclusive Society*, International Perspectives in Geography 12,
https://doi.org/10.1007/978-981-33-4473-0_10

care system by fulfilling welfare services for elderly people to allow for sustainable living by elderly people. Findings of future residential estates are discussed based on these cases.

Keywords Community-based integrated care system · Elderly care · Housing renovation · Kashiwa toyoshikidai · Suburban housing estate · Tama new town

1 Introduction

1.1 Background

Residential estates supplied to the market in large numbers in suburban areas after the 1960 s are now facing rapid aging (Kaneko 2017) because residents of the estate have similar life-stages, age composition, and social positions. Copious amounts of houses without diversity and vitality were supplied in a given sales price range and rent value range within a short period. After more than 30 years from the start of move-in, an increase in the number of elderly people because of aging of the husband and wife (absolute aging) and an increase in the ratio of elderly people because of a decrease in the younger generation attributable to employment and marriage (relative aging) advanced in parallel, thereby resulting in remarkable aging of society. Therefore, abandoned homes have resulted from the death of the residents and relocation of facilities for elderly people to suburban residential areas. This abandonment constitutes an important regional issue (Yui et al. 2016).

At the beginning of development of suburban housing estates, many estates anticipated their popularity with families with small children. Therefore, child-oriented facilities such as kindergartens, children's playgrounds, educational facilities such as elementary schools, and commercial facilities for the purchase of food and daily necessities and hospitals were included in the basic design, although, in contrast, medical facilities and nursing care facilities for elderly people were few. Many middle-rise apartments in estates were not equipped with elevators, and many steep sloping roads and steps cause hindrances today for the daily life of elderly people.

According to the Ministry of Land, Infrastructure, Transport and Tourism, 1,551 estates out of 4,970 estates nationwide are more than 35 years old after construction. The aging of buildings, installations and public facilities in the housing estate, aging of residents, and devitalization of neighborhood centers all have arisen together. Necessary urban functions such as welfare functions are not maintained; living environments have greatly deteriorated. Regeneration of housing estates is now cited as a social necessity. Given stresses imposed by population decline and aging population combined with the diminishing number of children, promoting regeneration of suburban housing estate where the number of abandoned houses is increasing and deviation from demand for land utilization has arisen, in January 2017, Ministry of Land, Infrastructure, Transport and Tourism established "Housing estate regeneration" liaison conference where related parties such as local governments and private enterprises exchange opinions. As described, the aging of buildings and the aging of residents in suburban housing estates are becoming severe. Now, regeneration and renewal have become one of important national concerns.

1.2 Purpose of This Study

Devoting attention specifically to housing renewal projects and particularly addressing welfare viewpoints, the aim of this study is to identify the meaning and task of regeneration of housing estates for activation of the community by integrated streamlining of the welfare service facilities including welfare facilities for elderly people instead of mere housing rehabilitation. From examples of regeneration projects of the housing estate, which introduced a community-based integrated care system by which welfare services are delivered to elderly people, outcomes and tasks of future housing estate regeneration projects are discussed.

As described in an earlier chapter, community-based integrated care systems are a regional comprehensive support and service delivery system aimed at support for preserving elderly people's dignity and independent life so that they can maintain their own lifestyle until the end of life in the home area. This should be established until 2025, as a goal, when baby boomers will be older than 75 years old. Five factors of community-based integrated care systems (house, medical care, nursing care, prevention, life support) are maintaining an organic relation while working together. The basic factors of community-based integrated care are life support, medical care, and welfare services for stable daily life in their house presented to them. Introduction of community-based integrated care systems in the regeneration project in the housing estate where aging of people and of housing units advanced are regarded as contributing to the formation of residential areas where aged people are able to continue to live by improvement of housing units and housing facilities such as installation of elevators and gentle slopes as well as fulfillment of welfare services.

2 Outline of Study Areas

The author chose following two study areas. One is Nagayama estate in Tama new town in the western suburbs of Tokyo and another is Kashiwa Toyoshikidai estate in the eastern suburbs of Tokyo (Fig. 1). These housing estates are picked up as the study target for the following reasons: they are old estates developed from the 1960s to the first half of the 1970s; aging of housing units and people advanced seriously and housing renewal projects are taking place, aimed at the introduction of community-based integrated care systems for life support and medical support for elderly people.

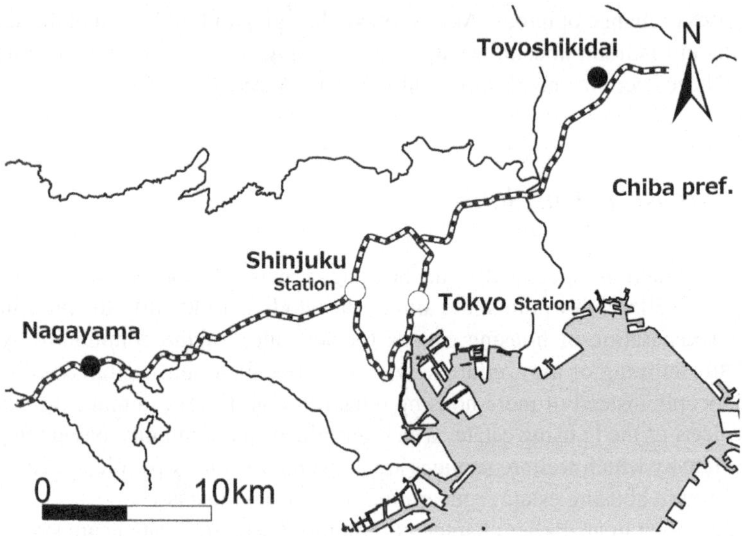

Fig. 1 Study areas

2.1 Nagayama Estate in Tama City

The total area of Tama new town developed by Japan Housing Corporation (later changed from Housing and Urban Development Corporation to the present Urban Renaissance Agency (UR) by clearing Tama hills was about 30 ha, making it the largest in Japan. Tama new town is located about 25–40 km southwest of Tokyo, involving Machida City, Hachioji City, Tama City, and Inagi City. The residential population reaches about 224,000 (in 2018). The first residents of Nagayama estate moved to the area in 1971: the earliest among the residences of Tama new town.

The population of Tama new town increased to 30,000 in as little as three years after the start of moving in. However, this short move-in period caused a marked lack of balance in the age-group of the residents. Today, super-aging is advancing in the estates. In the commercial areas in Suwa estate and Nagayama estate, almost all merchant stores have closed because of the influence of aged residents. They are now called "Shutter streets" (Fig. 2).

In the Suwa estate which located in adjacent Nagayama estate, various projects for estate regeneration are progressing at present. Housing rehabilitation of Suwa 2-chome, which is said to be "the greatest rehabilitation in Japan" is noteworthy. This is a large-scale project to replace aged Suwa 2-chome housing units (five-story, 23 buildings, 640 housing units without elevators) together. Large-scale housing rehabilitation was undertaken in 2010, 40 years after construction and in 2013, Suwa 2-chome housing units were regenerated as "Brillia Tama new town" with support by the local government. In all, 23 five-story buildings in the estate were regenerated as seven 11-story or 14-story high-rise buildings. The number of housing units roughly

Fig. 2 Commercial area in Nagayama housing estate (Photograph taken by Yoshimichi Yui in March 2019)

doubled, from 640 to 1,249. The additional housing units increased (609 houses) were sold as condominiums, with expenses of rehabilitation covered by profits from sales. The method of construction and finance attracted attention to all people concerned.

2.2 Kashiwa Toyoshikidai Estate

Another study area is Kashiwa Toyoshikidai estate, which is located at about 40 min from Tokyo station to Kashiwa station by JR Joban line. It takes slightly less than 5 min from the nearest Kashiwa station to the center of housing estate by bus and slightly less than 15 min even on foot. Consequently, the convenience of conditions of the site is good among suburban residential areas. Development of Toyoshikidai estate started in 1964 by large-scale estate creation by Japan Housing Corporation: a total of 103 buildings and 4,666 housing units were provided, with move-in started from April of the same year. This caused an increase of the population by about 15,000. The population of Kashiwa City exceeded 100,000 by the end of the same year. Most of the buildings supply medium-rise housing, although two buildings constructed in 1987 are high-rise buildings (Toyoshikidai No. 2 estate, 105 and 106 buildings).

No elevator is provided to medium-rise complex housing units in the Toyoshikidai estate, making it very troublesome for aged residents to ascend and descend stairs.

Old small-scale supermarkets and private shops that used to operate in the estate commercial area have been replaced by nationwide chain store type supermarkets.

3 Regeneration Project of Nagayama Estate in Tama City

3.1 *Outline of Project in Nagayama Housing Estate*

The Nagayama housing estate comprises condominiums for sale and rental apartment houses, which were supplied by Japan Housing Corporation (renamed Urban Renaissance Agency) and housing renewal have not been launched yet. In Nagayama housing estate, the aging of society poses a severe difficulty (Fig. 3) and the establishment of life support system for elderly people by community-based integrated care system is an urgent issue.

The "Nagayama model" by Tama City is well known as the project which introduced the community-based integrated care system. The "Nagayama model" is a challenge for regeneration of housing estates planned by Tama City and promoted mainly by residents. For Nagayama district, where aging is advancing rapidly, Tama City started a workshop with participation by the residents' associations, NPOs, citizen groups, and the like to create a method for deploying observation, places to stay, life support, and preventive care.

Tama municipal government held the "First Nagayama district workshop" in May 2016 and discussed watching activity, care prevention, exchange of many generations, and life support. About 40 people attended, representing diverse stakeholders

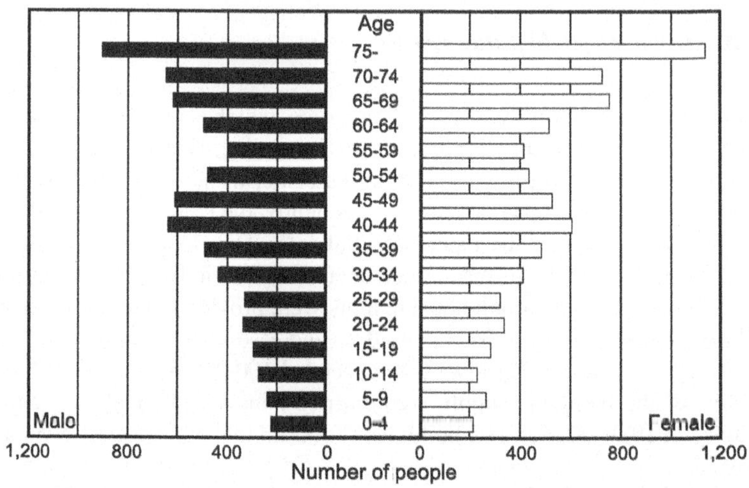

Fig. 3 Population composition by age of Nagayama housing estate (2015) (Prepared based on Population Census of Japan)

such as the Nagayama estate resident association, Nagayama estate commercial avenue, caseworkers, child committee members, welfare network Nagayama, day nursery, kindergarten, primary school, middle high school, medical association, UR, and comprehensive community support center.

Themes discussed by "Nagayama district workshop" included the following: (1) Watching activity (discovery of non-related people, relationship, "Choibora (devote a little attention to surroundings)"), (2) Scene for commutation for care prevention (approach to exit from the entrance, calling in cooperation with universities, increase in the place of assembly), (3) Life support ("Chokotto service" at reasonable price, life support coordinator, exchange between elderly people and children at nursery provided lunch), (4) Exchange of many generations (children are invited to volunteer activities and installation of a scene in which regional inhabitants converse together, new residence of many generations to the housing estate, and utilization of vacant school rooms). Tama city participated only in the setting of a scene of exchange of opinions while discussions were conducted mainly by inhabitants such as residents' associations.

It is noteworthy that enthusiastic discussion was made for "What we can do" and "Such a thing would be nice" for each assignment. In other words, regional assignment is identified by workshops, with possible measures selected by the residents themselves, with aged people support and care prevention discussed through resident participation, and administration creating a mechanism to support them.

3.2 *Outcome of the "Nagayama Model"*

As a preceding case, Tama City Central Area Comprehensive Community Support Center of Tama central area was moved to the UR rental facility in the avenue of well-known stores of Nagayama estate on October 24, 2016. A consultation counter for elderly people was newly opened to construct the "Tama City style community-based integrated care system" including persons with disabilities as well as elderly people. For construction of a community-based integrated care system, Tama City is performing "Consultation of all personal problems" and watching activities by supporters for elderly people at the comprehensive community support center, which used to be vacant stores.

Of those, elderly people support activities by "Welfare pavilion (Fukushitei)" are unique (Fig. 4). "Fukushitei" introduced by Miyazawa (2006) and Ueno and Matsumoto (2012) is a community dining place using vacant stores operated by elderly people where eating and distribution of meals (50 meals daily without delivery charge) are provided. Both the operation and customers of the dining places are elderly people of the community. According to the host, the "Fukushitei" is a multi-purpose facility serving as a salon, coffee house, and restaurant that serves only set meals, a GO parlor, and a Japanese style pub.

"Fukushitei" started in April 2001 when the committee of social participation enhancement for elderly people was established in Tama City and opened in February

Fig. 4 "Fukushitei" opened in Nagayama housing estate (Photograph taken by Yoshimichi Yui in March 2019)

2002 on commission. In April 2003, it became a major project of a specified nonprofit corporation: Fukushitei. The range of business operations is diverse, extending in scope from elderly people support project (formation of a scene of exchange of elderly people) and home living support (meal preparation, furniture movement, cleaning, taking them to the hospital, acting as a mediator between elderly people and healthcare centers, caseworkers, regional groups, care managers, and a home living support group "living support party"), urban development (renovation of pedestrian paths, publication and distribution of a living assistance book "Placed right beside" (Osobanioite), "Ribbon activity" for supervision of elderly people, and "mini-mini international" for supporting residents of foreign nationality. Consequently, their activities have been expanding.

As elderly people support, "Nekosapo station" was started by a major door-to-door delivery company (Fig. 5). "Nekosapo station" presents household duty support services such as support for shopping and cleaning, screen replacement at discount rates for registered members, avocational activity and health enlightenment activity to which points are added by a "Nekosapo station" reward card. Although the degree of recognition of the inhabitants for "Nekosapo station" is not high, it has been increasing gradually. The number of users has also been increasing.

Fig. 5 Nekosapo station (Photograph taken by Yoshimichi Yui in March 2019)

4 Regeneration Project by Kashiwa Toyoshikidai Estate

4.1 Outline of Project in Kashiwa Toyoshikidai Estate

As stated above, development of Toyoshikidai started in 1964. They are now faced with a super-aging society with buildings becoming increasingly older and society becoming 50 years older since its initial construction (Fig. 6). Aging of residents is reflected in the composition of the size of households and the number of households, with the parent generation increasing rapidly only after independence of children. As a result, the rate of one-person households exceed 40%; two-person households also exceed 35%, whereas about 80% of total households are so small as to have only one or two residents.

Housing redevelopment projects of Toyoshikidai as a countermeasure for severe aging and deterioration of buildings were commenced in 2004 by UR (Fig. 7). The first-phase redevelopment of buildings was completed in November 2008, when moving in of the residents started. This place of redevelopment was designated as the condominium which was named “Comfort Kashiwa Toyoshikidai.” Second-phase redevelopment buildings were built as high-rise condominium by a private real estate and sold as condominiums. Subsequently, third-phase high-rise condominium were built; then movement from UR rented housing to private high-rise housing were promoted.

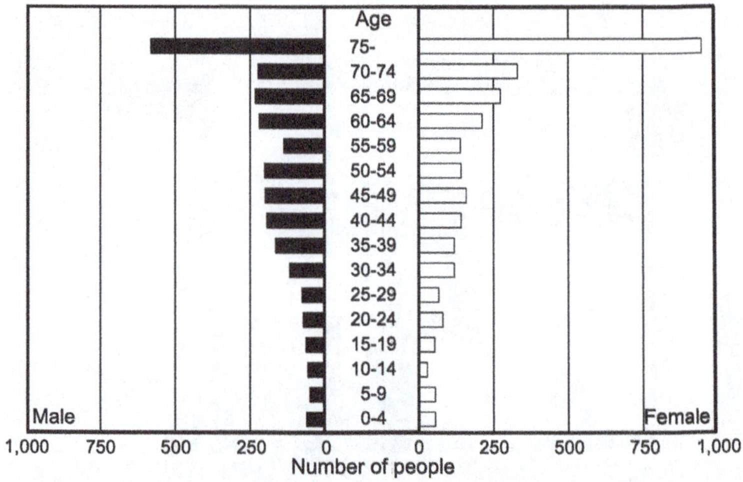

Fig. 6 Population composition by age of Toyoshikidai housing estate (2015) (prepared based on Population Census of Japan)

Fig. 7 Renewal of buildings in Toyoshikidai (Photograph taken by Yoshimichi Yui in March 2019)

As for the Toyoshikidai estate regeneration project, town development corresponding to rapid aging has been promoted by Kashiwa City, Institute of Gerontology of The University of Tokyo and UR (Institute of Gerontology, The University of Tokyo 2014). The purpose of this project is to promote "Town development for sustainable society" with basic policies of "Town where we can lead safe life in own house" and "Sustainable town where we can stay well continuously" by Kashiwa City. Specifically, "Realization of community-based integrated care system" and "Creation of meaningful working by elderly people" are set as the two pillars for concrete measures. A housing rehabilitation project by UR is now undertaken.

4.2 *Outline of Community-Based Integrated Care System in Toyoshikidai Estate*

The community-based integrated care systems in Toyoshikidai estate are attracting attention as a model project promoted cooperatively by local government, UR, and the university. This project is promoted as a part of urban redevelopment towards a long-lived society by cooperation of Kashiwa City, The University of Tokyo, and UR aiming at "Town where we can lead safety life in own house forever" and "Town where we can stay well continuously." As concrete measures, (1) Establishment of backup system to reduce burdens of home medical care (by cooperation with home medical care hub) project, increase in medical doctors in charge of home medical care and promotion of cooperation by various occupations, construction of information sharing system, enlightenment of citizens, consultation and support, installation of central hub (Kashiwa district medical cooperation center) that realizes projects described above; and (2) Creation of motivation in life by working, working in agriculture, livelihood support, nurturing, dining place and welfare, are described.

Challenges to Kashiwa City are promotion of home care and home medical care by promoting the pursuit of health of elderly people and by reducing excessive dependence on facility type nursing care, and are called the "Kashiwa project." Features of the "Kashiwa project" are attracting attention as "Urban development by long-lived society." Kashiwa municipal government is attempting realization of a society in which everyone can get older like oneself in their home area until the last moment while standing up to realize an "Aging in Place" society. After the "Kashiwa district medical care cooperation center" is streamlined, the city office and medical association were able to use the latest information and communication technology (ICT) and training program to promote home medical care. It is noteworthy that a framework to challenge home medical care by a team consisting of many job categories including medical doctors, dentists, pharmacists, nurses, and care workers is streamlined. This is regarded as being attributable to enhanced corporations with medical associations for construction of community-based integrated care systems. Installation of hub type housing units with services in the center of housing estate is also notable. A trend was apparent by which nursing homes for elderly people are

Fig. 8 Building accommodating aged people nursing home, clinic and comprehensive community support center in Toyoshikidai (Day service for aged people is also provided) (Photograph taken by Yoshimichi Yui in March 2019)

rarely provided at the center part, at such a place where users in the housing hesitate to visit easily. In Toyoshikidai estate, facilities that provide services of various types such as medical facilities, care facilities for elderly people, day nurseries, and certified centers for early childhood education are located at the center area, ensuring effective and highly convenient arrangements (Fig. 8).

According to the "Kashiwa Project", consideration has been given to (1) Motivation in life working (realization of affluent second life with abundant motivation in life), (2) The community-based integrated care systems (care system coordinated with home medical care (coordination with medical care)), and (3) Communities were established in which everything is attainable within walking distance (not isolating oneself from society, enjoying meeting with others outside of one's home (development of affluent community)) and such a system that ensures home care and medical care in a healthy condition without dependence on nursing homes.

The municipal government played a leading role in the operation of workshops and conferences aimed at construction of a community-based integrated care system in Toyoshikidai. Those related to medical care and welfare for elderly people were key members. Although the workshop members are those engaged with various jobs according to the City's official resources, in reality, their jobs are mostly medical care and welfare services for elderly people and for the general populace, participants with diverse attributes are not included, although social or welfare workers are included. However, it is considered that this was necessary for construction of care systems coordinated with home medical care, at which "Kashiwa project" had been aimed, and to plan and execute construction of various facilities and maintenance of facilities. It was considered that to perform coordination of large-scale facility maintenance with medical system, projects should be promoted mainly by local governments, and so-called government methods are judged as an effective method.

5 Conclusions

"Nagayama model" is a bottom-up type, intending regeneration of residential property by cooperation of local government, local residents, and related groups attracted attention as community activities for regional regeneration promoted mainly by elderly people. It might be regarded as a "Governance type by local government" community-based integrated care system. If sustainability of an aged society is considered, resident activities by residents other than aged people are not included; this feature is regarded as somewhat controversial.

A bottom-up type "Nagayama model" aimed at regeneration of residential property by cooperation of local governments, local residents, and related groups as housing site regeneration project in Nagayama area in Tama new town is attracting attention as a regional regeneration community activity promoted mainly by elderly people. However, if sustainability of a super-aging society is considered, then it is worrisome that resident activity by those other than aged people is not known.

In contrast, in the case of regeneration of housing site of Toyoshikidai estate, which is led by local administration and UR and which is a part of "Kashiwa project," medical care and nursing care related institutions constructed a framework for deep cooperation to support promotion of home medical care. Therefore, the respective facilities such as elderly nursing facilities and medical facilities are fulfilled, although aggressive participation is not apparent for labor cooperation involving residents, as observed in the Nagayama district of Tama new town.

These two housing renewal projects are presented in Table 1. Housing renewal projects were conducted with incorporation of a community-based integrated care system. In Tama City, establishment of this system was performed mainly by residents and local government acted only as a coordinator. Labor "cooperative" challenge was undertaken mainly by residents and diverse groups and organizations while the local government performed only a coordinating role.

In contrast, as a community-based integrated care system of Kashiwa Toyoshikidai estate, local government showed strong leadership, research institutes and UR challenged care prevention together with medical associations, and institutions cooperated functionally to promote labor cooperative projects. This might be a government-leading type project according to the definition presented by Nishiyama and Nishiyama (2008). A government-leading type project might be designated as a top-down type governance project in which government bureaucrats and specialists

Table 1 Comparison of community-based integrated care system between Tama new town Nagayama housing estate and Kashiwa Toyoshikidai

Features	Nagayama housing estate, Tama new town	Kashiwa Toyoshikidai
Year of development	1971	1964
Developed by	Japan Housing Corporation	Japan Housing Corporation
Redeveloped by	* Suwa housing estate by private developer	UR · Kashiwa City
Project name	Nagayama model	Kashiwa project
Core institute for construction of community-based integrated care system	Tama City	UR • Kashiwa City • The University of Tokyo
Participants of working for construction of community-based integrated care system	About 40 people from diverse stakeholder such as Nagayama housing estate residents association, Nagayama housing estate avenue of well-known stores, Case worker and child committee member, Welfare network Nagayama, Nursery school, Kindergarten, Primary school, Middle high school, Medical association, UR and Comprehensive community support center	Medical doctor (hospital, clinic), Dentist, Dental hygienist, Pharmacist, Nurse (home nursing, hospital • clinic), Hospital area cooperation staff, Care manager, Comprehensive community support center staff, National registered dietitian (at home, hospital), Physical therapist • Occupational therapist • Speech therapist, Nursing service operator, Health facility for elderly people receiving care • Staff of health facility for elderly people receiving care Furusato conference • Citizen and city office staff such as case worker and child committee member
Frequency of holding workshop	Monthly	General assembly held 2 or 3 times yearly and area meeting held once or twice yearly
Features	Nagayama housing estate, Tama new town	Kashiwa Toyoshikidai
Facilities newly constructed by streamlining	Comprehensive community support center using vacant stores	Kashiwa district medical cooperation center, Welfare facility for elderly people (Resident care facility) • Day service • Comprehensive community support center, Clinic in the facility), Certified center for childhood, etc.

(continued)

Table 1 (continued)

Features	Nagayama housing estate, Tama new town	Kashiwa Toyoshikidai
Services constructed	Elderly people and disabled people watching support team, Creation of place to stay, Life support, Nursing care prevention, Meal service by Fukushitei (Dining room and meal distribution), Watching service by Nekosapo, etc.	Home medical care for elderly people, Diverse cooperation for elderly people medical care, Childcare service for childhood center, Support for meaningful life of elderly people, etc.
Style of project promotion	Bottom-up style governance type	Top-down style government type

play a central role, whereas local governments and business institution control the projects with cooperation by related institutions.

On the other hand, with bottom-up type governance, local government only reminds related institutions to participate and simply acts as a supporter for progression of the meeting. The local government neither starts nor operates any project, but simply supports smooth operation as one member of labor cooperative operator. Although this is effective to remind local residents to participate and to activate resident activity, hardware aspects of facility maintenance such as medical care and nursing care are weak.

As described above, bottom-up type governance projects and top-down leading type government projects have their respective benefits and shortcomings. It is difficult to say which is better because thorough investigations of local circumstances should be made to know which is suited for the project intended. Therefore, the actual status of medical care of the region and details of community activities should be grasped; an appropriate community-based integrated care system reflecting local circumstances should be promoted urgently.

References

Institute of Gerontology, the University of Tokyo (2014) Implementing “integrated community care system” practical guidelines of multi-professional teamwork for home medical care. University of Tokyo Press, Tokyo (in Japanese)

Kaneko A (2017) Society history of new town. Seikyusya, Tokyo (in Japanese)

Miyazawa H (2006) The changing face of suburban new towns in large metropolitan areas: the case of Tama New Town. Ann Jpn Assoc Econ Geogr 52:236–250 (in Japanese with English abstruct)

Nishiyama Y, Nishiyama Y (2008) Governance type town planning in UK: urban regeneration by social enterprises. Gakugei Shuppannkai, Tokyo (in Japanese)

Ueno A, Matsumoto M (2012) Tama new town story: never all old town. Kajima Shuppankai, Tokyo (in Japanese)

Yui Y, Kubo T, Nishiyama H (eds) (2015) An increase in problematic housing vacancies in Japanese cities: geographical strategies to make better solution. Kokon Shoin, Tokyo (in Japanese)
Yui Y, Miyazawa H, Wakabayashi Y (2018) The regeneration of housing estates by the community-based integrated care system. Ann Bull Jpn Soc Urbanol 52:65–71 (in Japanese)

Integrated Care Systems Established to Strengthen Community Disaster Resilience

Hitoshi Miyazawa and Haruko Kikuchi

Abstract Building a resilient society is an important policy issue of Japan, a country where disasters occur frequently. Establishing resilient medical care and social welfare systems is stated as a major task in the Fundamental Plan for National/Regional Resilience formulated by the national government and local governments. Community-based integrated care systems, i.e., a network of support for medical care, nursing care, social welfare, various living support, and housing security are aimed at supporting residents' continued life in familiar areas. This system is also expected to contribute to the strengthening of disaster resilience of the community. This chapter consists mainly of the following two topics. First, we clarify the damage to medical care and nursing care in areas affected by the Great East Japan Earthquake and recovery processes that began immediately after the disaster, continuing to the present. Second, this chapter describes efforts of municipalities that have built community-based integrated care systems for support of affected persons and "better" reconstruction. This chapter presents arguments for the necessity of building resilient community-based integrated care systems by strengthening medical care and social welfare, reciprocal networks, and other factors, as well as mutually complementary relations within and outside the region.

Keywords Community-based integrated care system · Great East Japan Earthquake · Recovery and reconstruction · Resilience · Victim support

H. Miyazawa (✉)
Ochanomizu University, Tokyo, Japan
e-mail: miyazawa.hitoshi@ocha.ac.jp

H. Kikuchi
Kahoku Shimpo Publishing Co, Sendai, Japan

H. Miyazawa and T. Hatakeyama (eds.), *Community-Based Integrated Care and the Inclusive Society*, International Perspectives in Geography 12,
https://doi.org/10.1007/978-981-33-4473-0_11

1 Increased Concern for Resilience in Medical Care and Social Welfare

Japanese society emphasizes 'resilience' recently. In general, this concept refers to the ability of materials, people, systems and such to adapt to disruptive change and to continue to exist. For example, resilience has been used to overcome stress and adversity in psychology. Resilience has been used especially within Japanese policy-making to characterize a capability of communities to recover quickly to a former condition after becoming disordered by some event. The Hyogo Framework for Action adopted in the second UN World Conference on Disaster Reduction in 2005 defined 'disaster resilience' in the meaning above (UNISDR 2005). Furthermore, after the 2011 Great East Japan Earthquake, the national government and municipalities in Japan have been engaged positively to establish resilient nations and communities.

The national government of Japan decided on a Fundamental Plan for National Resilience in 2014. This plan specifies 15 individual and cross-sectoral fields for vulnerability assessment and management. One field is 'medical care and social welfare'. Even when a disaster occurs, it is necessary to provide medical care and social welfare services for people without interruption. If these are interrupted, they must be restored as soon as possible. Medical care is indispensable for the treatment of emergency patients and for the prevention of damage to the health of disaster area residents. Many people requiring social welfare services are vulnerable to disasters and need help in all aspects of life or support for evacuation. Moreover, these are fundamentally important for the reconstruction and revitalization of communities in the wake of a disaster.

Medical care and social welfare are also regarded as important fields in the Fundamental Plan for Regional Resilience formulated by local governments (Miyazawa 2017). Through vulnerability assessments, local governments have formulated measures for resilient medical care and social welfare such as those shown in Table 1. Because medical care and social welfare are closely related to housing and community measures, cross-sectoral measures are also required for toughening them. The establishment of integrated care systems is fundamentally important for resilient communities.

In this chapter, we review the provision of medical care and social welfare immediately after the Great East Japan Earthquake and subsequent attempts to establish community-based integrated care systems in the affected areas. A review of those days is valuable for the following reasons. The Great East Japan Earthquake brought great attention to the resilience of Japanese society and provided us with valuable experience as a foundation for efforts to establish resilient communities.

Table 1 Main measures for the establishment of resilient medical care and social welfare in the Fundamental Plan for Regional Resilience

Reinforcement of buildings and architectural structures
• Promotion of measures against seismic resistance and aging of medical and welfare facilities
Support for evacuation activities and securing of disaster shelters
• Support for evacuation activities and maintenance of evacuation shelters
• Creation and utilization of a list of persons requiring assistance for evacuation activities
• Designation of welfare facilities as welfare evacuation centers and promotion of concluding agreements
• Provision of medical care and welfare at evacuation centers for those who need assistance
• Promotion of safety measures and development of evacuation plans at welfare facilities
Strengthening medical and welfare systems at the time of a disaster
• Strengthening the disaster base medical centers
• Securing fuel, food, etc. at medical and welfare facilities
• Securing medical and welfare workers with intra-regional collaboration and inter-regional collaboration
• Formation of disaster medical assistance and disaster care assistance teams
• Securing of medicines at the time of disaster
• Preparation of business continuity plans related to medical and welfare facilities
• Training personnel engaged in medical, emergency relief and welfare support at the time of a disaster
Other measures related to resilient medical care and welfare
• Promotion of earthquake resistance of houses and countermeasures against aging
• Improvement of functions of emergency routes and promotion of measures against aging of road facilities
• Strengthening of local communities to promote the mutual aid of residents

Reprinted from Miyazawa (2017) with permission of the Japan Association of Surveyors

2 Medical Care Collapse Attributed to the Great East Japan Earthquake and Its Emergency Response

The decline in medical care and social welfare services in the aftermath of the Great East Japan Earthquake was greater than expected. In three affected prefectures (Iwate, Miyagi, and Fukushima), 300 hospitals (79.0%) were damaged by the disaster, 11 of

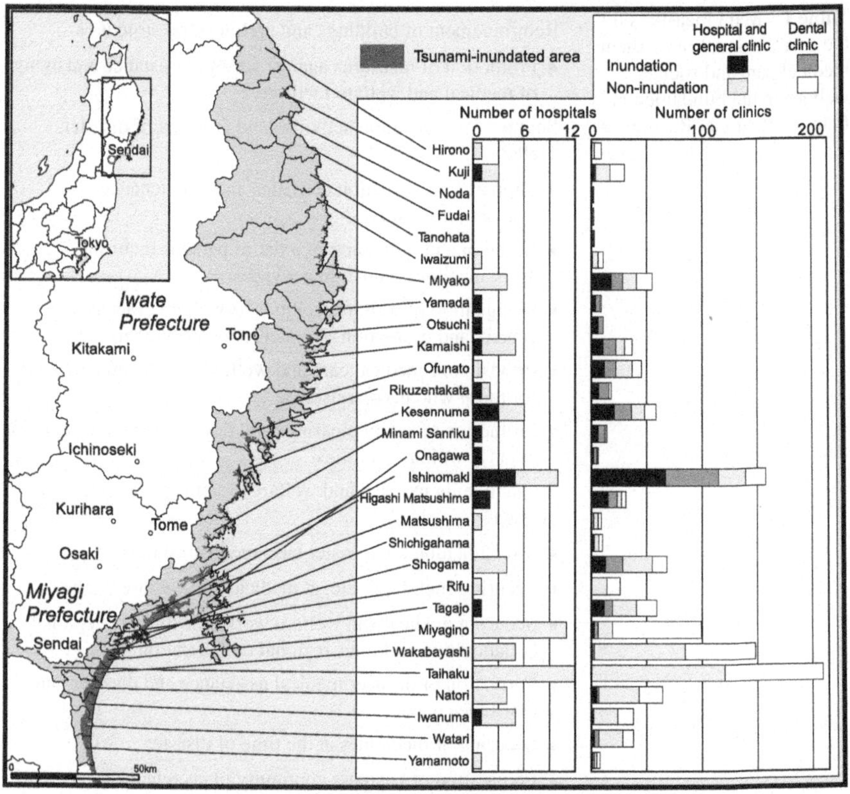

Fig. 1 Numbers of hospitals and clinics affected by flood damage caused by the tsunami (Reprinted from Miyazawa 2017 with permission of the Japan Association of Surveyors)

which were destroyed completely (Ministry of Health, Labour and Welfare 2011). There were 205 hospitals that restricted or refused acceptance of outpatients. In fact, 191 hospitals restricted or refused hospitalization. The damaged clinics also exceeded one thousand, 167 of which were destroyed completely. Similarly, social welfare facilities in the above three prefectures were heavily damaged by the disaster. The number of social welfare facilities damaged was 875, 59 of which were destroyed completely.

Because of the tsunami wave effects, the damage to medical and social welfare facilities was extensive in coastal municipalities. Regarding the medical facilities, Fig. 1 presents the number of hospitals and clinics that sustained flood damage from the tsunami waves in Iwate Prefecture and Miyagi Prefecture.[1] These were

[1]Fukushima Prefecture, which was affected by the Fukushima Daiichi Nuclear Power Plant accident, was excluded from these analyses. Data related to tsunami-inundated areas were used from the "Map of the area hit by the tsunami of 11 March 2011" prepared by the Tsunami Damage Mapping Team, the Headquarters for Disaster Response, Association of Japanese Geographers. However, for some

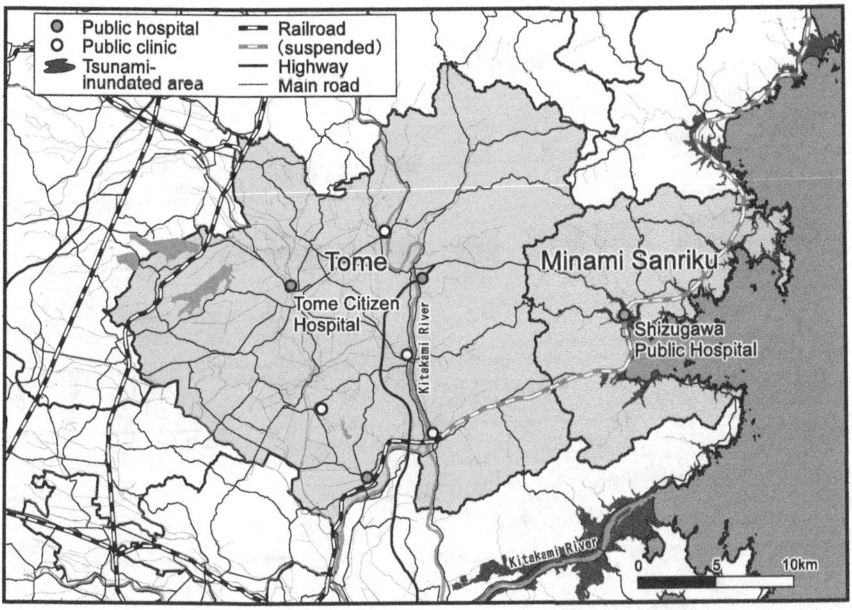

Fig. 2 Minami Sanriku Town and Tome City (as of April 2011)

estimated using GIS by overlaying the locations of hospitals and clinics with the area of inundation by the tsunami waves. This figure clarifies that municipalities with a large proportion of inundated hospitals and clinics are distributed from the south of the Sanriku area to the northern part of the Sendai Bay area.

Among them, one municipality in which the local medical care system collapsed because of the Great East Japan Earthquake was Minami Sanriku Town, Miyagi Prefecture (Fig. 2). Minami Sanriku Town, located in the southern part of the Sanriku area, had a population greater than 17,000 before the disaster; now it is about 13,000. This town sustained severe damage from the Great East Japan Earthquake. Beyond the human toll of 620 dead and 211 missing, 3,143 buildings were destroyed completely; 61.9% of all houses were damaged. We review the emergency correspondence related to medical care in Minami Sanriku Town when the disaster occurred.

In Minami Sanriku Town, one hospital and 14 clinics were destroyed completely by the tsunami waves. The complete destruction of Shizugawa Public Hospital, the only hospital and also the core hospital as a designated emergency hospital, led to the collapse of medical care in Minami Sanriku Town (Fig. 3). Before the disaster occurred, this hospital accepted patients from Tome City adjacent to the inland side of Minami Sanriku Town (Fig. 2), but many patients were transported from Minami Sanriku Town to Tome City when the disaster occurred (Niinuma and Miyazawa 2012).

---

inundation areas, the authors complemented the data with Google satellite images of the disaster area.

Fig. 3 Tsunami-destroyed Shizugawa Public Hospital (Photograph taken by Shiori Niinuma in May 2011)

According to data the authors obtained, the municipal hospitals and clinics in Tome City received 181 hospitalized patients and 4,475 outpatients from outside the city during the 8 weeks following the disaster. Patients from Minami Sanriku Town accounted for 61.3% and 74.0% of all patients mentioned above, respectively. With the support of doctors, public hospitals and clinics were able to accept numerous patients from outside the city. However, among the doctors who came in support of the area, the local practitioners withdrew when medical treatment was resumed at their clinics after the lifeline was restored during the third week following the disaster. In addition, the number of beds operated with the temporary increase of doctors was limited to around 96 beds which were unused because of the shortage of doctors. The number of hospitalized patients accepted from inside and outside the city exceeded the above number of beds during the first week of the disaster. To accept hospitalized patients continuously, it was necessary to promote discharge from the hospital. In some cases, patients in the city were transferred to hospitals outside the city. Patients who were unacceptable at the hospital in Tome City were transferred to hospitals in distant areas such as Kurihara City, Osaki City, and Sendai City (Fig. 1).

It has been assumed that the number of patients increases when disasters occur and that the demand, exceeding the capacity of medical facilities cannot be accommodated. However, in the affected areas of the Great East Japan Earthquake, the medical systems collapsed because of the tremendous damage to medical institutions. As a result, it became necessary to transfer patients among municipalities and prefectures. The inland medical facilities were compelled to accept patients from affected coastal areas, but they accepted local patients such as injured, chronically ill, and home care patients. The occurrence of the medical collapse in the area had enormous effects over a wide area.

3 Recovery of Medical Care and Social Welfare Services After a Disaster

This section specifically examines the process of restoring medical and welfare facilities that collapsed because of the Great East Japan Earthquake. In addition to Minami Sanriku Town, one area that sustained tremendous damage to medical care and welfare systems was Yamada Town in Iwate Prefecture (Fig. 4, Kikuchi 2013), located in the middle of the Sanriku area (Fig. 1). The population was approximately 19,000 before the disaster. Today, it is about 16,000. The human toll in Yamada Town was 816 dead and 2 missing. In addition, 3,346 buildings were damaged, of which 2,762 were destroyed completely. This damage was caused mainly by tsunami waves and the fires that followed.

Because of the tsunami, the Yamada Prefectural Hospital, Yamada Town's only hospital, was partially destroyed (Figs. 4 and 5). The damage to the first floor was especially severe. Utility services were also inconsistent. In addition, three of the four general clinics in the town were destroyed completely. Nevertheless, because doctors and nurses were safe at Yamada Prefectural Hospital, they started to visit patients at evacuation centers for medical examination from the fourth day after the disaster occurred. From around the tenth day, they resumed visits for medical treatment to patients who stayed at home.

Four months after the disaster occurred, medical treatment of outpatients resumed at a temporary clinic set up on land owned by the town. Patients hospitalized at Yamada Prefectural Hospital at the time of the earthquake were either transferred to hospitals and nursing homes outside the town that had remained unaffected or were discharged from the hospital. Resumption of inpatient hospital care was achieved in September 2016 when the hospital was rebuilt in an upland area of the town (Fig. 4). Until then, the Miyako Prefectural Hospital located in the adjacent municipality accepted hospitalized patients living in Yamada Town instead.

Regarding general clinics, one out of three clinics that had been destroyed completely was closed because of the death of a doctor. One other clinic resumed medical treatment using vacant facilities one month after the disaster, but it was closed in 2016. Another clinic reopened medical treatment one month after the disaster by borrowing a public building and transforming it into a temporary clinic. The clinic restoration was achieved about three years after the disaster when the clinic was rebuilt with nursing care facilities in an upland area of the town.

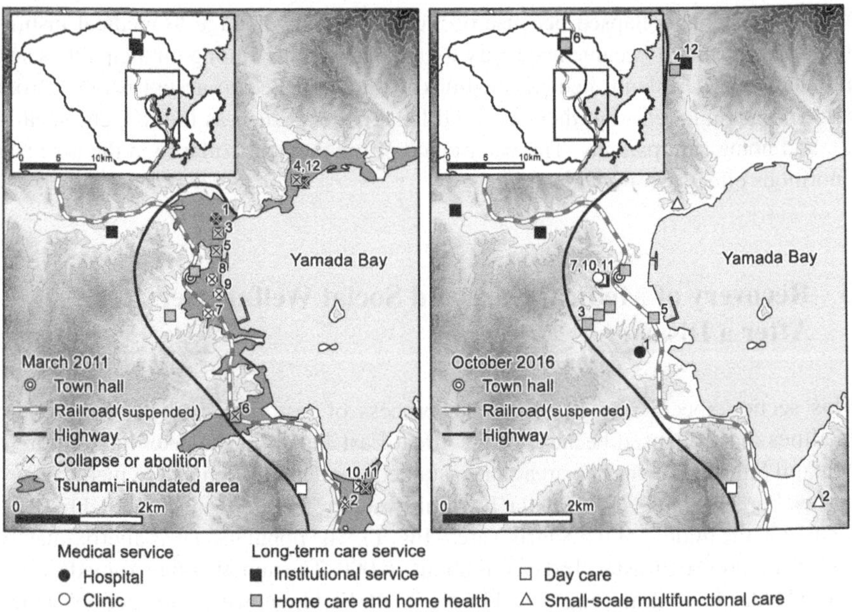

Fig. 4 Locations of medical and nursing care facilities in Yamada Town, Iwate Prefecture. Numerals in the figure correspond to the numerals in Fig. 5

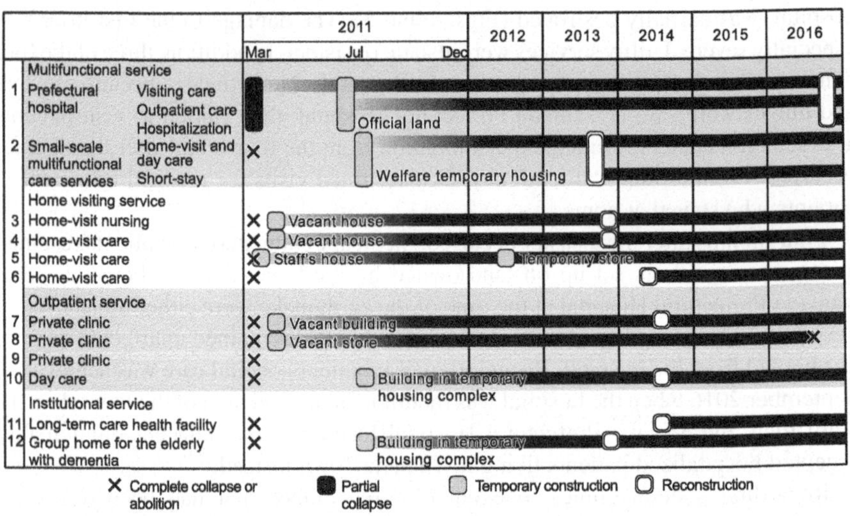

Fig. 5 Recovery process of medical and nursing care facilities after the Great East Japan Earthquake in Yamada Town, Iwate Prefecture (Reprinted from Miyazawa 2017 with permission of the Japan Association of Surveyors)

About half of the long-term care facilities in Yamada Town were destroyed completely. Of these, a health facility for the elderly and a day care facility were eventually restored after being installed in the clinic as described above, but only the latter was temporarily restored in the temporary housing complex four months after the disaster. In addition, using the temporary facility maintained using the prefecture government, a facility for small-scale multifunctional home-based care was resumed five months after the disaster. However, because its site became temporary welfare housing, the operational establishment restarted only with home care and day care services arranged in advance. The restart of the short-term stay service was achieved two and a half years after the disaster at a facility rebuilt in an upland area of the town.

At least visiting type services can resume if office space is secured, so the three operational establishments that were able to secure temporary space resumed before May 2011. However, because these spaces were secured in general homes, many inconveniences were encountered in carrying out service businesses. Therefore, one operational establishment moved to a temporary commercial facility opened in the center of the town. The other two operational establishments continued to work even in inconvenient temporary offices until they were rebuilt in upland or inland areas of the town in 2013. One effort was restarting a group home for people with dementia in the temporary housing complex half a year after the disaster. However, because it was small, visiting care nursing continued at the temporary office. The restart of services at Yamada Town in another operational establishment was delayed for three years after the disaster. Although the service operations continued at satellite facilities in Miyako City, the establishment was able to carry out the service business in Yamada Town only to a slight degree until the office was rebuilt in the inland area of Yamada Town.

A high probability exists that visited-type services and private practice clinics can resume quickly even after a disaster because, in those cases, the equipment that must be restored or rebuilt is small. Therefore, although self-help efforts are likely to restore those, it is expected that they will support medical care and nursing care in affected areas until other services are restored. Furthermore, at the temporary restoration stage, it is apparent that public buildings in temporary housing complexes play a major role in restoring medical care and welfare. However, the rebuilding of facilities requires several years after the disaster. Especially in the afflicted areas of the Great East Japan Earthquake, it was difficult to secure upland sites suitable for rebuilding facilities as a preparation for the tsunami waves again. Additionally, it was difficult to rebuild facilities quickly because of shortages of contractors and materials. Competition with other industries in securing labor, coupled with human injury accompanying the disaster further strengthened the shortage of care workers that existed before the disaster. This is a factor restricting the rehabilitation of medical care and welfare services in affected areas today.

4 Community-Based Integrated Care Systems for Reconstruction of Disaster-Affected Areas

4.1 Social Isolation Among Displaced Residents

The Great East Japan Earthquake caused enormous damage to housing. For residents who lost their houses, more than 50,000 units of public temporary housing were built. Moreover, 80,000 houses were secured for houses rented as private rental homes, arranged as emergency temporary housing. In addition, more than 30,000 units of disaster resolution public housing were planned. By the end of 2018, approximately 29,500 units had been completed (Reconstruction Agency 2019). Securing residence is the first step for recovery of livelihood. However, regarding temporary housing and resolution public housing, residents' isolation and solitary death have been regarded as difficulties faced after the Great Hanshin-Awaji Earthquake (Tanaka et al. 2009, 2010). Maintaining close relationships that prevailed in the previous community has been pointed out as difficult because a lottery was used as the method of choosing who would move into those houses. In temporary housing and the disaster resolution public housing, relationships among residents become less robust. Thereby mutual opportunities for observation and interaction functions are encountered less. Furthermore, the number of residents who withdraw to their own residences will increase. Those residents will continue to lose their physical and mental capabilities.

Therefore, at the time of the Great East Japan Earthquake, the Ministry of Health, Labour and Welfare asked municipalities to set up facilities and welfare temporary housing to support displaced persons such as elderly people in the temporary housing complex. Nursing care facilities were temporarily restored using the buildings in the temporary housing complex at Yamada Town described above corresponded to these measures. However, it is said that even in temporary housing built in the disaster area of the Great East Japan Earthquake, more than 200 people have already died solitary deaths. Residents of temporary houses are now moving to disaster resolution public housing. Reorganization of temporary housing has begun. Residents occupying disaster resolution public housing and those who continue to remain in temporary housing remain strongly apprehensive about the occurrence of isolation and solitary death.

Under these circumstances, some municipalities in affected areas assign importance to the construction of community-based integrated care systems as part of reconstruction projects. As an example of these municipalities, we specifically examine the efforts of Ishinomaki City, as explained hereinafter.

4.2 Disaster-Affected Residents Supported in Ishinomaki City's Community-Based Integrated Care System

Ishinomaki City, located at the mouth of Kitakami River, is the second-largest city in Miyagi Prefecture (Fig. 6). Ishinomaki City was established in 2005 via the merger of the former Ishinomaki City with six neighboring towns. Before the Great Eastern Japan Earthquake, Ishinomaki City had a population of about 160,000. However, the current population is less than 150,000. Damage sustained by Ishinomaki City from the disaster was the most severe among affected municipalities. Human casualties in Ishinomaki City included more than 3,500 deaths and about 400 people missing.

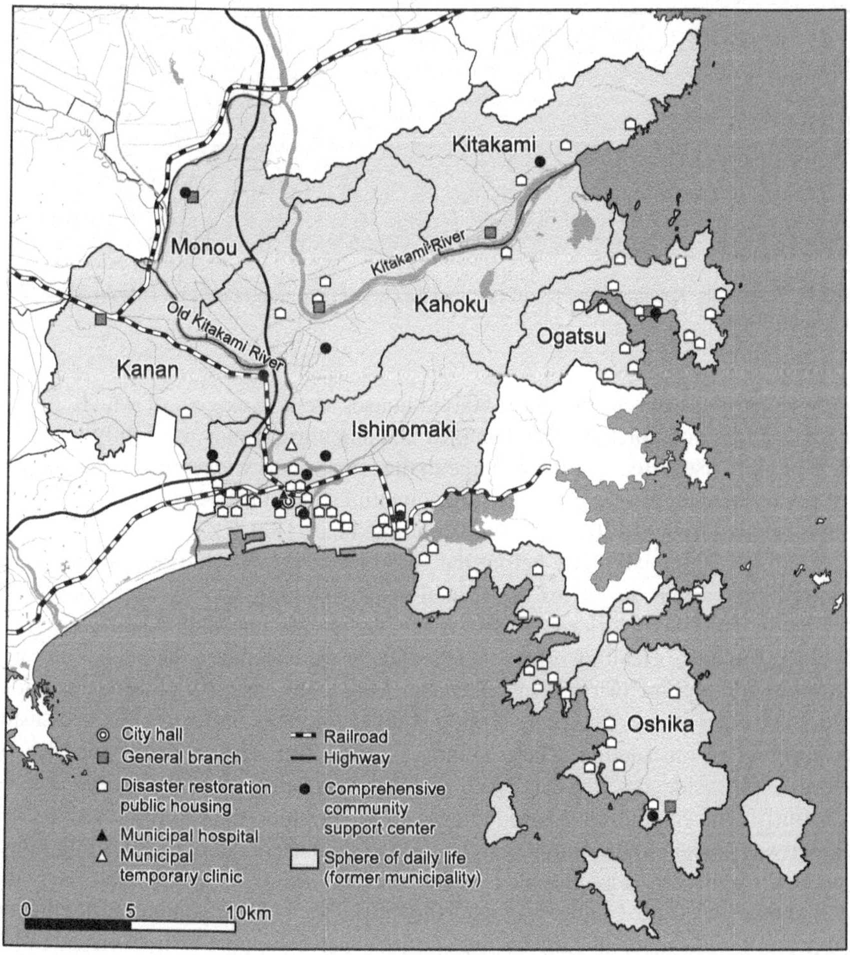

Fig. 6 Location of comprehensive community support center in Ishinomaki City, Miyagi Prefecture

Fig. 7 Temporary housing complex in Ishinomaki City (Photograph taken by Hitoshi Miyazawa in March 2019)

Even in terms of structures, about 50,000 houses were damaged, of which 20,000 were destroyed completely. The damaged housing corresponds to slightly less than 80% of all houses. Because the damage to the houses was widespread and extreme, temporary housing was supplied in large quantities. About 7,000 temporary houses were supplied to over 130 districts in Ishinomaki City (Fig. 7).

Although Ishinomaki City before the Great Eastern Japan Earthquake had been confronting various problems associated with an aging society and population decline, conditions deteriorated further because of the enormous damage caused by the earthquake. Health workers and caretakers were tested to their limits in their activities for support of residents. Then, in May 2013, the city chose to promote community-based integrated care aiming at relief for people affected by the disaster and prompt reconstruction. Three months later, a temporary housing complex of about 2,000 residences, the largest in the city, was selected as the model area of construction of a community-based integrated care system; the Comprehensive Care Center was opened in the complex to be used as the base. A temporary health clinic operated by the city was provided in the housing complex. Six months later, the Comprehensive Care Center was relocated next to it. This change strengthened cooperation with medical care for the comprehensive support of housing complex residents. The Ishinomaki Municipal Hospital, which had suspended operations because of the tsunami disaster, was reconstructed in front of JR Ishinomaki Station in September 2016. Home-care-supporting functions, to which a high value was attached in community-based integrated care, were reinforced.

According to the basic concept of community-based integrated care system by Ishinomaki City established in 2014, the system should be built in seven areas including one old city and six towns before the merger considering characteristics of each area (Ishinomaki City Community-based Integrated Care Promotion Committee 2014, Fig. 6). In the old town area, comprehensive community support center was established for every old town to supervise elderly citizens while maintaining cooperation with each branch office. In the old town, based on this system, they aim to build a community-based integrated care system through coordination of various job categories. Many medical institutions and nursing care facilities are located in the old Ishinomaki City area, whereas comprehensive community support centers are provided at six sites. According to the project concept, the Mutual Support Center (tentative name)[2] estimated to be completed by 2020, will be constructed in front of Ishinomaki Station next to the city office and city hospital; this center will coordinate various job coordination activities by cooperating with six comprehensive community support centers.

For major victim assistance with community-based integrated care, Ishinomaki City has attached great importance to support for living places and motivation in life as well as support for medical care and nursing care as described above (Ishinomaki City Community-based Integrated Care Promotion Committee 2015). Initially, a high value was attached to support for the prevention of social isolation in the temporary housing complex. Almost all temporary housing has an assembly room and a lounge, which are used as support centers by various professionals for life consultation, lifestyle support for residents, and for promotion of exchanges of residents. A base support center stationed by office staff is provided in large temporary housing complex to support patrol of inhabited houses, overall consultation, and promotion of car sharing. Visits by life consultation officers of the social welfare council and by welfare support coordinators are important attempts. A card by which contact information can be confirmed easily is distributed to residents of temporary housing. Lending of facilities to volunteers and NPOs engaged in support for people affected by the disaster and reconstruction assistance are comprehensive community activities. Even with such a sequence of activities, incidents of solitary death in the temporary housing continued to occur.

After completion of disaster resolution public housing, emphasis was placed on people affected by the disaster who moved from temporary housing to disaster resolution public housing. In Ishinomaki City, the supply of disaster resolution public housing started from 2012. Supply of as many as 4,456 housing units, the planned number of housing units, was completed by fiscal 2018. To the present day, relocation of people affected by the disaster from temporary housing to the disaster restoration

[2]Management function of Comprehensive Care Center provided in the temporary housing complex is transferred to Mutual Support Center and the target of this attempt includes the old Ishinomaki urban area and the six old towns.

Fig. 8 Disaster restoration public housing in Ishinomaki City (Photograph taken by Hitoshi Miyazawa in March 2019)

public housing achieved a major aim. However, it is considered that social isolation of residents occurs easily because of the following reasons: People affected by the disaster living in the disaster restoration public housing came from temporary housing at various locations, which implies that human relationships among residents are very weak. Units of disaster restoration public housing are mostly medium height and high-rise condominium buildings (Fig. 8). Therefore, contact with people next door is less likely to occur than in row-house type temporary houses. Residents are likely home-bound, even if they need some help in the abnormal state, unlocking them from the outside might not be possible in some cases. Residents of disaster resolution public housing include many single persons and aged people (Table 2). In many cases, they have mental and physical difficulties and a sedentary lifestyle. They might have nobody to turn to and voice their concerns. Continued observation of activity and the creation of community for residents of disaster resolution public housing are urgent issues. Residential areas in the new urban area developed by reconstruction present similar difficulties.

Then, as one activity of community-based integrated care, Ishinomaki City is promoting the observation of activities in the disaster resolution public housing and in a new urban area developed by restoration by case workers, child committee members, community welfare coordinators, and livelihood support staff. For example, regular home visits of disaster resolution public housing by life support

Table 2 Residents of disaster restoration public housing in Ishinomaki City

Number of target households	2,612
Number of households responding	1,592
Number of people responding	2,694
One-person household	762 (47.9%)
Two-person household	553 (34.7%)
More than 65 years old	1,373 (51.0%)
Solitary aged household	487 (30.6%)
15–64 years old without regular occupation	269 (24.0%)
More than 65 years old certified as needing long-term care	273 (19.9%)
People carrying physical disability certification	250 (9.3%)
People in bad health	576 (22.1%)
People with an illness	1,606 (64.0%)
People who possibly have mood disorder or anxiety disorder	172 (7.8%)
People who have reduced opportunities for physical activity	937 (37.6%)
People who have no person to consult	499 (20.4%)
People who do not participate in events	1,573 (61.3%)

Prepared based on Ishinomaki City (2018)

officers of the social welfare council are conducted every 10 days. For those judged as needing support, professionals such as health nurses support them individually. In addition to supporting for formation of a new community, support is provided for human resource development, formation of a base, lounge activity, fitness, autonomous activities by residents to organize exchange events, and volunteer activities. Mutual assistance type housing equipped with a space for increasing chances of contacts by the residents themselves is supplied in part (Fig. 9).[3] There are many low-income earners. Those of productive age without regular occupations exist among residents of the disaster restoration public housing. Then, Ishinomaki City extended the rent reduction period of the disaster restoration public housing from 10 years to 20 years. This is a peculiar doctrine undertaken by Ishinomaki City.

[3]Such a consideration from building construction viewpoints is provided that the housing area is reached by entry via a common entrance and passage through a common multi-purpose space without fail. Housing is arranged face-to-face, fronting on an inner garden.

Fig. 9 Mutual assistance type disaster restoration public housing in Ishinomaki City (Photograph taken by Hitoshi Miyazawa in March 2019)

5 Conclusions

This chapter introduces concrete examples of support of people affected by the disaster based on damage to medical care and welfare facilities, restoration, reconstruction and establishment of community-based integrated care system in the disaster-stricken area from immediately after the occurrence of the Great East Japan Earthquake up to the present day. The community-based integrated care system is a system for supporting welfare in a broad sense of continuing to live in one's homeland. Therefore, the establishment of resilient community-based integrated care system is expected to contribute to resilience of the community. In the Japanese society of the future, disasters might occur in the midst of population decline, aging, and attenuation of community. Therefore, the establishment of a resilient community-based integrated care system is an important task. It is often said that resilience has a narrower concept of robustness, redundancy, substitution, reactivity and recovery. For resilience of the community-based integrated care system, investigation based on these concrete viewpoints and efforts is required. The first step to do this might be strengthening the compositional units of the community-based integrated care system, including medical care and welfare according to the measures presented below.

In the event of occurrence of a wide-scale disaster, medical care, and welfare facilities and service providers might be damaged severely. According to the Fundamental Plan for Regional Resilience, many local governments consider the importance of earthquake resistance to increase the robustness of the buildings. However, to reduce the risk of falling into a null state of medical care and welfare and to avoid the worst case, which should never occur, further robustness should be provided. In addition to base hospitals, for core hospitals acting as the fundamental level of medical care under disaster occurrence and welfare facilities able to cope with broad needs, higher robustness should be provided as important social infrastructure. Considering disaster characteristics of every region, resistance against disasters other than earthquakes, relocation of facilities located at disaster dangerous area to other area with lower disaster risk, improvement of robustness of related infrastructure for continuation of functions, and securing human resources are necessary. From experience gained from The Great East Japan Earthquake, it is known that if important facilities are damaged catastrophically, considerable time is necessary before they can be restored to a normal level. Final restoration might be strongly affected by delays in incremental restoration.

From viewpoints of widening the range of medical care and welfare at disaster occurrence, small and medium-sized medical institutions and welfare facilities should establish a system that allows continuation of service business and rapid recovery when damaged. Even if (temporary) restoration is possible at an earlier stage, institutions have important roles to fulfill with the base facilities at disaster occurrence and provide medical care and welfare in lieu of them. Therefore, a business continuation plan should be designed and necessary preparations should be made ready.

For the local community, to improve disaster response, mutual support relations should be strengthened at all opportunities. This reinforcement might result in the preservation of local communities from rehabilitation to reconstruction time periods. After a wide-scale disaster affecting wide areas, such as The Great East Japan Earthquake, large numbers of people are forced to move. Even if such relocation is reduced to the greatest degree possible, local communities are invariably altered to a drastic degree. Maintaining a spirit of mutual support is important as grounds for forming a new community.

However, preparing for damage beyond one's imagination is an important lesson learned from The Great East Japan Earthquake. Construction of regional coalitions is greatly anticipated to continue care for sick and injured people when medical and welfare facilities were stricken and became ineffective. Although disaster assistance system is institutionalized in the medical care field as represented by Disaster Medical Assistance Team[4] and Japan Medical Association Team,[5] in the welfare

---

[4]Medical care teams consist of trained medical doctors and nurses who are ready to act immediately after disaster occurrence. Medical care activity is started within 48 h, as a goal, after disaster occurrence.

[5]Medical care teams are dispatched to stricken areas by the Japan Medical Association. They visit the stricken area about 3 days after disaster occurrence to support local healthcare until the local medical system has recovered.

field, organization of Disaster Care Assistance Team[6] just started after The Great East Japan Earthquake. Therefore, the implementation of measures is sluggish.

For a regional coalition to develop sufficient substitution capabilities, geographical characteristics should first be considered for establishment of coalitions. In the case of a tsunami disaster, it is highly probable that wider coastal areas might be damaged simultaneously. Therefore, establishing coalitions with inland areas that are not likely to be damaged should be sought. This preparation can construct a horizontal type coalition by regions having entirely different characters. Regions acting as partners have already been realized. Therefore, preparations for emergency situations in areas with different geographical backgrounds will become possible by prior instruction.

Second, it is important that the system has redundancy capable of responding to medical and welfare needs that increase temporarily after a disaster. Areas with sufficient medical care and welfare workers are difficult to maintain even under normal times. Reduction in the number of sickbeds has been advanced as a means of suppressing medical care costs. Since Tome City experienced The Great East Japan Earthquake, some areas are in a reduction and contraction state where functions are not fully developed, even if medical personnel are added at times of disaster occurrence. It is noteworthy that, according to resilience viewpoints, optimum efficiency improvement might lose redundancy and increase vulnerability for the emergency state.

As described above, the establishment of a resilient community-based integrated care system presents numerous challenges. One example used in the geography that contributes to the establishment will be introduced here. The GIS used in part in this chapter enables the distribution and spatial relation of various elements related to community-based integrated care systems, whereas the scale is changed in a perspective manner, from local thinking to more circumspect consideration. Utilization of GIS is helpful for evaluating vulnerability for resilience improvement if the status of various elements of the integrated care system in every region is grasped. Furthermore, it will develop effectiveness for discussing efforts such as interregional coalition where regional coalition is considered flexibly. Strengthening of medical care and welfare and community-based integrated care systems can be done.

References

Ishinomaki City (2018) Results of health survey investigation of residents of Ishinomaki City restoration public housing in fiscal 2017. http://www.city.ishinomaki.lg.jp/cont/10351000/2500/20180704132611.html. Accessed 1 Nov 2018 (in Japanese)

Ishinomaki City Community-based Integrated Care System Promotion Committee (2014) Basic concept of Ishinomaki City community-based integrated care system promotion plan. https://www.city.ishinomaki.lg.jp/cont/10355000/1111/kihonkousou.pdf Accessed 1 Nov 2018 (in Japanese)

[6]Teams of professionals being dispatched to the stricken area immediately after disaster occurrence to support elderly people and people with disabilities at a care facility and evacuation center.

Ishinomaki City Community-based Integrated Care System Promotion Committee (2015) Ishinomaki City community-based integrated care system promotion execution plan. https://www.city.ishinomaki.lg.jp/cont/10355000/1111/270316jisshikeikaku.pdf. Accessed 1 Nov 2018 (in Japanese)

Kikuchi H (2013) Impacts on medical, nursing care service system and changes in the lives of survivors from the Great East Japan Earthquake of 2011: a case study of Yamada Town, Iwate Prefecture. In: Proceedings of the general meeting of the Association of Japanese Geographers 2013s, 132 (in Japanese)

Ministry of Health, Labour and Welfare (2011) Annual health, labour and welfare report 2011. Nikkei Printing Inc., Tokyo (in Japanese)

Miyazawa H (2017) Increased resilience of medical care and welfare and community-based integrated care systems. J Surv 67(10):12–17 (in Japanese)

Niinuma S, Miyazawa H (2012) Tsunami damage on medical institutions by the Great East Japan Earthquake and the acceptance of affected patients by inland medical institutions: a case study of Minami-sanriku Town and Tome City, Miyagi Prefecture. Q J Geogr 63:214–226 (in Japanese with English abstract)

Reconstruction Agency (2019) Status of supplying public housing for disaster victims and private ground after The Great East Japan Earthquake (end of December, 2018). http://www.reconstruction.go.jp/topics/main-cat1/sub-cat1-12/20190131_jutakukyokyu.pdf. Accessed 1 April 2019 (in Japanese)

Tanaka M, Takahashi C, Ueno Y (2009) The relationship between the actual conditions of "isolated death" occurrences and residential environments in disaster restoration public housing: case of the Great Hanshin-Awaji Earthquake. J Architect Plann (Trans AIJ) 74:1813–1820 (in Japanese with English abstract)

Tanaka M, Takahashi C, Ueno Y (2010) The conditions of "isolated death" occurrences and its background in public temporary housing for disaster: case of the Great Hanshin-Awaji Earthquake. J Architect Plann (Trans AIJ) 75:1815–1823 (in Japanese with English abstract)

UNISDR (2005) Hyogo framework for action 2005–2015: Building the resilience of nations and communities to disasters. https://www.unisdr.org/2005/wcdr/intergover/official-doc/L-docs/Hyogo-framework-for-action-english.pdf. Accessed 1 Nov 2018

A Prospect of Community-Based Inclusive Society in Japan

Establishing Community-Based Integrated Support Systems for Pregnancy, Childbirth, and Childcare in Japan: Focusing on Regional differences

Hitoshi Miyazawa and Kanoko Tada

Abstract Various difficulties have arisen as to childbirth and parenting in Japan when the trend toward the nuclear family is progressing and the local community is weakening. For that reason, more local governments are building integrated support systems that provide health, medical care, and welfare in the fields of maternal and child health and parenting from the gestation period to the parenting period in a seamless and integrated manner, just as with the field of elderly care. As described in this paper, we selected Wako City, Saitama Prefecture and Hokuto City, Yamanashi Prefecture as study areas because they are addressing the building of community-based integrated support systems for pregnancy, childbirth, and childcare ahead of others. Devoting attention to how interprofessional cooperation and local residents' organizations work in collaboration, we compared the characteristics of their care systems and examined the regional background that made a difference between the two care systems. Results revealed that Wako City and Hokuto City had respectively different regional characteristics as an urban area and a rural area. For example, professions that have experience and know-how of interprofessional cooperation differ and also there are differences in local resident organizations as cooperative partners. Similarly to community-based integrated care systems in the field of elderly care, to build effective community-based integrated support systems in the fields of maternal and child health as well as and parenting, it is necessary to ascertain regional resources that contribute to the purpose and form networks between them.

Keywords Comprehensive support centers for childrearing generation · Community-based integrated support systems for pregnancy · Childbirth and childcare · Maternal and child health · User support program (maternal and child health-type) · Regional difference

H. Miyazawa (✉)
Ochanomizu University, Tokyo, Japan
e-mail: miyazawa.hitoshi@ocha.ac.jp

K. Tada
Bridgestone Corporation, Tokyo, Japan

H. Miyazawa and T. Hatakeyama (eds.), *Community-Based Integrated Care and the Inclusive Society*, International Perspectives in Geography 12,
https://doi.org/10.1007/978-981-33-4473-0_12

1 Introduction

Maternal and child health programs in post-war Japan have emphasized the provision of services to support the physical health of mothers and children, including health guidance, health checkups, and medical assistance. In recent years, however, difficulties such as parents' anxiety about childbirth and childcare have arisen, in addition to child abuse. Consequently, the requirements for maternal and child health have become increasingly complicated in terms of including needs for parenting support. For that reason, similarly to community-based integrated care systems in the field of elderly care, it has become necessary in the field of maternal and child health to build systems that provide health, medical care, and welfare in familiar communities in a seamless and integrated manner.

To this end, it is necessary for specialists in various fields, including health, medical care, social welfare, and education, to support pregnancy, childbirth, and childcare, working beyond the frameworks of their respective professions (Sasakawa 2014). However, cooperation across professions remains insufficient in activities to support pregnancy, childbirth, and childcare, and engender appropriate support to only a slight degree. It has been pointed out that building interprofessional systems is difficult because specialists are overburdened by their own duties. Therefore, they lack understanding of each other's duties and lack time to share information among professions (Kurihara and Okuyama 2012; Mitsuhashi et al. 2008). In addition, at present, cooperation with nonprofit organizations and community organizations are regarded as necessary to provide meticulous support for maternal and child health and parenting.

Therefore, the Ministry of Health, Labour and Welfare implemented the model program of integrated support for pregnancy and childbirth in 2014 to promote integrated support systems for pregnancy, childbirth, and childcare in local governments. The program consists of the maternal and child health consultation support program, the prenatal and postpartum support program,[1] and the postnatal care program.[2] In 2015, the maternal and child health consultation support program was enshrined into law, becoming an obligatory program by making the shift to the maternal and child health-type user support program based on the Act on Child and Childcare Support enforced in that year.

"User support programs" are of three types: the basic type, the specific type, and the maternal and child health type. The basic-type program is a program in which parenting support specialists provide various consultation programs for parenting households and undertake coordination with related organizations. Similarly, the

---

[1]In the prenatal and postpartum support program, specialists such as public health nurses and midwives, as well as those who experienced parenting and the senior generation give counsel on various matters before and after childbirth to support pregnant women and nursing mothers and parenting households who are isolated in their community.

[2]The postnatal care program is aimed at giving support for mother and child health immediately after leaving maternity hospitals. It is intended to alleviate worries and anxieties through counseling by midwives and other specialists providing meticulous professional support and health guidance through overnight stays or one-day stays as well as home visits.

specific-type program is a program in which parenting support specialists provide information about childcare services and support the use of the services. The maternal and child health-type program is a program in which pregnant women and nursing mothers are supported by maternal and child health coordinators whom municipalities place in health centers (undertaken by public health nurses and midwives who have specialized knowledge of maternal and child health). After seamlessly grasping the conditions of pregnant women and nursing mothers based on information obtained from pregnancy notification, individual consultation, and health checkups, the coordinators select maternal and child health services available for pregnant women and nursing mothers. They build cooperative systems with related organizations. Maternal and child health services are provided by various bodies from public institutions to nonprofit organizations and community organizations. Therefore, forming networks for cooperation with organizations concerned is regarded as an important objective in the maternal and child health-type program, just as with other types. Each municipality is establishing comprehensive support centers for child-rearing generation as a one-stop center. Such centers deal seamlessly with information provision and consultation support. Such centers also provide maternal and child health-type user support programs, which are services needed for pregnant women and nursing mothers and parenting households through the gestation period to the parenting period.

As described above, the national government has urged each local government to build mechanisms for mutual understanding and cooperation between parties concerned, such as administration, public institutions, and local organizations, to bolster support spanning from the gestation period to the parenting period and for a community as a whole to provide seamless support. Nevertheless, the national government has presented no concrete method for that and has stipulated the provision of programs according to the actual situation of each area. Consequently, the reality is that municipalities are seeking to make use of resources that each area possesses, such as related facilities and talented persons, are seeking to build cooperative systems, and are investigating integrated support programs considering the actual circumstances of their respective areas.

Researchers in nursing science and sociology have long argued the need to build integrated support systems for pregnancy, childbirth, and childcare (Furukawa 2008; Sasakawa 2014). However, although many articles have introduced examples of integrated support systems for pregnancy, childbirth, and childcare, few reports of the relevant literature have clarified the actual situations of the building of the systems according to regional characteristics. Therefore, this chapter examines integrated support systems for pregnancy, childbirth, and childcare that have been built in areas with different characteristics as examples. Devoting attention to the building of cooperative systems with interprofessional and local organization activities, we aimed at consideration of factors of regional differences observed there. It has not been long since build-up programs of integrated support systems for pregnancy, childbirth, and childcare have begun to be implemented. Local governments implementing the systems are still few. When the spread of integrated support systems for pregnancy, childbirth and childcare to all parts of Japan becomes a goal, we would like to offer suggestions on how the system should be constructed from a geographical perspective.

2 Study Areas and Methods

Twenty-nine municipalities were approved as model areas for the model program of integrated support for pregnancy and childbirth implemented in 2014 (Fig. 1). The maternal and child health consultation support program in the model program was made as a shift to the maternal and child health-type user support program in 2015, as explained above. In 2018, the program was implemented in 942 municipalities (54.2%), of which 799 municipalities implemented as the maternal and child health-type. More than half of the municipalities were carrying out the maternal and child health-type program together with other type programs. In addition, comprehensive support centers for childrearing generation were established at 1,436 locations in 761 municipalities (43.8%) as of April 2018.

As described, although the number of local governments that have implemented the program is increasing, the number is still about half, with large regional differences in terms of the status of implementation.

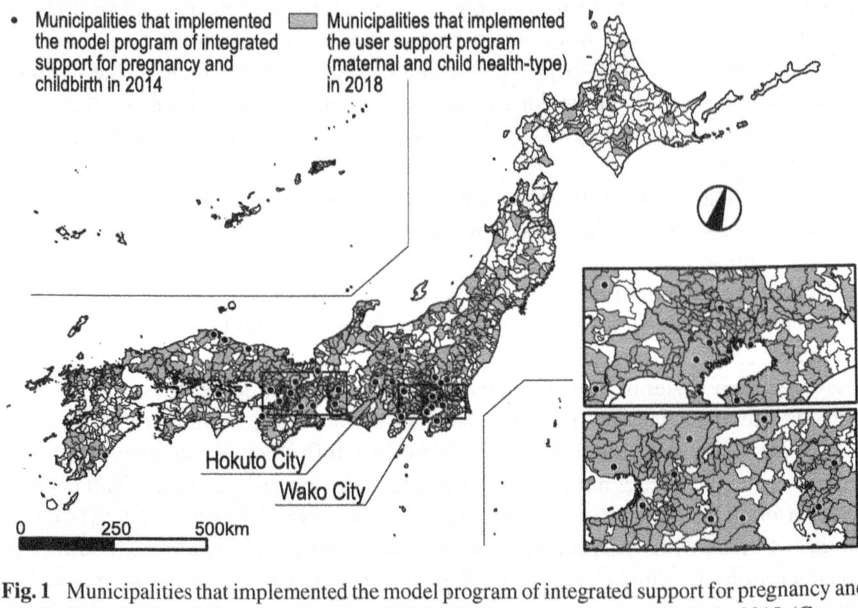

Fig. 1 Municipalities that implemented the model program of integrated support for pregnancy and childbirth in 2014 and the maternal and child health-type user support program in 2018 (Created based on materials provided by the Ministry of Health, Labour and Welfare)

Particularly, few local governments have addressed the program from the beginning of the model program of integrated support for pregnancy and childbirth and accumulated achievements. The breakdown of 29 municipalities that have addressed the building of integrated support systems from the beginning is 11 local governments in metropolitan areas, 16 provincial cities, and 2 small towns and villages in provincial areas. As presented above, local governments in metropolitan areas and municipal governments in provincial areas account for the majority. As described herein, we selected Wako City, Saitama Prefecture as an example of local governments in metropolitan areas and Hokuto City, Yamanashi Prefecture as an example of the provincial cities (Fig. 1).

Wako City, Saitama Prefecture is located about 20 min by train away from Ikebukuro, a sub-center of Tokyo. The population of Wako City has continued to increase as a bedroom town in the suburb of the Tokyo metropolitan area (Table 1). The in-migration and out-migration rates are also high. According to a Population Census of Japan, the population was 80,826 people in 2015, many of which live in young households. The preschool population was 4,929 people (6.1%). The birthrate based on vital statistics was 10.9‰ in 2016. The population increased naturally. Of households with children younger than six years old, 96.9% are nuclear family households.

Hokuto City, Yamanashi Prefecture, another example area, is located in the northwestern part of the prefecture, about 170 km distant from Tokyo. It is within about 30 km from Kofu City, the prefectural capital of Yamanashi Prefecture. Hokuto City was

Table 1 Overview of study areas

	Wako City	Hokuto City
Location	Metropolitan area	Rural provincial area
Area	11.0 km^2	602.5 km^2
Total population (2015)	80,826	45,111
Population under 6 years old (2015)	4,929 (6.1%)	1,506 (3.3%)
Total households (2015)	36,898	18,408
Households with under 6 years old (2015)	3,729 (10.1%)	1,017 (5.5%)
Three-generation households with under 6 years old (2015)	98 (0.2%)	171 (0.9%)
Live births (2016)	884 (10.9‰)	202 (4.2‰)
Deaths (2016)	441 (5.4‰)	667 (14.0‰)
In-migrants (2016)	7,623 (93.7‰)	1,728 (36.2‰)
Out-migrants(2016)	7,279 (89.5‰)	1,518 (31.8‰)
Employed persons 15 years old and over (2015)	40,963	22,520
• Primary industry (2015)	317 (0.8%)	3,597 (16.0%)
• Secondary Industry (2015)	6,155 (15.0%)	5,571 (24.7%)
• Tertiary Industry (2015)	29,827 (72.8%)	13,028 (57.9%)

Created by Population Census and Vital Statistics of Japan

established as a result of merging eight towns and villages, which was done during 2004–2006. Because it is surrounded by mountains, the city has the characteristics of rural areas. In recent years, it has been developed as a tourist site mainly in the highlands of the city's surrounding areas. The number of workers is increasing in service industries (Table 1). The population of the city was about 48,000 people at the time of the merger. Subsequently, it has shown a declining trend, decreasing to fewer than 45,111 people in 2015. However, the recent population change shows that in-migration exceeds out-migration. This phenomenon reflects the increased number of migrants to this city from metropolitan areas. Additionally as political measures, the city puts effort into programs to promote in-migration and settlement of the parenting generation. The administration has developed housing with equipment to support parenting at several locations in the city. However, population aging is progressing rapidly in Hokuto City. The rate of population aging has risen to 36.5%. The population under six years old is small, accounting for 3.3%. Of households with children under six years old, 83.0% are nuclear family households.

Both cities have a few three-generation households who can get cooperation for parenting from grandparents within their households. Furthermore, because Wako City has many people moving in and out, many families do parenting without building neighborhood relationships and friendships. Hokuto City, by its nature, has a few parenting households in the large city area. Under the circumstances, each family does parenting with a few parenting households around. Furthermore, newly moved-in parenting families have not fully formed neighborhood relationships and friendships.

In the following section, we present specific examination of the two cities and clarify the characteristics of integrated support systems for pregnancy, childbirth, and childcare that have been built in each area, devoting particular attention to the building of cooperative systems with interprofessional and local organization activities. To this end, we referred to documents, including the minutes of administrative meetings, and conducted a hearing survey of employees in the divisions in charge of administration and employees at comprehensive support centers for childrearing generation. The hearing survey was implemented from 2016 to 2018.

3 Integrated Support Systems for Pregnancy, Childbirth and Childcare in Wako City: An Example of Local Governments in Metropolitan Areas

3.1 Overview of Integrated Support Systems in Wako City

Wako City is divided into three districts based on the lower secondary school districts. It has established a total of five comprehensive support centers for childrearing generation, one center in the south district, and two centers in each of the other district (Fig. 2). It has made them into bases for integrated support programs for pregnancy,

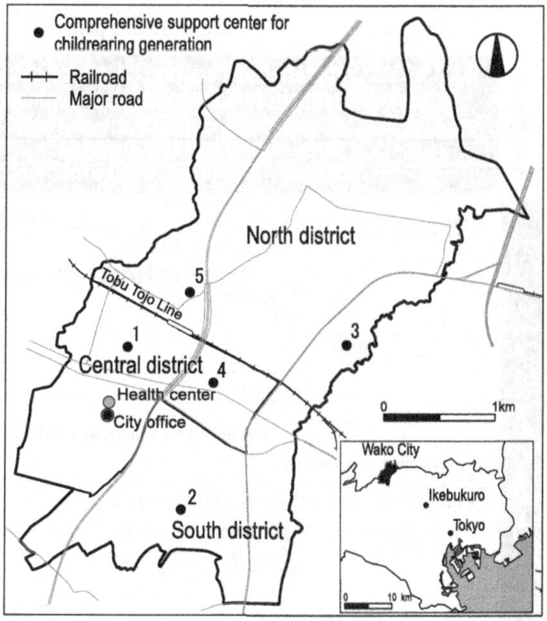

Fig. 2 Distribution of comprehensive support centers for childrearing generation in Wako City

childbirth, and childcare. In Wako City, comprehensive support centers for childrearing generation are mainly established in existing nursery. The operation of the centers is entrusted mainly to local medical corporations and social welfare corporations. Each center permanently stations specialists who have a qualification of midwife or nurse as maternal and child health care managers, as well as licensed social workers and nursery teachers as parenting support care managers. User support programs in comprehensive support centers for childrearing generation in Wako City are, except some centers, are provided in the form of combining the maternal and child health-type programs mainly by maternal and child health care managers and the basic-type program mainly by parenting support care managers within the same center.

Integrated support programs for pregnancy, childbirth, and childcare in Wako City are shown in Fig. 3. Comprehensive support centers for childrearing generation in Wako City carry out the following programs. Besides the city office, the Maternal and Child Health Handbook is issued at four comprehensive support centers for childrearing generation in which maternal and child health care managers are permanently stationed. In issuing a Maternal and Child Health Handbook, a maternal and child health care manager conducts an interview and assesses risks that the household faces. Having assessed needs for cooperation with other specialists or organizations, such as high risks that are of deep concern, the manager creates a care plan at a community care conference, as described later, and uses it as a guideline for support.

Along with these, each comprehensive support center for childrearing generation makes the following efforts as prenatal and postpartum support programs. Parents' classes and parturient women's classes that had been held at the municipal health

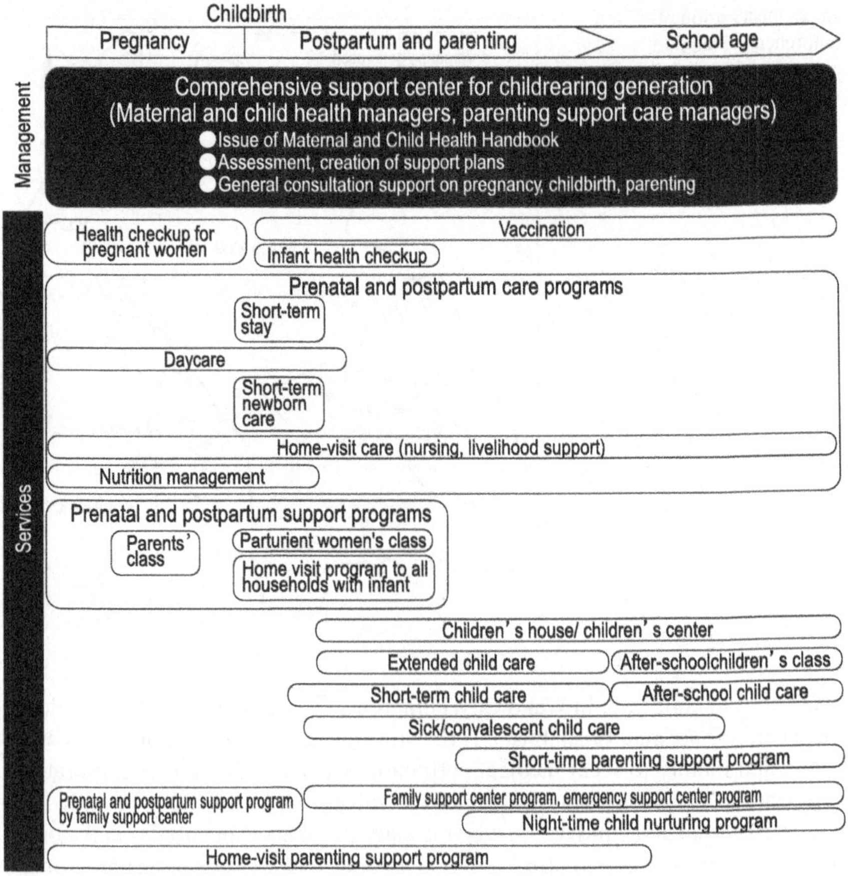

Fig. 3 Integrated support programs for pregnancy, childbirth, and childcare in Wako City (Created by editing "*Neuvola* Guide in Wako City")

center are now held at each comprehensive support center for childrearing generation. The content of the classes varies among centers. Some centers have gatherings for people who experienced late childbirth and hold parents' classes for foreign residents. Some host monthly birthday parties. In addition, maternal and child health care managers and parenting support care managers accept consultation at any time and try to relieve anxieties related to pregnancy, childbirth, and parenting.

Furthermore, mothers who are in poor health after childbirth or who have strong anxiety about parenting, whose baby's health is unstable, or who cannot receive support from their families can access short-term stay and daycare services at a maternity hospital[3] in the city as postnatal care programs. With regard to households

[3]The maternity hospital was invited by the administration of Wako City and was established in 2011.

with members who are unable to go out, home-visit support is provided as an outreach postnatal care program. For high-risk households, public health nurses, midwives, and other specialists are dispatched from the administration. Also provided is an outreach postnatal care program available to everyone, at which specialists visit parenting households and give support in cooperation with NPOs in the city, as described later.

3.2 Building of Interprofessional Systems Led by Administration

The background by which Wako City addresses integrated support systems for pregnancy, childbirth, and childcare is that existing administrative organizations became unable to cope with complex difficulties that parenting households must confront. Another important point is that know-how, organizational strength, and human resources accumulated in building community-based integrated care systems in the field of elderly care can be shared among city employees. Wako City is highly regarded as an advanced local government that developed community-based integrated care systems with emphasis on preventive care for elderly people. Mr. Tonai Kyoichi, Director of Health and Welfare Department, Wako City, who contributed greatly to promote community-based integrated care systems for elderly care considered that the experience might be applied to the field of parenting, and chose to address integrated support programs in the field of pregnancy, childbirth, and childcare. For this reason, Wako City has built integrated support systems for pregnancy, childbirth, and childcare using integrated care systems in the field of elderly care as a reference.

To build integrated support systems, Wako City reorganized the administrative system. In April 2014, the Welfare Policy Division was newly established in the Health and Welfare Department responsible for health and welfare in the city office. Vertical sectioning within the department was solved by assigning one staff member from each division in the department to the Welfare Policy Division. With this, consulting service desks for citizens that had been provided separately in each division were consolidated into the desk in the Welfare Policy Division. Then, the city created mechanisms by which the persons in charge of coordinating general consulting support within the Welfare Policy Division gather information and have a discussion at a community care conference in an interprofessional manner, as needed. Furthermore, in April 2017, the Child Safety Department was established to improve children's welfare and parenting support (Fig. 4). The *Neuvola* Division dealing with administrative affairs of maternal and child health and child allowance was newly established within the department. In addition, to build a system for liaison and coordination of measures between the Health and Welfare Department and the Child Safety Department, the Welfare Policy Division changed its name to the Community-based Integrated Care Division. The renamed division was to continue being responsible for the community care conference.

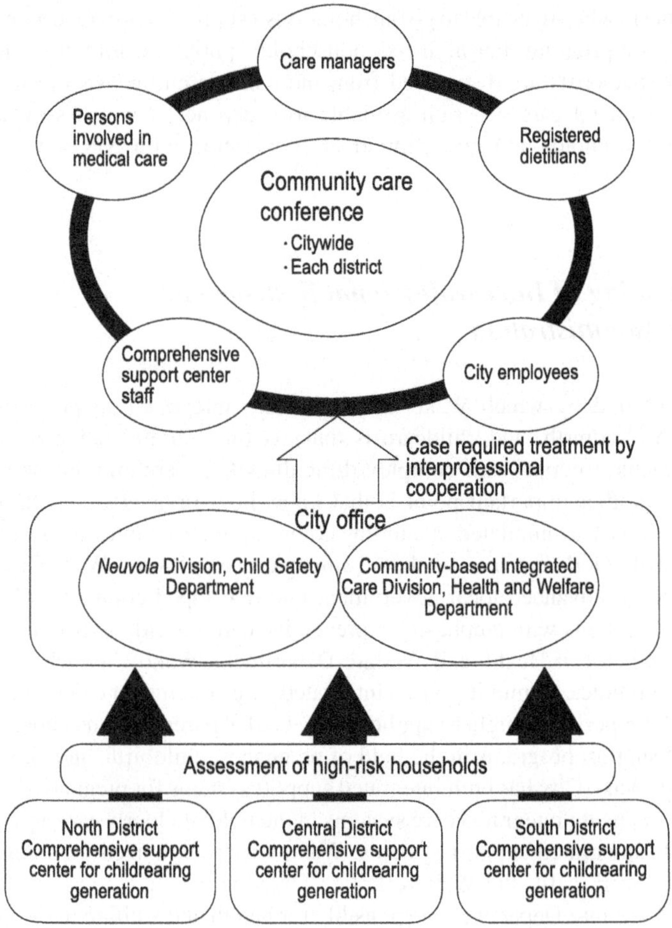

Fig. 4 Interprofessional system in integrated support systems for pregnancy, childbirth, and childcare in Wako City (Created based on a hearing survey)

The community care conference is a meeting in which various professionals and organizations gather and create individual care plans for cases that require support with cooperation and connection with many professionals. This technique has been incorporated into community-based integrated care systems in the field of elderly care. In Wako City, the community care conference in the field of pregnancy, childbirth, and childcare is held regularly three times a month. Participants in the conference are mainly maternal and child health care managers, parenting support care managers, city employees, staff of comprehensive community support centers, registered dietitians, occupational therapists, pharmacists, and persons involved in medical care. In the community care conference, various specialists and related persons create care plans in cooperation with households confronting difficulties extracted in issuing the Maternal and Child Health Handbook or in health checkups, who were determined as high risk, or who were reported to the city office and who determined that the creation of a care plan is needed.

In addition, to enhance cooperation with respective specialists and organizations, Wako City uses information and communications technology (ICT) to share information about high-risk households, using it for comprehensive care. The city office manages electronic medical charts that include information such as whether they have a past medical history and receive support from their parents. When the need arises, concerned persons can access the charts and build a support team.

3.3 Cooperative Systems with Theme-Type Local Organization Activities

In promoting integrated support systems for pregnancy, childbirth, and childcare, Wako City cooperates with the Wako Kosodate Network, a NPO corporation in the city. The NPO provides a range of Wako City's parenting support services. For example, it runs a website that includes parenting information and operates one of comprehensive support centers for childrearing generation. As reflected in the experience of the corporate representative who raised children in Wako City, it has created face-to-face connections within Wako City communities. In this city, interpersonal relationships in the local community are weak. The improved face-to-face connections support parenting households, with the aim of forming networks. As a result, parenting families in the networks become less isolated. A society is created in which everyone supports parenting.

Home-visit parenting support services provided by the NPO play an important role as part of integrated support systems for pregnancy, childbirth, and childcare in Wako City. As noted previously, the service is provided mainly to households other than those confronting high risk. In the service, pregnant women and nursing mothers as well as parenting households are visited once to twice a week upon request. The home visitors listen closely to their worries and concerns, and do house chores or childcare together. The activity is not intended for doing house chores or childcare for the parent, but doing house chores or childcare together. The NPO offers training to become home visitors for local residents who experienced parenting and equips them with preparedness, attitudes, and skills as supporters through taking an eight-day training course.

Wako City also tries to prevent mothers and children from being isolated from the community after childbirth in cooperation with parenting circles in the city. The background is that because there are few places to give birth in Wako City, there are many pregnant women who want to give birth in their hometown, many pregnant women give birth at medical facilities outside the city. An important difficulty was that many mothers and children were isolated when they returned to Wako City after becoming stable after delivery. Wako City had 19 parenting circles to conduct their activities based on comprehensive support centers for childrearing generation. A care manager in a comprehensive support center for childrearing generation confirms whether a woman will give birth at a facility outside the city during an interview with the pregnant woman conducted when issuing a Maternal and Child Health Handbook. By introducing pregnant women who plan to give birth in facilities outside the city to parenting circles, they make the women establish a connection with circles from before childbirth and become acquainted with members of the circles. In doing so, it will be easier for pregnant women to join parenting circles when returning to Wako City after childbirth, thereby preventing them from being isolated from the community.

4 Integrated Support Systems for Pregnancy, Childbirth and Childcare in Hokuto City: An Example of Local Governments in Rural Provincial Areas

4.1 Overview of Integrated Support Systems in Hokuto City

Pregnant women in Hokuto City feel inconvenienced because the city has few places at which to give birth. To accommodate the needs of those pregnant women and to promote in-migration and settlement of the parenting generation, Hokuto City has worked on building integrated support systems for pregnancy, childbirth, and childcare to create an environment to ease parenting in the city. For this reason, Hokuto City established the Hokutokko Genki Division as the division in charge of administration with authority over countermeasures against the declining birthrate as well as programs to support pregnancy through parenting. The Hokutokko Genki Division was established as a comprehensive support center for childrearing generation at the Hokuto City Health Center. The city has only one comprehensive support center for childrearing generation (Fig. 5). The comprehensive support center for childrearing generation is a base for user support programs. It also has an annex with the Tsudoi no Hiroba, which is a place for parents and children to interact with other parents and children during parenting, as well as a matchmaking service desk.

Integrated support programs for pregnancy, childbirth, and childcare in Hokuto City are shown in Fig. 6. The city implements user support programs at the comprehensive support center for childrearing generation combining the basic-type program and the maternal and child health-type program in an integrated manner. Specialist staff members in charge of the programs are two user support specialists, four public health nurses, two dietitians, and a midwife, all permanently stationed, in addition to seven transit professionals. In Hokuto City, the Maternal and Child Health Handbook is issued at the Hokutokko Genki Division. In issuing a Maternal and Child Health Handbook, pregnant women are interviewed by a public health nurse who is responsible for the district in which she lives. If a pregnant woman is assessed as having difficulties, then the responsible public health nurse creates a care plan in cooperation

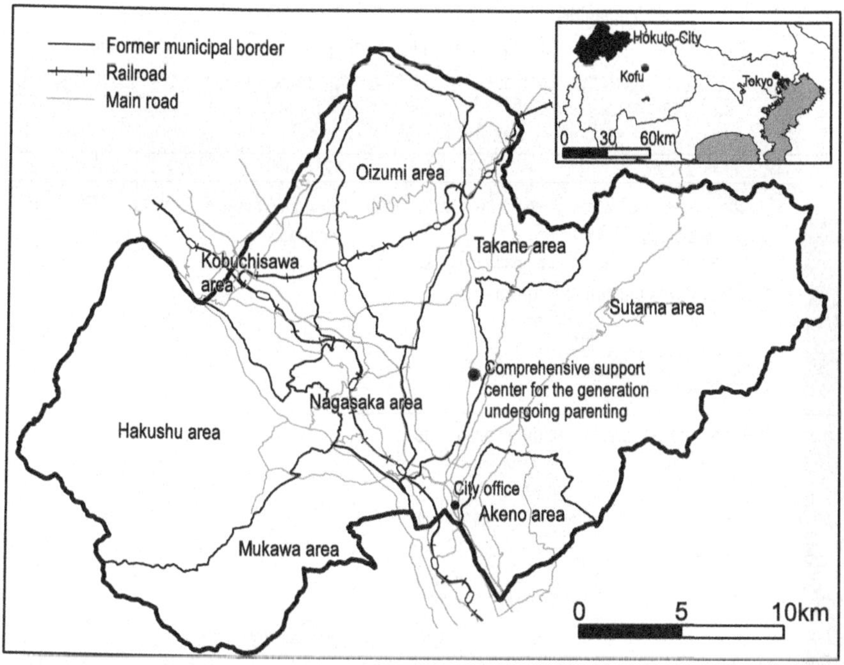

Fig. 5 Location of the comprehensive support center for childrearing generation in Hokuto City

with concerned parties such as other professions and organizations. Additionally after childbirth, responsible public health nurses create individual care plans according to child development based on the conditions of infant health checkups. As described, in Hokuto City, public health nurses continuously ascertain the situations of the people for whom they are responsible during pregnancy through parenting until a child enters elementary school: pregnant women, nursing mothers, and parenting household members. If the need arises, then concerned parties form networks and build a support system.

Furthermore, at the comprehensive support center for childrearing generation, midwives consult with parents about worries and concerns before and after childbirth, while applying for prenatal and postpartum support programs. They conduct consultations on various concerns such as childbirth, childcare, breast milk, and weight. Additionally, they hold baby-care and prenatal and postpartum self-care programs twice a month. Not only does the center provide highly professional learning activities that only midwives would be capable of. The center also tries to create places where mothers mutually interact in a classroom style. The center also offers postnatal care programs. It provides overnight stay care programs and dispatches helpers for infants and mothers who are in poor health and who have anxieties and concerns related to childcare.

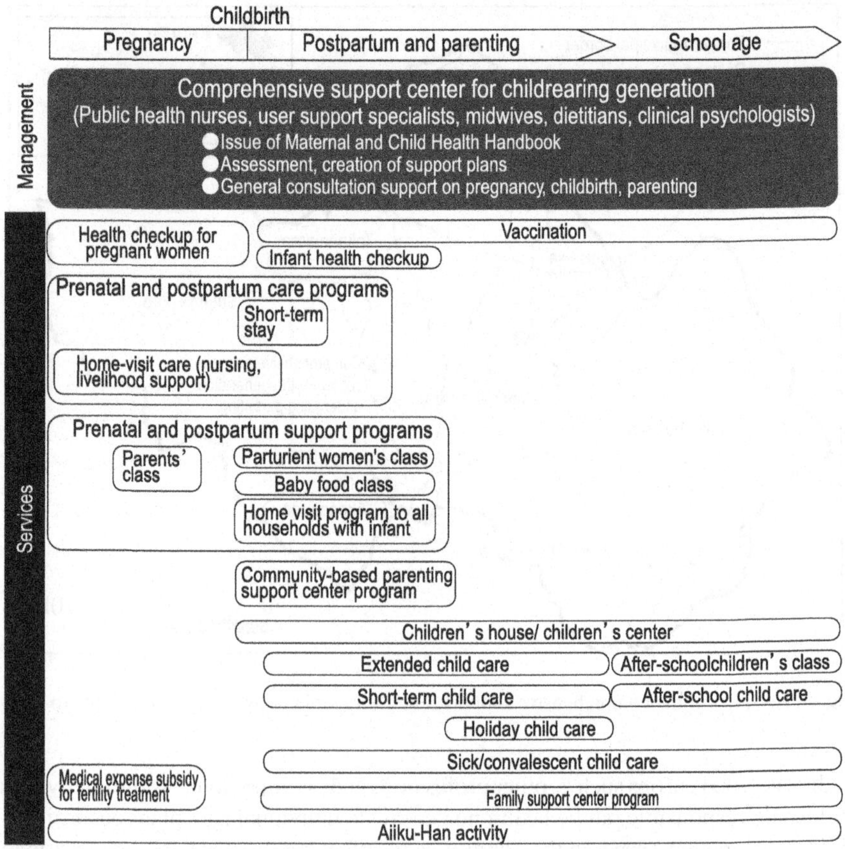

Fig. 6 Integrated support programs for pregnancy, childbirth, and childcare in Hokuto City (Created based on materials provided by Hokuto City)

4.2 *Building of Interprofessional Systems Led Mainly by Public Health Nurses*

Hokuto City had 25 public health nurses as of 2008, which was after the merger of local governments, of which 21 nurses belonged to the same department, performing duties together. Subsequently, the city changed the placement of public health nurses. Through the period in which eight public health nurses took charge of maternal and child health programs, the Hokutokko Genki Division was established in 2017. Since then, four public health nurses have taken charge of maternal and child health. At present, one public health nurse undertakes maternal and child health programs in one to three districts based on the unit of a town or village before the merger.

Public health nurses must continually ascertain the situations of pregnant women and nursing mothers and parenting households in the districts for which they are responsible. As described above, this is accomplished starting with an interview with a pregnant woman in issuing a Maternal and Child Health Handbook, and after childbirth, through mainly infant health checkups. Whenever identifying a difficulty of pregnant women and nursing mothers or parenting households in districts for which they are responsible, they communicate each time with public health nurses working in other divisions or sectors, city employees, and further specialized agencies. If necessary, they gather concerned parties and hold a meeting (Fig. 7). However, public health nurses neither hold regular meetings nor do they set up places for sharing information. When the need arises, they create a care plan for the household and build support systems every time.

Public health nurses in Hokuto City also go to their districts of jurisdiction by themselves and make home visits. They assign importance on directly ascertaining how each household or resident is progressing and strongly recognize the importance of maintaining face-to-face relationships with residents. Public health nurses even assess personal relationships among district residents: what household lives in which district and who is a leader-like figure in each district. When public health nurses take a leading role in creating a care plan, they build a support system suitable for each household, considering not only support from specialized agencies, but also informal support from such as neighbors. The relationship between parenting households and public health nurses is extremely strong in Hokuto City, as known from responsible staff members saying, "One public health nurse for a single house."

Furthermore, in Hokuto City, the placement of public health nurses in administrative organizations has a strong effect on the cooperative systems described above. As described earlier, in Hokuto City, 21 public health nurses performed duties in the Health Promotion Division at the time of the merger. Because the placement of public health nurses was changed later, the 21 public health nurses were assigned, several each, to multiple divisions. Their work began to be conducted with a combination of

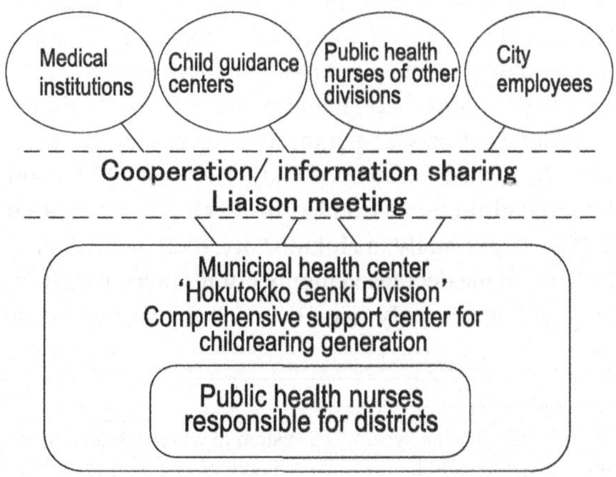

Fig. 7 Interprofessional system in integrated care systems for pregnancy, childbirth, and childcare in Hokuto City (Created based on a hearing survey)

the district sharing system and the work sharing system.[4] Even for different departments within the same city office, the difficulty in mutual cooperation tends to present additional difficulties. However, in the case of Hokuto City, because the public health nurses used to perform duties together in the Health Promotion Division, they are mutually communicating and working as a team still now, even after being assigned to different divisions. This arrangement serves as a basis for cooperation among departments.

4.3 Cooperative Systems with the Territorial Bonding-type Local Organization Activity "Aiiku-Han"

In integrated support systems for pregnancy, childbirth, and childcare in Hokuto City, special mention must be made of the fact that cooperative systems with Aiiku-Han have been built through public health nurses. Aiiku-Han is a territorial bonding-type local organization that the Imperial Gift Foundation Boshi-Aiiku-Kai, established in 1933, launched to disseminate knowledge and skills related to health and hygiene of pregnant women and nursing mothers as well as mothers and children to local communities in rural areas and to improve circumstances thereafter (Ohkawa and Hirayama 2001). Aiiku-Han activities started before World War II and expanded nationwide after the war. Aiiku-Han activities are continuing today with the aim of improving the health of pregnant women and nursing mothers as well as mothers and children.

Aiiku-Han activities in Hokuto City are conducted mainly by women in their 50s to 70s in each district, in addition to a small number of men, based on the unit of a town or village before the merger. The main activities are to promote intergenerational exchange through friendly greetings to residents, home visits, and presentation of gifts to pregnant women and nursing mothers as well as parenting households, issues of information magazines, festivals, and parenting exchanges. In this manner, today's Aiiku-Han activities in Hokuto City emphasize that group members become acquainted with parenting households.

From a nationwide perspective, there have been areas where Aiiku-Han activities were abolished because of budgetary constraints and a shortage of supporters (the Imperial Gift Foundation Boshi-Aiiku-Kai 70-year History Compilation Committee 2005). Additionally in Hokuto City, there were districts where activities were ceased because of the decreased number of mothers and children. In some districts, activities were conducted only once or twice a year and became merely nominal. However,

---

[4]The district sharing system is a system in which the area in the jurisdiction is divided into districts and one public health nurse takes charge of one district and undertakes the work comprehensively. The work sharing system is a system in which public health nurses undertakes work across all districts according to the work (mother and child, mental health, etc.). In some cases, public health nurses are assigned combining the district sharing system and the work sharing system.

public health nurses in Yamanashi Prefecture and Hokuto City who had considered that Aiiku-Han activities were indispensable for future maternal and child health programs revitalized Aiiku-Han activities in each district. They went around explaining to residents the necessity of Aiiku-Han activities while holding lectures and study meetings. The administration supported Aiiku-Han activities financially. Residents reaffirmed the necessity of Aiiku-Han activities through public health nurses' advocacy. In some districts, residents who agreed to the call became group members and restarted Aiiku-Han activities. For example, Aiiku-Han in the Takane district resumed activities in May 2014 because they were able to secure 70 members. Aiiku-Han activities are currently being conducted in five districts.

The background by which public health nurses needed Aiiku-Han activities is described as follows. Trends toward nuclear family lifestyles are progressing even in Hokuto City, which is located in a rural area. Also, the increases of households isolated from local communities occur. In Hokuto City, one concern was that the relationships between public health nurses and residents might become tenuous because the number of public health nurses responsible for one district was decreased compared to before. As prescribed previously, public health nurses in Hokuto City tried to conduct face-to-face activities with local residents. Public health nurses still ascertain the situations of their districts, but the actual circumstances of local communities are increasingly less apparent. For that reason, they strove to ascertain the actual circumstancces of local communities through cooperation between public health nurses and Aiiku-Han. In Hokuto City, infant health checkups were done in each former town or village before the merger of local governments; the consultation rate was nearly 100% in all the former towns and villages. However, after the current health center became the only place for infant health checkups after the merger, the consultation rate dropped to nearly 80%. They thought that the call to parents to go for checkups by Aiiku-Han was needed as a method to raise the consultation rate.

Aiiku-Han and public health nurses share information properly. In Hokuto City, cooperation between public health nurses and Aiiku-Han makes it possible to ascertain the situations of parenting households and to identify problems in their early stages on a district-by-district basis.

5 Regional Differences in Integrated Support Systems for Pregnancy, Childbirth and Childcare: Comparison Between an Urban Area and a Rural Area

In previous sections, we have clarified the characteristics of integrated support systems for pregnancy, childbirth, and childcare in two local governments, studying Wako City, Saitama Prefecture as an example of metropolitan areas and Hokuto City, Yamanashi Prefecture as an example of rural provincial areas. Herein, we examine how regional characteristics of urban and rural areas are reflected in integrated support systems for pregnancy, childbirth, and childcare in the respective local governments.

In Wako City, the administration took the initiative to produce efficient and systematic interprofessional systems, making use of know-how and organizational strength cultivated to build community-based integrated care systems in the field of elderly care. The background against which the city adopted those systems is that parenting professionals in Wako City have limitations that each parenting household in the city must be faced directly and supported independently because of numerous annual births and their daily duties. In fact, professionals in comprehensive support centers for childrearing generation cannot necessarily conduct interviews with all households with time constraints. Other staff members are asked to conduct interviews on a temporary basis as a result of a shortage of professionals. Providing sufficient follow-up support for all households is also difficult. Consequently, professionals conduct repeated interviews only with households that have difficulties. It seems difficult to build interprofessional systems that rely on each professional person's independence and ability in Wako City that were apparent in Hokuto City. We think that the administration must demonstrate strong leadership to build interprofessional systems efficiently through the introduction of ICT and other measures, in addition to holding regular community conference.

In contrast, cooperative systems with other professions and those with local residents have been created mainly by public health nurses in Hokuto City. As expressed in a phrase used by city employees, "One public health nurse for a single house," the connection between parenting households and public health nurses responsible for the district is strong in Hokuto City. Public health nurses have long passed down the importance of the activities in Hokuto City that they themselves go to the actual places. Behind that, there is a regional background peculiar to rural areas in which the population is aging and the birthrate is declining. Insufficient obstetric care systems resulting from a shortage of obstetricians and gynecologists now constitutes a nationwide problem. Particularly, the problem has become aggravated in areas with increasingly aging populations and declining birthrates. This has produced a situation that hinders the safety and comfort of pregnancy and childbirth (Ohira et al. 2007; Miyazawa 2017). Not only are medical facilities dealing with pregnancy and childbirth a few in number; means of public transportation are not well developed in Hokuto City. There is no easy access to those institutions. For that reason, public health nurses in the city have visited households to provide health guidance and have engaged in activities such as household situation assessment and disease prevention.

As depopulation continues and the annual number of births is not so high in rural areas, the newborns that one public health nurse must care are few. In Hokuto City, the annual number of births is about 200, buy the number of public health nurses is four; one public health nurse newly cares for 50 newborns a year, more or less. Public health nurses have engaged in activities, assigning importance to visitation of places where residents live by themselves. The number of newborns is a manageable size for the activities. Through the activities, they came to confront complex domestic problems while being involved deeply in each household. To resolve difficulties, they communicate independently with organizations, specialists, and city employees and build support systems by taking a leading role. However, the reduced number of public health nurses responsible for districts in recent years raises the concern that relationships between public health nurses and parenting households might weaken. As a measure against that, emphasis has been assigned to cooperation with Aiiku-Han, the territorial bonding-type local organization.

As explained above, cooperation with local organization activities is indispensable in both urban and rural areas to build integrated support systems for pregnancy, childbirth, and childcare. Nevertheless, we found differences between Wako City and Hokuto City in terms of local organizations to be partners working with them. In urban Wako City, although the territorial bond-like connection is weak, theme-type local organization activities such as parenting-related NPOs and parenting circles have been conducted actively. The administration made it possible to provide meticulous support for pregnant women and nursing mothers as well as parenting households through the building of cooperative systems using distinctive features of respective organizational activities. We learned from reports by NPO staff members with whom we conducted a hearing survey in this study that enhanced cooperation with administration made activities easier to conduct and that having credibility with administration increased staff members' willingness, which led to the vitalization of activities. Enhanced cooperation with administrators brings positive effects to both administration and theme-type local organization activities.

In rural Hokuto City, public health nurses and Aiiku-Han, a territorial bonding-type local organization, each of which had experience of working on maternal and child health programs, have built cooperative systems. Regional differences in Aiiku-Han activities were apparent in the city. In some districts, the activities were merely nominal. However, Aiiku-Han activities have begun to take place in many districts to date after public health nurses took a leading role in advocating the revitalization of Aiiku-Han. We think that the background is that residents highly trust public health nurses in Hokuto City and that there is a connection based on ties to the land that persists in rural areas. Continuous safeguarding of group members and continuous financial support by administration will be necessary to maintain Aiiku-Han activities.

6 Conclusions

As described in this chapter, we clarified the characteristics of community-based integrated support systems for pregnancy, childbirth, and childcare that have been created and maintained in respective example local governments in urban and rural areas, particularly addressing and comparing cooperative systems of interprofessional systems and local organization activities. We also examined the respective regional backgrounds. Findings from the examination can be described as follows.

In Wako City, integrated support systems for pregnancy, childbirth, and childcare have been created based on the techniques and experiences obtained from the building of community-based integrated care systems in the field of elderly care. The administration took a leading role in building cooperative systems with interprofessional and local organization activities. However, as for integrated support systems for pregnancy, childbirth, and childcare in Hokuto City, public health nurses took a leading role in building interprofessional systems and cooperative systems with local organization activities. Because Hokuto City is located in a rural provincial area, its population is small, with little access to medical facilities. Consequently, public health nurses there worked together independently with other professions, other organizations, and local residents' organizations while visiting places where residents lived and while conducting health activities. It was difficult for Wako City, which is located in an urban area, to build systems that emphasize individual public health nurses' activities, as conducted in Hokuto City. Consequently, Wako City built efficient systems, such as a community care conference and the introduction of ICT, by the administration.

Differences in regional conditions influence the difference in local residents' organizations that are selected as partners for cooperation to provide more meticulous support. Because of regional characteristics of urban areas, Wako City has built cooperative systems with theme-type local organization activities such as NPOs and parenting circles. However, Hokuto City revitalized Aiiku-Han, a territorial bonding organization, and has rebuilt cooperative systems with public health nurses.

Integrated support systems for pregnancy, childbirth, and childcare are currently being built nationwide. Local governments actively visit each other to learn how to build the systems. However, based on findings of this study, professionals who have experience and know-how for interprofessional cooperation differ among areas. Moreover, regional differences exist in the local organization activities that can make an effective approach to each household. Consequently, to build effective systems, we again indicate the need for each local government to ascertain the actual circumstances of respective local communities and establish integrated support systems for pregnancy, childbirth, and childcare according to the context, similarly to community-based integrated care systems in the field of elderly care.

References

Furukawa Y (2008) Constructing the Japanese support system for women in birthing and child-rearing stages. J Hum Nurs Stud 6:71–76 (in Japanese)

Kurihara M, Okuyama M (2012) The cooperation with the many types of job of public health nurses in maternal and child health activities in cities and towns. Journal of Mie Prefectural College of Nursing 16:35–43 (in Japanese)

Mitsuhashi M, Masumoto T, Fukumoto M (2008) Actual situation of parenting support activities of district welfare commissioners: from a viewpoint of the cooperation with maternal and child health services. Bull Schl Nurs Kyoto Prefect Univ Med 17:101–110 (in Japanese)

Miyazawa H (ed) (2017) Mapping health, medical care, and welfare in Japan. Akashi Shoten, Tokyo (in Japanese)

Ohira M, Imada Y, Nagami E, Muramoto I, Maehara S, Yoshikawa Y, Oi K, Nakamura Y, Shindo S, Sibutani H, Urano S, Fujita T (2007) Public-health-nurses' perspective on the relationship between public-health-nurses and midwifes [sic] for maternal care. J Mie Prefect Coll Nurs 11:9–19 (in Japanese)

Ohkawa K, Hirayama M (2001) Changes of Aiikuhan activities and expectations. Jpnese J Hum Sci Health-Soc Serv 7(2):53–59 (in Japanese)
Sasakawa T (2014) The present conditions and problems of the child care support in a community: About the change of the child care support system and the present conditions of the child-rearing family. Bull Kawasaki Coll Allied Health Prof 34:13–18 (in Japanese)
The Imperial Gift Foundation Boshi-Aiiku-Kai 70-year History Compilation Committee (2005) Boshi-Aiiku-Kai 70-year history. Imperial Gift Foundation Boshi-Aiiku-Kai (in Japanese)

Creation of Social Ties for Prevention of Isolation of Elderly Public Assistance Recipients: The Case of a Project for the Provision of "a Place of One's Own" in Nishinari Ward, Osaka City

Nanami Inada

Abstract This chapter elucidates the process through which elderly public assistance recipients create ties, using as an example the case of a project in Nishinari Ward, Osaka City, which provides elderly people with a place of their own. Then, this chapter presents consideration of the meaning of "a place of one's own" that eradicates "poverty in relationship." Results revealed that the functions of "a place of one's own" are to be a place to create new relationships, to implement indirect preventive care and livelihood support through participation in programs, and to coordinate ties with the local community. Among the functions, programs for community contribution activities can be the motivation for users to raise feelings of self-esteem and be effective to prevent social isolation. Results also demonstrate that new interactions are being developed between NPOs of different types and citizens' groups involved in "a place of one's own" and new local networks in which "a place of one's own" serves as a hub are being formed. We can expect that projects for the provision of "a place of one's own" will not only prevent social isolation, but will also become a new basis for local networks.

Keywords A place of one's own · Needy elderly person · Nishinari ward · Osaka city · Public assistance · Social isolation

1 Introduction

In recent years in Japan, there has been a growing interest in livelihood collapse and poverty among elderly people. In fact, the impoverishment of elderly people has increased rapidly since the mid-1990s in Japan. The number of elderly households receiving public assistance is still growing. At present, more than half of the households receiving public assistance are elderly households, with members aged 65 years or older, accounting for 52%, of which 90% are one-person households. Additionally, it has been pointed out that a lowering of pension benefits by the macroeconomic

N. Inada (✉)
Osaka City University, Osaka, Japan
e-mail: nanamii@zac.att.ne.jp

H. Miyazawa and T. Hatakeyama (eds.), *Community-Based Integrated Care and the Inclusive Society*, International Perspectives in Geography 12,
https://doi.org/10.1007/978-981-33-4473-0_13

slide system increases elderly households for which the level of pension benefits falls below the standard of public assistance. Furthermore, the percentage of elderly people aged 65 years or over among the homeless population is increasing year by year, rising from 21.0 to 42.8% period of 2007–2017.[1] There is growing awareness that poverty in Japan is typified by elderly poverty as well as child poverty (Iwata 2016).

We often think that such elderly poverty becomes apparent mainly as a result of economic factors. However, some point out that poverty of single elderly persons with a high life risk is likely to occur not only because of economic difficulties, but also when ties and relationships are fragile (Kotsuji 2010; Kuroiwa 2010): the impoverishment of elderly people appears not only as economic disparity resulting from their working career. It also appears as a result of complex factors that "poverty in relationship" (Matsumiya 2013) seen in the weakening of family relationships and ties with the local community lie on top of one another. Furthermore, poverty caused by these complex factors is related to the issue of social isolation of elderly people.

Efforts have been undertaken across Japan to prevent elderly people from social isolation and seclusion. Most are projects that provide elderly people with a place where they can casually drop by and interact with others. According to the results of a survey conducted by the Cabinet Office (2011), current projects to provide a place for elderly people assign importance in health benefits from the perspective of preventive care. Having said that, the definition of a place for elderly people and its meaning are not confined to preventive care: such a place depends on the field and subject of study (Nakashima et al. 2007). Nonetheless, research examining places for elderly people from the perspective of poverty issues is insufficient. Only a few cases can be found in which a local government official in charge makes a report on the activities to provide a place for elderly public assistance recipients, including Abe (2014). The most important reason why research has not been accumulated sufficiently is that the issue of social isolation is not uniform throughout the country; it emerges in a regionally concentrated manner (Kawai 2013). Consequently, only a few local governments and communities recognize both elderly poverty and social isolation as regional challenges. Other reasons include the following. Because preventive measures for isolation of elderly people receiving public assistance are mainly addressed in casework for public assistance recipients, few examples exist of concrete efforts. Even if a project to provide a place for elderly people has already started, it is still new and garners little social recognition.

The Revised Social Welfare Act (promulgated on June 2, 2017 and came into effect on April 1, 2018) clearly states the strengthening of integrated support systems in local communities in terms of needy people with complex challenges. It also illustrates the importance of the strength of local communities as the basis for support from a perspective of community-based inclusive society. In community-based integrated care systems that are being built by the respective local governments, an important challenge to respond to elderly poverty and social isolation has been how to build

[1]Figures are based on The 2017 National Survey on the Actual Conditions of the Homeless (Estimates) by the Ministry of Health, Labour and Welfare.

overwatch systems and provide seamless and smooth support. For that reason, this chapter devotes attention to local support systems, mainly local government projects that provide a place for elderly people, and considers the process to eradicate "poverty in relationship" confronted by elderly people who are having economic difficulties.

The region to be studied is Nishinari Ward, Osaka City, where the public assistance rate is the highest among local governments in large cities in Japan. The percentage of single-elderly person households is also the highest in the population of households with elderly people. In Nishinari Ward, a project to provide a new place for elderly people was launched in 2013. The official name of the project is "Nishinari ku tanshin korei seikatsuhogo jukyusha no shakaiteki tsunagarizukuri jigyo (The Project for the Creation of Social Ties of Single Elderly Public Assistance Recipients in Nishinari Ward)," which is commonly called the "Hito-hana (One Flower) Project." The base facility of the project is the "Hito-hana Center." For this study, we conducted two surveys of users of the Hito-hana Center and a survey of neighboring local residents during 2014–2015 (Inada 2015). The first survey of users addresses the use of the center. The second one assesses the living conditions of center users. This chapter clarifies the functions of projects to provide a place for elderly people that prevent elderly public assistance recipients from social isolation, based on the results of analysis in the three surveys, with consideration of how "a place of one's own" is related to ending social isolation.

2 Elderly Public Assistance Recipients in Nishinari Ward, Osaka City

2.1 Aging of Precarious Workers in Nishinari Ward

Nishinari Ward is located in an inner-city area of Osaka City (Fig. 1). The percentage of the aged population in the ward is 38.9%, which is outstandingly high among administrative wards in Osaka City. In Nishinari Ward, a day labor market has existed since before the World War II. A discriminated community called the "Buraku" and concentration areas of ethnic minorities have been formed, including Korean residents in Japan. The ward has been recognized as a place with many regional challenges such as poverty and discrimination. Consequently, various regional measures and social services have been undertaken in Nishinari Ward as measures against poverty since before the war. In recent years, great efforts have been launched under Nishinari Tokku Kousou,[2] including measures against child poverty and support for public assistance recipients and needy people.

[2]Nishinari Tokku Kousou is a plan by which Osaka City designates Nishinari Ward as a special zone to promote measures directed at solving various issues that the ward has. The goal is to promote the following measures: support for needy people, improvement of public order, measures against tuberculosis, attraction of tourists, and support for parenting households.

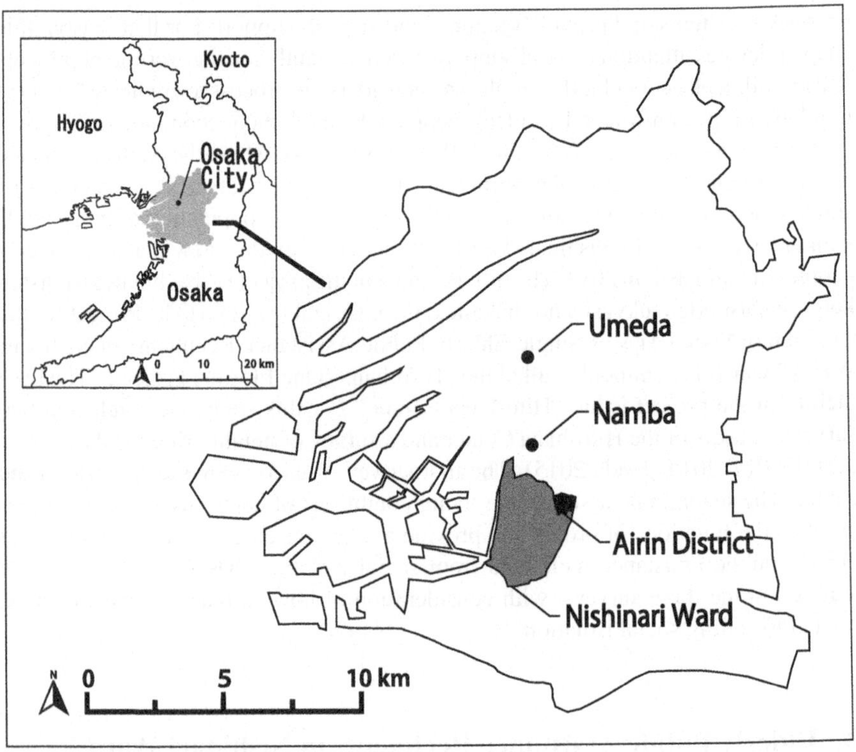

Fig. 1 Nishinari Ward, Osaka City

It is the "Airin District" (commonly called Kamagasaki) where issues of poverty and aging are particularly apparent in Nishinari Ward. The district has supplied cheap labor that underpins local economic activities as the largest day labor market in Japan. The Airin District has a unique mechanism to reproduce day labor. It is the function to support day laborers' living and work, as well as to supply day labor stably. The district has day labor placement services, functions related to food, clothing, and shelter (single room occupancy housing, diners, public baths, etc.), free or low-cost clinics where people without health insurance can see a doctor, and a welfare office dedicated specifically to day laborers.

However, after the mid-1990s, many of the day laborers who lost their jobs in the aftermath of the collapse of the bubble economy were forced to live on the street. They became homeless. Laborers are aging and are increasingly being excluded from day labor markets. Most day laborers have no fixed place to live because single room occupancy housing is the base for their lives. Therefore, because persons with no fixed address cannot be registered as residents, they are excluded from every system of social security in Japan. When excluded from day labor markets, many day laborers lose their places to live and are thrown out on the street without a pension plan or health insurance. Most such day laborers are single men. Because day labor is a precarious form of employment that demands sudden movement from place to place, they have little prospect of marriage or having a family of their own. Consequently, they cannot receive family support when some difficulty befalls them.

2.2 *Elderly Homeless People and Public Assistance*

Homeless people increased at the beginning of the 2000s, not only in precarious worker groups, such as day laborers, but also among regular employees who had lost their jobs. Homelessness has become a common urban problem in large cities throughout Japan. In response to the pressing situation, the "Law on Special Measures for Self-Sufficiency Support for the Homeless" was enacted in 2002; self-sufficiency support projects for homeless people have started in main cities across Japan. In addition to helping enable homeless people with no fixed address to receive public assistance, the requirements of a place of residence and age for the receipt of public assistance were to be corrected.[3]

As described above, by the improvement of support for homeless people at the national level, Nishinari Ward came to provide active self-sufficiency support for homeless people. There are two ways of homeless self-sufficiency support. One is self-sufficiency support projects aimed at employment for working age individuals under 65 years old. Another is efforts using public assistance systems aimed at social independence of those who cannot work because of age or because of illness or injury.[4]

The most important matter related to the latter, efforts using public assistance, is the determination of address for resident registration, i.e., support to secure housing. Most housing prepared in Nishinari Ward for support is single-room occupancy housing in the Airin District, where day laborers have resided to date. When day laborers become homeless, the occupancy rate for single room occupancy housing dropped; management became concerned. For that reason, single room occupancy housing was converted to housing for formerly homeless people with support. Such housing came to accept many public assistance recipients. Furthermore, neighboring private rental housing came to accept public assistance recipients actively, even people who had been avoided in the past. The rent paid by public assistance recipients

---

[3]In the past, when a person who did not have an address, such as homeless people and day laborers, receive public assistance, entering a public assistance facility was a requirement. However, lawsuits seeking to receive public assistance in general housing in addition to the facilities accumulated at the beginning the 2000s. Additionally, the official notice to permit the receipt of public assistance in residential houses was released from the Ministry of Health, Labour and Welfare. These made it possible to receive public assistance in residential housing.

[4]The mechanism of self-sufficiency support, the latter, is the following: First, a homeless person who is eligible for support is helped to find a residential house and signs a lease. Second, the person applies for resident registration with a supporter at a city office. At this time, the lease agreement is a proof of the place of residence. After resident registration, the person applies for public assistance. Through such a process, a homeless person can receive public assistance.

is transferred directly to landlords from the ward office. Public assistance recipients became desirable tenants for landlords because they do not fail to pay rent.

Consequently, former day laborers were to return to the Airin District as public assistance recipients through being homeless. In the Airin District, activities of NPO corporations and citizen groups to support housing and living of public assistance recipients became active in a combined way. A rapidly growing number of needy elderly people who have had no connection to Nishinari Ward to date moved to areas around the Airin District, seeking the receipt of public assistance and support. However, with this, another problem arose: some public assistance recipients were able to receive housing support, but were unable to receive sufficient livelihood support and counseling later on. They became isolated.[5]

3 Outline of the Hito-Hana Project

Because of the regional background presented above, livelihood support for elderly public assistance recipients became an urgent challenge in Nishinari Ward. Some public assistance recipients maintain ties with supporters involved in housing support. Some live in housing where support staff are stationed permanently. However, health deterioration and lonely deaths have increasingly occurred among people who receive no support and who are isolated in their rooms. The isolation and lonely death of elderly public assistance recipients became a regional problem.

To respond to the regional challenges, Nishinari Ward started the Hito-hana Project as an Osaka City's supplementary project based on Nishinari Tokku Kousou. The Hito-hana Center was established in the Airin District as the base of the project. Figure 2 portrays the position of the Hito-hana Center in the regional care systems for elderly public assistance recipients in Nishinari Ward. The operator of the Hito-hana Center is a consortium consisting of five NPO and social welfare corporations that have worked for many years based in Nishinari Ward. Their main activities are nursing care services, community planning, expressive activities, and support for day laborers and public assistance recipients. The Hito-hana Center aims to realize "a place of one's own" into which the corporations' seamless support is integrated.

The project of "a place of one's own" in the Hito-hana Center aims to create a reason for living for single elderly public assistance recipients and to prevent their social isolation through the implementation of various programs for livelihood support. The center has four permanently stationed staff members. The staff deal with daily life consultation related to "overwatch" supervision such as the operation of specialized programs and financial management. Additionally, they do external

[5]With the increase of public assistance recipients, one caseworker in Nishinari Ward deals with more than 300 public assistance recipients. This far exceeds 80 people, which is the average number of people in urban areas. It is increasingly difficult to maintain sufficient consulting systems.

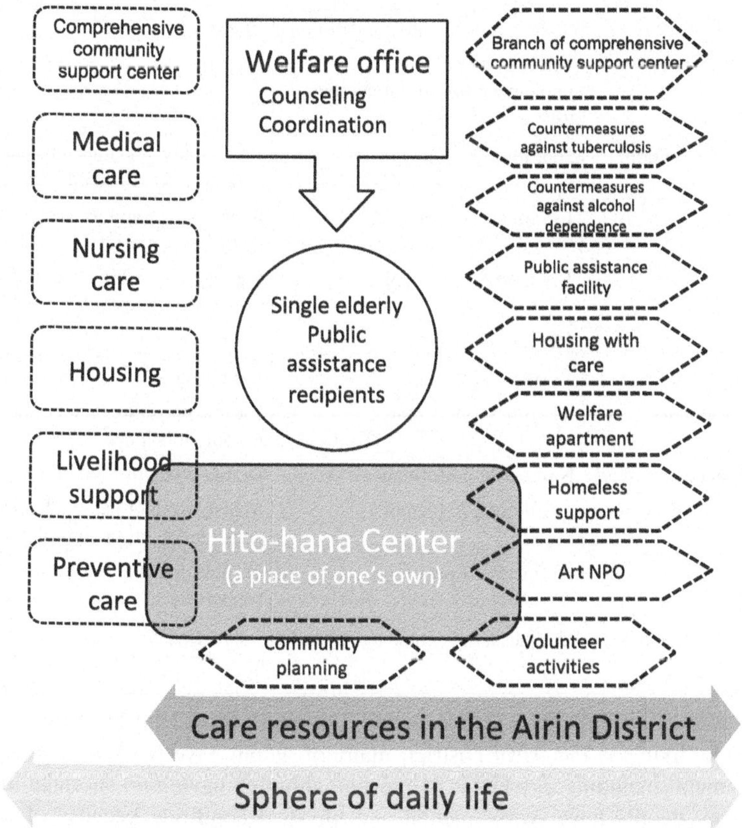

Fig. 2 Community-based integrated care system for single elderly public assistance recipients in Nishinari Ward (Created based on a field survey conducted in the Airin District, Nishinari Ward, Osaka City)

coordination work, including the arrangement of lecturers for programs, information sharing with welfare office caseworkers, consultation with neighborhood associations, social welfare councils, NPO groups, and other neighboring organizations. Table 1 presents programs for social participation and livelihood support provided by the center. Each program invites professional lecturers from outside. The five corporations of the consortium invite various unique individuals, making full use of their own networks: highly original programs are implemented.

People eligible to use the Hito-hana Center are single elderly public assistance recipients in Nishinari Ward who have undergone screening by a welfare office.[6] As of December 2014, the number of people who have registered to use the Hito-hana

[6]Registration for use is determined after meeting the requirements (e.g., aged 65 years or over; living alone; not being in a condition requiring nursing care) and undergoing an interview by a welfare office caseworker.

Table 1 Outline of programs implemented by the Hito-hana Center

Name of program	Total number of implementations/total number of users	Content
Social participation	265 times/2474 people	Community activities (cleanup volunteer activities, mowing, etc.)
	135 times/918 people	Expressive programs (dance, play, calligraphy, music)
	154 times/1070 people	Recreation (cooking, haiku, watching movies)
	150 times/925 people	Farm work
	13 times/122 people	Health class
Work experience	221 times/249 people	Bicycle recycling factory help, restaurant help, etc.
Independent activities	Number of performances: six	Drama group of users
Financial management	135 times/172 people	Management of passbook and seal, consultation on savings plan, etc.
Other	63 times/301 people	Trouble consultation, sewing class, support for taking medicine, etc.

Created by the author based on a survey conducted at the Hito-hana Center

Center was 129 people in all (122 men and 7 women). Most Hito-hana Center users are single, living in the Airin District, many of whom have worked in precarious employment, including day labor. At present, although having no shortage in daily life, many people have worse mental and physical condition because of regular health problems, medical history of alcohol or gambling dependence, and homeless experience. Difficulties and challenges that people with such a life history have when living in a community are often not subject to social welfare. This leaves them left out from support. For that reason, support programs of the Hito-hana Center are to provide the following livelihood support, even if they can live independently and are not subject to nursing care services or welfare services for persons with disabilities: financial management, "overwatch," drug administration support, and opportunities to participate in society through farm work, cleanup work, and work experience.

4 Use of the Hito-Hana Center and User Evaluation

This section clarifies the users' participation in Hito-hana Center programs and presents their evaluations based on a survey of Hito-hana Center users. The survey was administered in December 2014 in the form of a questionnaire survey to elucidate the use of the Hito-hana Center. Of 78 registrants who gave their consent to cooperate with the survey, 66 registrants responded (response rate 84.6%). The respondents comprised 65 men (98.5%) and 1 woman (1.5%); their average age was 71.3 years. All respondents are single elderly persons who received public assistance in Nishinari Ward.

4.1 Purpose of Visit

Half of those registered to use the Hito-hana Center use the center on a regular basis. The remaining half merely registered and use the only slightly center for reasons such as deterioration in health conditions. Of regular users, those who use the center between twice and five times a week account for about 70%. Those who only visit the center when undergoing an interview by a caseworker or a staff member account for about 30%. Regular users are likely to visit the center in accordance to the dates of programs in which they want to participate.

Table 2 shows the top five ranking of purposes to visit the center. "To participate in programs" has the highest percentage of 22.7%. The next common response is "To meet up with people" (19.7%). It is noteworthy that they use the center just to meet up with people even when not having any plan to participate in programs. For example, they visit the center when wanting to chat with someone over tea or to talk with lecturers and volunteers involved in the center. Then, "To contribute to community and society" (9.1%) follows. There are many people who not only participate in provided programs, but who also aim at community contribution through the programs. A user responded: "Even I have a feeling that I want to do something to help people or society, I do not know what to start with. However, I come here; I can think of many different ideas of volunteer activities and put them into practice."

Among the top five items to which users look forward at Hito-hana Center shown in Table 3, recreation is ranked the highest, accounting for 43.9%. According to

Table 2 Purpose of users' program participation (Top five items)

Purpose	Percentage (%)	Specific purpose
To participate in programs	22.7	To have a good time; I want to do something with someone; So I can feel a sense of unity
To meet up with people	19.7	To see friends and acquaintances found in the center; To see staff members, volunteers, and lecturers at the center
To contribute to community	9.1	When coming here, I can learn activities where I can be helpful; Because I receive public assistance, I want to be any help to people
To read books	6.0	I can borrow books and DVDs; I read books at the center, or borrow books and read them at home
For financial management	3.0	As a program, I use it once a month or so

Created by the author based on a survey conducted at the Hito-hana Center

Table 3 Fun for Hito-hana Center users

Content	Percentage (%)	Specific content
Recreation	43.9	Watching movies; Cooking; Participating in a haiku gathering
Farm work	39.0	Farm work in empty land near the center
Community cleanup activities	36.6	Mowing of nearby parks; Cleanup volunteer activities after community events and gatherings; Arrangement of bicycles parking on the street
Expressive program	29.3	Drama group practice, calligraphy, dance, music, etc.
Use of a place of their own	26.8	Reading; Chats with users and staff members; Helping do chores at the center; Taking care of flowers in the garden

Created by the author based on a survey conducted at the Hito-hana Center

the results of a center staff interview, recreation is popular because users can not only participate in them casually, but can also participate in them easily irrespective of whether they prefer indoor activities or outdoor activities. Following recreation, physical activities are popular: farm work (39.0%) and mowing in parks and community cleanup activities (36.6%). Many users cook for themselves to make ends meet. They sometimes cook together using crops harvested on farms as ingredients and sit around the table. The Hito-hana Center functions as a place to spend time to do something with other people in the same space. It is similar to expressive programs that are implemented in cooperation with other participants and to the use of "a place of one's own" in which each can spend time freely.

4.2 *Evaluation as "a Place of One's Own"*

Next, we clarify how users recognize and evaluate the Hito-hana Center. To direct attention to differences in the frequency of use of the center, we designated users who use the center on a daily basis as "regulars" (about 70% of all users) and those who only slightly use the center as "non-regulars" (about 30%). Based on the respective evaluations, we clarify what function the center is expected as "a place of one's own."

Figure 3 shows evaluations of the center by regulars and non-regulars, respectively. In the case of regulars, "A place where I feel safe" (37.5%) and "A place where I feel ties with others" (37.5%) hold high rankings, followed by "A place where people understand me" (31.3%). In the case of non-regulars, "A place where I have fun" (32.4%) is followed by "A place where I need not care about others" (26.5%) and "A place where I feel safe" (23.5%).

We can read the case of non-regulars as short-term use in which users enjoy a once-in-a-lifetime interaction in temporary use. However, regular users appear to

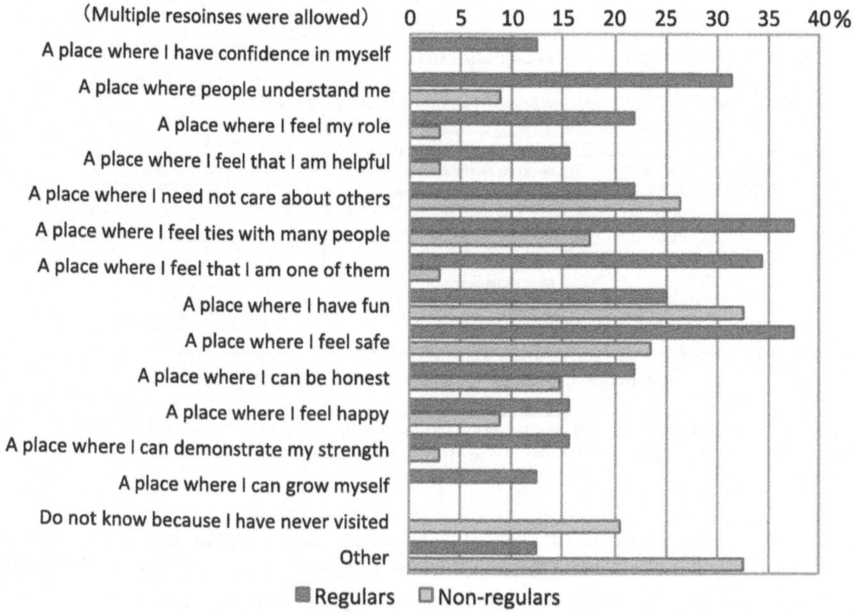

Fig. 3 What kind of place of one's own is the Hito-hana Center for users? (Created based on a survey on the use of the Hito-hana Center)

understand each other while having ties with many people, which makes the center a place to feel safe. We can see the process through which regular users have created a place of their own by building relationships among people involved in the Hito-hana Center over time and developing a sense of belonging and a sense of camaraderie.

4.3 Changes in Perception of Living Using Hito-Hana Center

Regulars and non-regulars differ in terms of changes in perception about living after they started using the Hito-hana Center. In the survey, we asked 24 item questions with a five-scale response, rating from good to bad, about changes that occurred after they started using the Hito-hana Center. The trends of items to which they commonly responded "Better" and "Much better" differ between regulars and non-regulars (Fig. 4). In the case of regular users, it is apparent that the following items are ranked high: "Mental calmness," "Desire to help people and to contribute to society," "New friends and acquaintances," "An opportunity to have communication," "Willingness to work and volunteer activities," and "Participation in community volunteer activities." Because regular users gained mental calmness, positive changes are becoming apparent in their involvement in people and in society.

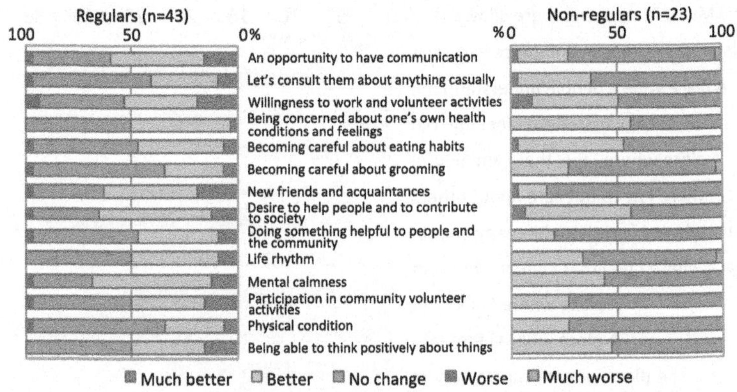

Fig. 4 Changes in users (regulars and non-regulars) after starting to use the Hito-hana Center (Created based on a survey on the use of the Hito-hana Center)

In the case of non-regular users, the following items are ranked high: "Being concerned about one's own health conditions and feelings," "Becoming careful about eating habits," and "Being able to think positively about things." They are more likely to feel that positive changes occurred in their own inter-self. Although regular users feel external changes such as ties with people and the community, non-regular users feel internal changes, such as facing their own inter-self.

4.4 Roles of the Hito-Hana Center

The survey results presented above suggest that the Hito-hana Center has the following three functions.

First, the center is a place for a user to create new relationships with other users, staff members, lecturers, and volunteers. The Hito-hana Center is a place where users can drop by casually as a place to meet up with people and also as a place of their own to spend some time. For single elderly persons, having a place of their own like this motivates them to go outside and to have interactions with others, which helps in preventing their seclusion.

Secondly, the center provides indirect preventive care through programs. It is significant that expressive activities such as painting, music, and dance in expressive programs and recreation, are enjoyable and favorite things. Visiting a place of activities for such reasons marks the first step to vitalizing users' mental and physical health conditions. Additionally, continuing participation in programs such as health promotion, cooking class, and trouble consultation improves users' skills for life and health maintenance, which engenders preventive care.

Thirdly, the center has a function of coordinating ties with people and the local community. It is difficult to obtain information related to community activities individually. However, if coming to the center, then they can obtain detailed information about activities and can conduct activities with multiple members. Gathering people and information engenders community activities. Community activities consist mainly of volunteer activities such as cleaning up. This reflects users' willingness to help people and the local community.

5 Building of Social Ties Through the Use of the Hito-Hana Center

5.1 Characteristics of Center Users Seen from a Survey on Living Conditions

This section considers the prevention of social isolation through the creation of ties by clarifying users' relationships with people and the local community based on the results of a survey of living conditions conducted among users of the Hito-hana Center. The survey was administered during February and March 2015. Subjects were 66 people who responded the survey on the use of the center. A face-to-face questionnaire and interview were conducted individually.

Characteristics of center users became more evident through a survey of living conditions. Results demonstrated that users of the center are classifiable into two types: those who have long lived in the Airin District as laborers and those who came to receive public assistance.

The former is those who made a living through precarious employment such as day labor during their working career and who came to receive public assistance in their old age after experiencing difficult conditions such as homeless life. A once-in-a-lifetime style of working, day labor, increased laborers' anonymity and made it difficult for them to have deep ties individually. Consequently, it is rare that former day laborers have mutual ties that continue long into their old age.

The latter is those who had nothing to do with precarious work or poverty during their working career, but who were suddenly reduced to poverty when entering old age because of unemployment, lack of pension, or other reasons. When being reduced to poverty, they were unable to receive support from their family and relatives and came to receive public assistance in Nishinari. Therefore, they are already in an isolated situation, having been disconnected from every relationship.

Health conditions of the surveyed subjects can be described as follows. Although the average age of the subjects is over 70 years old, they can use daytime facilities. Their health conditions are generally good. However, about 80% have regular health problems and see a doctor regularly. Of them, 30% have symptoms of chronic disease and 10% have mental illness or dependency problems, such as alcohol. In addition, about half feel physically impaired, although they are not given such a diagnosis by their doctor. They feel numbness and muscle weakness in the limbs, decreased physical strength, frequent urination, and failing eyesight. Because many users are elderly people, signs of deteriorating mental and physical functions are emerging along with aging.

5.2 *Building of Ties with People and the Community Through Use of the Center*

Although center users are classifiable into two types, as described above, both are in a condition of social isolation. However, as explained below, they have strong ties with support of NPOs or volunteer organizations who conduct support activities for day laborers and homeless people in the local community and who do further work in support of the receipt of public assistance and housing support.

Table 4 shows whether the surveyed subjects have ties with people and the community. Those who have friends sharing common topics and interests account for 59.1%. Most are friends and acquaintances found in the Hito-hana Center and fellows found when they are homeless; they have ties mainly with people who became close recently. Those who have interactions with people in the neighborhood account for 63.6%. The interactions are shallow, merely greeting each other or standing and talking. Those who responded that they had someone with whom to consult about difficulties account for 65.2%. They consult with friends or acquaintances, welfare office caseworkers, or Hito-hana Center staff members; they are more likely to use

Table 4 Whether users have ties

Item	Percent "Yes" (%)	Percent "No" (%)	Examples of "Yes"
Friends with whom share common topics or interests	59.1	40.9	Friends found in the Hito-hana Center; Fellows when I was homeless
Socializing with neighbors	63.6	38.4	Just exchanging greetings; Just simple standing talking; Not going into details if we meet
Advisers in times of need	65.2	34.8	Friends and acquaintances; Welfare office caseworkers; staff members at the Hito-hana Center; Counselors of support groups
Participation in local events and activities	56.1	43.9	Community cleanup activities; Mowing; Program activities (drama performance activities)

Created by the author based on a survey conducted at the Hito-hana Center

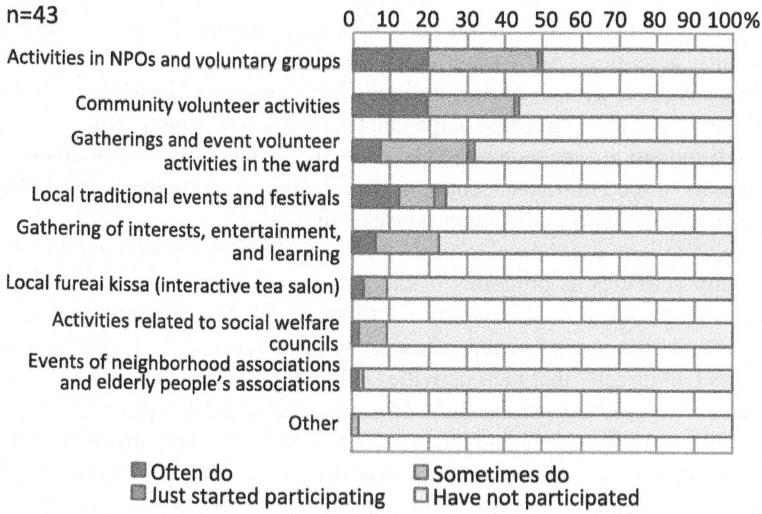

Fig. 5 Users' participation in community activities and the frequency (Created based on a survey on the use of the Hito-hana Center)

public consultation functions. Those who responded that they participate in community events and activities are 56.1%. They mainly participate in community volunteer activities.

In this manner, recent relationships are strong and ties with blood relatives or those to the land are weak in terms of socializing and ties of users. Then, how do users who are acquiring new relationships in recent years have ties with the community and how are they involved in it?

Figure 5 presents the frequency of participation in community activities by content. It is apparent that 50% participate in activities in NPOs and voluntary groups. This result is followed by community volunteer activities (43.9%) and gatherings and volunteer activity events (32.8%). The percentage of participation in activities hosted by neighborhood associations and the public sector, such as social welfare councils, is low, accounting for less than 10%.[7] The following can be regarded as the main factors in the high frequency of participation in NPO activities and volunteer activities among registrants to the Hito-hana Center: community activities are active in Nishinari Ward where registrants have their residence; the Airin District is abundant in activity resources; and furthermore opportunities for those activities play a role to mediate between registrants and the community just as the Hito-hana Center does.

[7]According to the "Hitorigurashi koreisha ni kansuru ishiki chosa (An attitude survey on elderly people living alone)" conducted by the Cabinet Office in 2014, with regard to experience of participating in local group activities among elderly people living alone, gatherings for interests or sports are the most common, accounting for 28.6%. The next is activities in residents' associations and neighborhood associations, accounting for 28.2%. However, the percentage of social activities in NPOs and volunteer groups is low of 5.6%, which shows the exact opposite trend to that of our survey.

Regarding motivators for starting participating in activities (Fig. 6), 33.5% learned community activities "As a program of the Hito-hana Center." The second is "Finding posters or flyers" (21.8%), followed by "Obtaining information at the Hito-hana Center" (12.7%), "Because acquaintances or friends participated" (8.7%), and "Through support groups such as NPOs" (8.4%). The tendency to participate as part of programs of the Hito-hana Center, or the tendency to participate after obtaining information at the Hito-hana Center is outstanding.

As shown there, users develop ties with the community through participation in community activities as programs of the Hito-hana Center. They are more likely to want to participate in community contributions such as community and event volunteer activities among community activity programs. Through volunteer activities, users feel directly that their activities help people and the community. For that reason, their sense of accomplishment or sense of fulfillment is great. However, that is not their only reason for participation. Hito-hana Center staff members point out that underlying users' willingness to community contribution, they have a will to try to overcome "a sense of inferiority" resulting from the fact that they receive public assistance. Indeed, it has been understood as one stigma attached to public assistance that "a sense of inferiority" for the general public that public assistance recipients themselves have engenders social isolation (Matsuoka 2007). For that reason, community activity programs, mainly for community contribution, are regarded as being the motivation for users to raise their feelings of self-esteem.

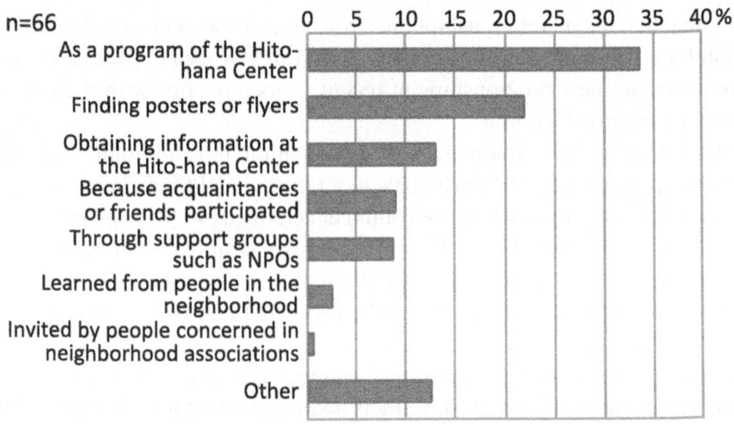

Fig. 6 User motivations for starting participating in community activities (Created based on a survey on the use of the Hito-hana Center)

6 Local Residents' Evaluation of the Hito-Hana Center and Attitude Toward Public Assistance Recipients

Here we consider the possibility of coexistence of single elderly public assistance recipients who have a high risk of isolation with local residents by approaching the attitude of local residents. The attitude survey of local residents conducted for the purpose is to clarify how the Hito-hana Center and single elderly public assistance recipients are perceived by the community and how they build relationships. The survey was administered in March 2015. We distributed a questionnaire to neighboring residents who are involved in the Hito-hana Center. Later, we collected the questionnaire after respondents had filled it out. The questionnaire consists of 18 item questions divided broadly into two themes. One theme is the evaluation of efforts made by the Hito-hana Center. Another is an attitude survey related to their impressions of single public assistance recipients and ties with them. The local residents surveyed include individuals who have experience of participating in events or gatherings of the Hito-hana Center, groups involved in the Hito-hana Center, and persons concerned with neighborhood associations. The respondents are 67 people (38 men, 27 women, 2 non-respondents) aged from their 10 s to their 80 s. Male respondents are 56.7%. Female respondents are 40.3%.

Because registration to use the Hito-hana Center is conducted through screenings by public assistance caseworkers, the establishment itself is not widely informed. Consequently, most respondents recognize the existence of the Hito-hana Center by hearsay, such as word of mouth and rumor. In addition, the Hito-hana Center came to have open days (called "Yume Hiroba (Dream Square") once a month. The center is now a place that is open to local residents, in addition to users, as a place of interactions among neighboring residents and local support groups. With the expansion of the contents of the activities, neighboring residents and the Hito-hana Center have developed relationships through participation in gatherings (open days of the center and performance days of the drama group) of the Hito-hana Project, requests for help in work, such as cleanup and mowing, and requests for the participation in gatherings and events. Of individuals and groups who made work requests of the Hito-hana Center, 80% gave a good evaluation to its careful work. More than 90% made a work request again.

Then, when local residents are recognizing efforts made by the Hito-hana Center and are building relationships, how have their impressions of public assistance recipients and ties with them changed? We conducted an analysis after categorizing respondents' attributes into five groups: participants in events of the Hito-hana Center, local support groups, local support groups for children and parenting, groups related to neighborhood associations, and activities groups such as local circles.

Figure 7 presents neighboring residents' impressions of public assistance recipients using a five-scale evaluation. Overall, 44.3% responded that there were positive changes. No one felt that there were negative changes; 16.4% responded that there is no change. Also, 36.0% said that they do not know. Looking at the responses by group category, positive changes are more likely to have occurred in local circle

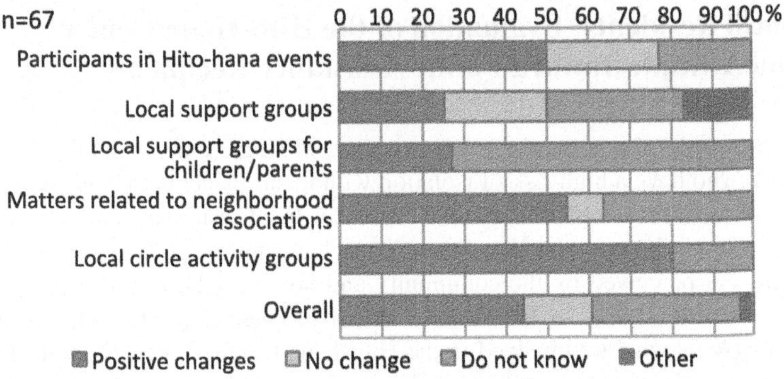

Fig. 7 Changes in the impressions of public assistance recipients that local residents have (Created based on a survey of attitudes toward the Hito-hana Center)

activity groups and groups related to neighborhood associations that have had no connection with public assistance recipients or support activities to date.

Figure 8 shows evaluations in the ties between single elderly public assistance recipients and the community. Overall, 55.7% felt that the ties were developed. Those who responded "No change" account for 9.8% and "Do not know" 34.4%. Looking at the responses by group, more than 50% in most groups responded that the ties were developed. However, only the groups related to neighborhood associations show a low percentage of 36.4%.

Figure 9 presents effects on the community brought by the establishment of the Hito-hana Center. To the question of whether the Hito-hana Center is a strength of the community, as a whole, 76.7% responded that they think so. In evaluations by the group, the groups related to neighborhood association show a low percentage compared to others.

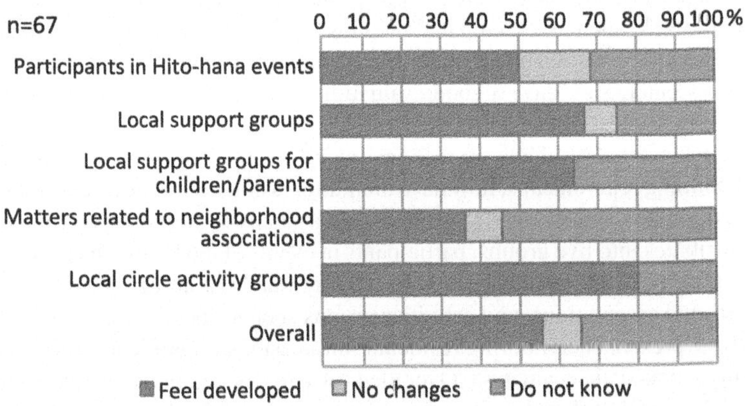

Fig. 8 Changes in ties between single public assistance recipients and the community that local residents feel (Created based on a survey of attitudes toward the Hito-hana Center)

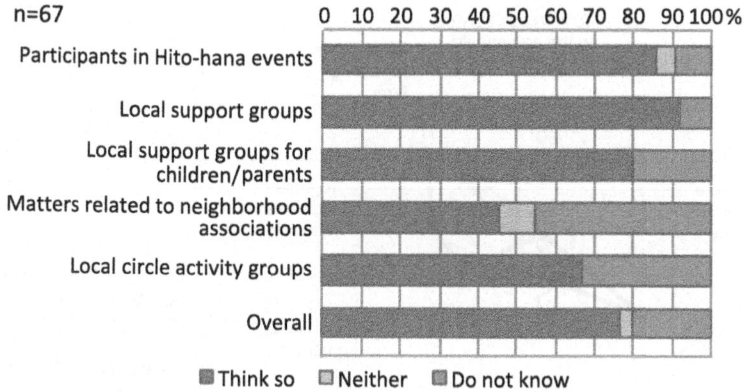

Fig. 9 Local residents consider whether the Hito-hana Center is a community strength (Created based on a survey of attitudes toward the Hito-hana Center)

Figure 10 exhibits things that they want the Hito-hana Center to do in the future. The most common response is "To continue for a long time" (66.0%). Next is "To ease the requirements for registration" (49.1%). At present, only elderly public assistance recipients can be registered. However, many people said that they want the center to be a place that more elderly people can use by easing the requirements for registration. From other responses, it is apparent that the center is expected to strengthen its ties with the community and help the community: "To develop face-to-face ties" (34.0%), "To increase help to the community" (24.5%), and "To increase the involvement in the community" (24.5%).

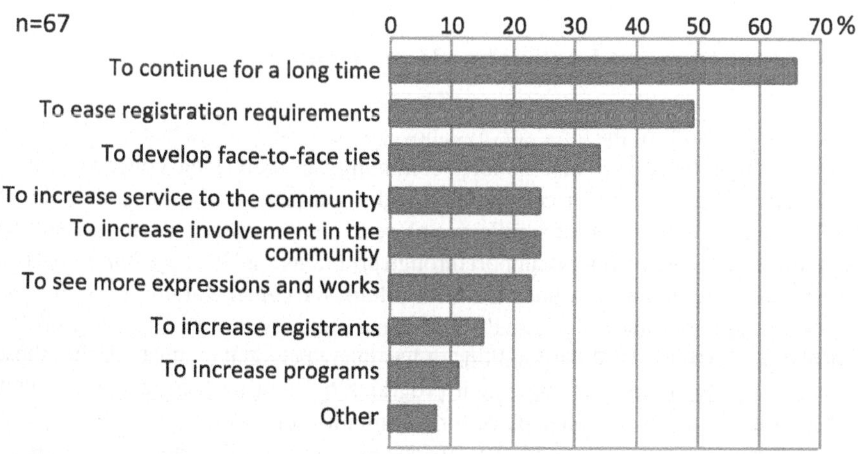

Fig. 10 Things that local residents want the Hito-hana Center to do (Created based on a survey of attitudes toward the Hito-hana Center)

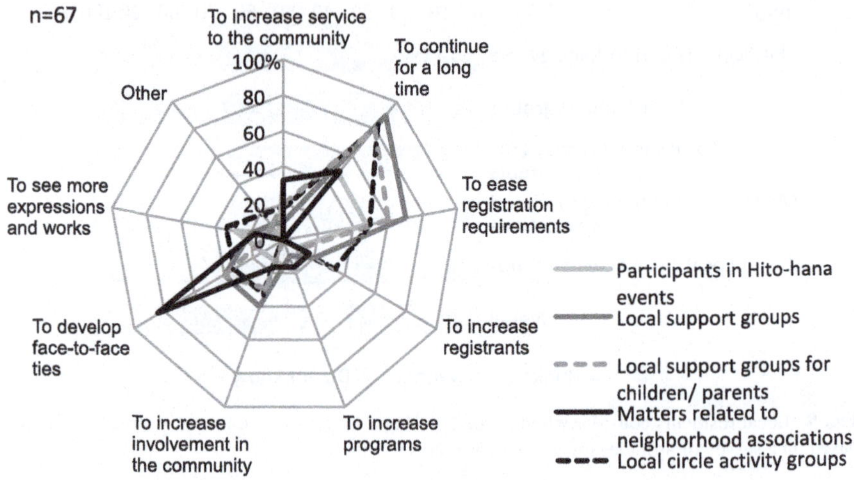

Fig. 11 Things that local residents want the Hito-hana Center to do (by attribute) (Created based on a survey of attitudes toward the Hito-hana Center)

Figure 11 shows an arrangement of the above according to the respondents' attributes. The request for the long-term continuation is strong among all attributes; it is particularly strong among local support-type groups and activities groups. However, people related to neighborhood associations show a different trend here, too. It is apparent that 83.3% make a request to increase face-to-face ties. Because they do not usually have strong connections in actual support sites or in places for community activities, including the Hito-hana Center, they want to start with building face-to-face ties.

7 Things Brought by the Hito-Hana Center

Based on the results of the three surveys above, we describe things that the Hito-hana Center brought to its users and the local community, as well as the consequences.

The first is an ability for coordination and a grasp of needs. The functions of the Hito-hana Center have three pillars: a place to create new relationships, indirect preventive care, and livelihood support through programs, and the coordination of ties with the local community. Among the three pillars, the center's ability to coordinate is of the most importance. Because the Hito-hana Center is operated by a consortium consisting of five fields of corporations, it has many personal contacts. Using these diverse ties, the center can implement programs by inviting lecturers from many different fields and can create a place for community activities.

Furthermore, not only ties with the outside world, but also the operation system by the consortium, engender an accurate grasp of user needs. Users have needs for livelihood support that the framework of conventional welfare systems cannot address.

The five corporations of the consortium have dealt with issues of those who have fallen down the gap separating conventional social welfare systems, including support for day laborers and homeless people and the issue of poverty among the younger generation. Consequently, based on the idea of support in various forms beyond the framework of conventional welfare services, they complement the creation of ties of single elderly public assistance recipients that lies in the gap separating welfare services and systems. There, not asymmetric relationships of supporting and being supported, but gentle support systems of "participation" are built, which engenders the provision of "a place of one's own" where anyone can have a good time.

The second is the raising of feelings of self-esteem and the prevention of social isolation. Many users strongly feel a stigma that they receive public assistance, or feel "a sense of inferiority" because they receive living allowances even though they are not working. Of course, public assistance recipients' rights to a minimum level of living and social participation should be guaranteed. However, when a strong stigma is attached to public assistance recipients, the awareness of their rights like these are likely to weaken. For that reason, their sense of inferiority toward the general public and the weakness of awareness of their rights engender even deeper isolation.

Programs of the Hito-hana Center ease users' sense of inferiority toward the general public and reduce their hesitation to participate in society. As a result of casual participation in community activities programs, users are able to feel that they are engaged in activities contributing to the community. This feeling suggests that efforts for various community contributions through community activity programs are effectively functioning as an instrument to improve self-evaluation and self-usefulness. As described, by helping to unravel negative feelings toward receiving public assistance that users themselves have and affirm their feelings of self-esteem, the programs prevent users from isolation.

The third is the possibility to create new bases and networks for community activities. The establishment of the Hito-hana Center led to the creation of an unconventional new base for community activities. In addition to the function as a place for single elderly public assistance recipients, it has a function as the base for community activities by elderly public assistance recipients. Various local organizations and groups make a request to the Hito-hana Center for cleanup, mowing, and other work. At the same time, activities of the Hito-hana Center are opportunities to produce interactions between organizations and groups. Local organizations that have conducted activities independently to date are beginning to cooperate actively.

Little cooperation or interaction exists between conventional self-governing organizations for long-time residents and networks of new local groups because of differences in their views on regional issues. Although the two have little mutual interaction, in addition to the networks of conventional self-governing organizations, the networks of new community activities for which the Hito-hana Center serves as the base are expected to engender strengthening of safety nets in the community.

8 Conclusions

Based on efforts for the provision of "a place of one's own" in Nishinari Ward, Osaka City, this chapter has presented consideration of the creation of ties of old and single public assistance recipients who have two difficulties, economic difficulty and poverty in relationship. Results revealed two points. The first is that to prevent social isolation, it is important to have an ability to coordinate the creation of relationships in the base of projects for the provision of "a place of one's own." The second is that program activities in projects for the provision of "a place of one's own" engender the vitalization of users' everyday life and raise feelings of self-esteem and function as an instrument to prevent social isolation.

However, the premise of these activities is that the user is not in a condition requiring nursing care. Most single elderly public assistance recipients have regular health problems and chronic diseases. Their social isolation risk is very high if their health conditions worsen and they become unable to go to a place of their own. Consequently, it will be important to take measures for health maintenance and for preventive care for persons with high risks of health deterioration. However, if people who have ties with the recipients can watch over them gently, such as keeping an eye on them, then their isolation and symptoms are preventable from becoming critical even when their health conditions get worse. Because the operation of the Hito-hana Center involves nursing care providers, one challenge for future efforts is expected to be the building of an overwatch or supervisory system through home-visit projects as comprehensive regional care through cooperation with the specialized agencies concerned.

Osaka City has had financial difficulties in recent years. In addition, views on public assistance recipients are increasingly critical. Even though the Hito-hana Project is a project based on Nishinari Tokku Kousou, various difficulties, mainly arising from the costs, will arise to maintain the project in the future. However, without spending money on places or functions such as these, public assistance recipients who have too much time on their hands are likely to seclude themselves. Their dependence, such as gambling and alcohol, will worsen, which might bring them economic collapse and health condition deterioration. This eventuality leads not only to the improper use of public assistance expenses, but also to increased medical expenses. Consequently, a downward spiral in which public assistance recipients deepen their isolation and the community falls into ruin will occur, which can be expected to generate different social costs.

Having made that observation, efforts of the Hito-hana Center have no small ripple effect on the community as a place of interaction for support, as surveys conducted for this study demonstrate clearly. Many people in the community say that they want it to continue for a long time. Relationships brought about in new exchanges between the Hito-hana Center and the community suggest a richness of ties that cannot be expressed by cost conversion and the possibility of regional coexistence brought by that. Efforts of projects undertaken for the provision of "a place of one's own" such as this have the possibility of drawing strength that the community has not only as a place to create ties of public assistance recipients, but also as a place to make networks of local organizations multilayered. For such positive results to engender an ending to social isolation, the key will be how long they can continue projects for the provision of "a place of one's own."

References

Abe R (2014) An effort of front-line public assistance "self-sufficiency support project for elderly people in public assistance households": the base of support and the provision of a social place of one's own through cooperation and collaboration with the community. Life Welfare 696:19–21 (in Japanese)

Cabinet Office (2011) Case research on a place for elderly people and their turn (in Japanese)

Cabinet Office (2014) Survey of the Elderly Living Alone (in Japanese)

Inada N (2015) The 2014 Hito-hana project survey report. In: Health and Welfare Section, Nishinari Ward Office (eds) The 2014 project for creation of social ties of single elderly public assistance recipients in Nishinari Ward: Business report of the Hito-hana Project in 2014 (in Japanese)

Iwata M (2016) Is the redesign of systems for "bright life after retirement" possible? La revue de la pensée d'aujourd'hui 44(3):98–113 (in Japanese)

Kawai K (2013) What is the issue of social isolation? In: Kawai K, Sugano M, Itakuda K (eds) Challenges for the issue of social isolation. Horitsu Bunka Sha, Kyoto (in Japanese)

Kotsuji H (2010) Escape from no-relationship society: An approach from citizens toward the issue of social isolation. Fukushi no Hiroba 128:26–27 (in Japanese)

Kuroiwa R (2010) Social isolation of the elderly in urban areas. J Poverty 4:88–97 (in Japanese)

Matsumiya A (2013) The poverty of social networks on the elderly and "Dying Alone": a case study in Aisai City. Ann Jpn Assoc Urban Sociol 30:15–28 (in Japanese)

Matsuoka Y (2007) A study on the relationship between policies to prevent the bloating of Japanese public assistance payments 'rankyu-boushi' and the attachment of a stigma to the welfare system. Bull Nayoro City Univ 1:69–89 (in Japanese)

Ministry of Health, Labour and Welfare (2017) The 2017 National Survey on the Actual Conditions of the Homeless (Estimates) (in Japanese)

Nakashima K, Hirode M, Konagai A (2007) A study on the concept of place. Bull Faculty Educ Mie Univ 58:77–97 (in Japanese)

Current Situation and Challenges of "Inclusive Care": An Investigation of the "Community-Based Inclusive Station Program" in Saga Prefecture

Mikoto Kukimoto

Abstract As described in this chapter, we examined the current situation and challenges of "inclusive care," which has gained attention as a mode of providing care for a wide scope of users, such as elderly people, children, and children and others with disabilities. Inclusive care initiated from a grassroots "old folks' home" movement led by the private sector after the 1980s. It became a policy as a national program through the support of independent programs by prefectural and other local governments. Saga Prefecture has promoted inclusive care as an independent local government program. However, challenges remain that hinder wider adoption of the idea at the field level, which provides care for widely diverse users in the same space. Particularly, providers have stabilized the operational foundation through long-term care insurance services, but now have "vague anxiety" that they might be in conflict with standards for long-term care insurance by introducing inclusive care. In addition, persons engaged in care shoulder a burden because they must provide different care for different people and because differences exist among local governments and among responsible departments in terms of the knowledge of exceptions spanning the systems separated according to the users and the stance toward promotion. Providers bear heavy psychological and procedural burdens. These problems might be improved to some degree through clarification of responsible departments through the progress of the national program and sharing of the idea. At the same time, guidance and supervision based on the idea of inclusive care and intangible support are required.

Keywords Care for disabled · Childcare · Elderly care · Inclusive care · Old folks' home

---

This chapter is a translated version of Kukimoto (2019).

M. Kukimoto (✉)
Senshu University, Kawasaki, Japan
e-mail: kukimoto@isc.senshu-u.ac.jp

H. Miyazawa and T. Hatakeyama (eds.), *Community-Based Integrated Care and the Inclusive Society*, International Perspectives in Geography 12,
https://doi.org/10.1007/978-981-33-4473-0_14

1 Background and Purpose

This chapter presents attempts to examine the current situation and challenges of so-called "inclusive care," which has gained attention as a mode of providing care to diverse users such as elderly people, children, and children and others with disabilities through investigation of the "Saga Prefecture Community-based Inclusive Station Program."

Governments have promoted "realization of a 'community-based inclusive society'" as one component of social security reform. A community-based inclusive society is defined as "A society in which local residents and various entities in the community participate as 'their own business' beyond 'vertical sectioning' in each system or field and the relationship between 'supporter' and 'recipient' and jointly create each resident's livelihood and reason for living along with the local community by connecting people to people and also people to resources as 'a whole' across generations and fields." It aims at the idea of a "shift from 'vertical sectioning' to the 'whole' in public support." "Inclusive services" are positioned in "the strengthening of community-based integrated support" in the road map for the realization of a community-based inclusive society and have been promoted as something "to generalize the idea of comprehensively securing necessary support," including support for children (persons) with disabilities, children, and needy people.[1]

Such "inclusive services" started with the practice of "old folks' homes and group homes" (hereinafter, uniformly designated as "old folks' homes")[2] led by the private sector engaged in elderly care. Now such services follow the idea of "inclusive care," which has developed in a grassroots manner. The services have become policies by practicing organizations approaching central and local governments. Hirano (2005) defines "inclusive care" as activity that (1) provides a small-scale place to live in the local community, taking it for granted; (2) has no restrictions on who makes a request for use, whether elderly people, children, or persons with disabilities; and (3) shapes various human relationships developed at the scene as a new community in which people live together.

It has been pointed out that inclusive care that accepts people without restrictions on the subjects not only has functions to take on needs and be "the last resort"; it might produce high-quality effects through interaction among users, which cannot be produced in places providing conventional care for homogeneous subjects (Hiroi 2000; Murayama et al. 2017). At the same time, even though sympathizing with the idea of inclusive care, it is sometimes difficult to spread and expand care practically during actual operation, except in some developed areas. In addition, systems that make it possible to provide inclusive care, including long-term care insurance

---

[1]Quotes were taken from "Toward the realization of a 'community-based inclusive society'" on the home page of the Ministry of Health, Labour and Welfare. https://www.mhlw.go.jp/stf/seisakunitsuite/bunya/0000184346.html. Accessed 14 December 2018.

[2]Hirano and Okuda (2005) see old folks' homes and group homes in a unified manner as a regional hub of small-scale care. For the descriptions presented herein, we use the term "old folks' home" as a type of facility, including group homes.

systems, and modifications have affected service continuity based on the idea of inclusive care led by the private sector.

As described in this chapter, we first confirm the idea of inclusive care based on details of the practice. On that basis, we organize the state of implementation and challenges through interviews with administration, representatives of networks, and providers practicing inclusive care, taking up the "Community-based Inclusive Station Program" in Saga Prefecture, which promotes inclusive services unique to the prefecture. The structure of the paper is described as follows. Section 2 organizes the development and idea of "Toyama-style day care services," which became the model for a national system as a typical practice of inclusive care, and the process by which the services became a policy, using existing materials. Section 3 presents the current situation and challenges of the program based on interviews with administration and providers, taking up the "Community-based Inclusive Station Program" and "Nukumoi Home Program" in Saga Prefecture. Finally, Sect. 4 considers future challenges, devoting attention to a gap separating the idea of inclusive services and actual operation.

2 Old Folks' Homes and the Idea and Development of Inclusive Care

2.1 Development of the Old Folks' Home Movement

As a practice evolving into inclusive care, the "old folks' home" movement developed after the mid-1980s. Hereinafter, we organize the history of old folks' homes and the development of the movement based on research conducted mainly by Hirano (2005), Hirano and Okuda (2005), and Ikeda (2010).

Old folks' homes appeared in the mid-1980s as "something responding to the thoughts or wishes of elderly people and their family that 'I want (my family) to continue to live in my house or the community in which we have lived for a long time,'" particularly as a place for elderly persons with dementia. At that time, the knowledge of how to address elderly persons with dementia was not widespread in intensive nursing homes for elderly people. Few places existed for elderly persons with dementia to stay and to receive care in the community. Under those circumstances, the "old folks' homes" drew attention as a place to look after elderly people who needed some sort of care, including dementia care, and as a place for widely diverse people to stay beyond the institutional framework. They spread rapidly throughout the country during the late 1990s. A survey conducted by Miyagi Prefecture in 1998 revealed 600 old folks' homes in operation nationwide.

Although old folks' homes had been run by residents' voluntarism before the long-term care insurance system was established, the continuous operation became possible by being designated as day care services with the enactment of the Long-Term Care Insurance Act. Hirano (2005) argues that the process of old folks' home practice can be organized into three periods. The first period is the pioneering start of old folks' homes mainly in the 1980s. The second period is the institutionalization of small-scale care and the launching of nationwide networks mainly in the 1990s. The third period is the application of insurance as small-scale multifunctional care based on the long-term care insurance system used during and after the 2000s.

Before the long-term care insurance system was established, no entity, except for local governments and social welfare corporations, was able to provide care services using public funds, apart from some pioneering local governments. For that reason, many old folks' homes have been run as unauthorized or unreported facilities in independent programs. Founders of old folks' homes at that time were mainly care and nursing professionals who realized the limits of care at facilities and hospitals, persons who used to care for their family members and were not satisfied with the care at facilities, and persons who wanted to create a place for care that they would want to use as they or their relatives become older. To establish and operate an old folks' home without public subsidies, most providers started services by opening their private homes to elderly people or using low-priced older rental housing. An old folks' home is a small-scale facility with the capacity of five to eight people and a staff of two to three people. An atmosphere differing from public facilities and hospitals gives a sense of security to elderly people with dementia. Successful cases originating in the private sector later became models for the national program.

Efforts for old folks' homes that started before the long-term care insurance system are classified into two types of care: small-scale multifunctional care for elderly people that provides multifunctional care for the elderly mainly in small-scale facilities, including temporary stays; and inclusive care that has widened the range of people eligible for care in addition to elderly people (Table 1). A movement was undertaken by local governments to provide public support for such pioneering practices. For example, against the background of the "Day Home Program for Elderly People Suffering from Dementia" established in 1989, Tochigi Prefecture developed a program in 1994 to grant old folks' homes an annual amount of 6,000,000 yen as operational subsidies. In 1996, a prefectural organization of old folks' homes was launched. In 1998, a national forum, which later became the "National Research Exchange Forum for Old Folks' Homes and Group Homes", was held. In 1999, the "Old Folks' Home and Group Home Network (National Network of Old Folks' Homes)" was inaugurated.

Table 1 Major pioneering old folks' homes

Type	Small-scale multifunctional care for elderly people	Inclusive care for elderly people, children, and children and others with disabilities
Typical facility	"Kotobuki-en" Shimane Pref., 1987 "Yoriai" Fukuoka Pref., 1991 "Nozomi home" Tochigi Pref., 1993	"Kono yubi tomare" Toyama Pref., 1993

Created based on Ikeda 2010

After the long-term care insurance system was started in 2000, long-term care insurance became applicable to some day care services as a result of facilities practicing inclusive care, the National Network of Old Folks' Homes, and other organizations approaching the Ministry of Health and Welfare (of the day). Even at old folks' homes using private houses, providers obtained a corporate status, such as an NPO corporation. The part of day care services forming the basis of the program was designated as day care services in the long-term care insurance system. Other parts were operated as an independent program. In doing so, users' burdens were eased. Old folks' homes became stable in economic terms. The government understood the accommodation function of old folks' homes, which made it possible for providers to provide short-term stays for personal care applicable to standards (short stay). However, the "applicable to standards" stipulation did not spread, except with some local governments. The main reasons include that many municipalities hesitated to carry out short-term stays for personal care applicable to standards at their discretion; moreover, local government employees believed strongly that such support was intended only for providers in hilly and mountainous areas.

Entering 2003, the central government came to recognize the importance of the base where seamless services in elderly care can be provided in an integrated and combined manner. Old folks' home providers were called to committees of the central government and were asked their opinions. As a consequence, "small-scale multifunctional home-based care services" were newly established in 2006.

As described, with no legal backing in the first place, old folks' homes were started. They have since developed as a grassroots practice led by the private sector. Their common features are that they are "a 'small-scale' place with the anything-goes 'multifunction' that offers 'community-based' family-like living space" (Ohara et al. 2018: 48). The capacity is 10 people, more or less. Operational entities are quite diverse, including NPO corporations, social welfare corporations, medical corporations, limited liability companies, and voluntary organizations.

A diverse and unrestricted practice such as this involves the following challenges. First, to continued practices of old folks' homes and inclusive care crossing institutional barriers, providers invariably face the challenge that "the business can be run relatively unrestrainedly. Furthermore, financial resources to cover the operational and management costs are necessary to develop the business continuously" (Ohara et al. 2018: 48). Because any operational entity can easily enter the business using a private home, the practices are likely to be a mixture of wheat and chaff in terms of quality.

Influences can alter national systems and revisions to laws. For example, the long-term care insurance system started in 2000 has played a major role in building the operational foundation for old folks' homes. Many old folks' homes have responded to diverse needs and have provided inclusive services, including those for children and children (persons) with disabilities as an independent program outside of the long-term care insurance system, particularly addressing the long-term care insurance reimbursement for day care services under the long-term care insurance system. However, a "prolonged stay" that continues for a long period of time by adding overnight stay to day care services came to be seen in the structure above; it has been pointed out that, in some cases, old folks' homes are unreported private elderly care homes in reality (Ikeda 2010). In the wake of fire incidents, including a fire that occurred at a group home in 2013, the Fire Service Act was revised in April 2015, which required small-scale group homes and old folks' homes to install sprinklers. This caused difficulty for them in continuing operations because of installation costs. Many providers scale down the of old folks' home operations.

2.2 *Idea and Practice of Inclusive Care: Toyama-Style Day Care Services as a Successful Case*

Next, taking up "Konoyubi Tomare" in Toyama Prefecture, which is a pioneering effort for inclusive care, we can organize the ideas and details of the practice.[3]

Ms. Kayoko Soman, one of the founding members and the present representative of "Konoyubi Tomare," used to work at hospitals in Toyama Prefecture as a nurse. She learned the fact that even when hospitalized patients expressing the wish that "I want to go home. I want to die a natural death," there was often a "family reason" that "if I am discharged from the hospital, I have no one to take care of me because my 'yome' (daughter-in-law) has a job." She experienced a situation at medical sites by which "Even when we saved patients' lives (at the hospital), patients were crying 'I want to die at my home' to the end." Ms. Soman started thinking that "If I care for grannies during the day, they can live at home and their 'yome' (daughter-in-law) can work outside home" (Soman 2009: 379). She left the hospital at which she had been working at the time with the consideration that local communities and home care might need her skills and experience as a nurse. In 1993, Ms. Soman and her fellow nurses opened a private day care house "Konoyubi Tomare" in Toyama City.

"Konoyubi Tomare" made it their motto from the beginning that "We will not refuse anyone" if there is a demand. The care house accepted not only elderly people, but also children with disabilities upon request. The opening leaflet says, "We make it our motto to provide services that anyone can use when needed and as much as needed. There is no restriction on age or area of residence. The procedure is simple, so that you can use the services from the day on which you applied." "By spending time together with widely various age groups, from children to senior citizens, irrespective of whether they are persons with or without disabilities, a warm rapport with others will be created. There are no terms or conditions of use" (Tanaka et al. 2003: 154). It is clear that they aimed at inclusive style care by which elderly people, children, and persons with disabilities spend time in the same space. In respect of the safety in spending time together with various users, "People say that 'if we leave babies and

[3] With regard to the details and idea of the inauguration, we mainly referred to written works by the corporation's representative (Soman 2009), Tanaka et al. (2003), and Health and Welfare Department of Toyama Prefecture (2013), as well as the following interview video. "We're All Different and All Wonderful: Ms. Soman Kayoko talking about Toyama-style day care services" (Video of ninchisho-forum.com) https://www.ninchisho-forum.com/movie/n_028_01.hmtl. Accessed 3 September 2018.

little children in a place where elderly persons with dementia stay, the elderly people might behave violently.' However, there might be a potentially dangerous situation even when gathering and caring for elderly people alone. We have never had such an elderly person who been violent toward babies." The care house assigns importance to the staff's response that they do not see the disease, but see the totality of the individual user, the building of relationships between staff members and users, and for the purposes, the placement of extra staff.

However, facilities caring for various people were unable to receive authorization and public subsidies under existing systems because public systems had different standards for day care services for elderly people, persons with (physical, intellectual, and mental) disabilities, and the age categories of under 18 and 18-and-older. The city officer and other people involved said that "You should specifically examine elderly people, or persons with disabilities." and "We cannot provide subsidies to you because that is not the law." There was neither a care system nor a political framework that went beyond the barriers of people to be cared for. Consequently, Ms. Soman and her fellows started Konoyubi Tomare "at a stretch" as an independent program of the private sector without public subsidies.

For five years from the time of its opening in 1993, the operation was financially unstable without public subsidies. In 1997, commissioned to undertake the "Day Care Program for Persons with Disabilities" by Toyama City, they came to be involved in administration for the first time. This originated from a signature campaign organized by guardians who wanted their children with disabilities to be cared for in "Konoyubi Tomare." The number of signatures collected by the guardians was nearly 130 in a week, which helped the administration make a decision to entrust the program to the private sector. Most of those who wanted their children to be cared for had children younger than three years old who had difficulty obtaining places in nursery schools and kindergartens and children with disabilities who needed tailored care. The main users were children aged from 3 months to around three years. They placed a value on functioning as a recipient of users who were in institutional gaps.

Entering 1998, facilities with a capacity of 10 persons with disabilities and elderly people in all became able to receive an annual amount of 3,600,000 yen subsidy. This is a "flexible subsidy" provided by Toyama Prefecture and Toyama City by removing the boundary between elderly people and persons with disabilities. Local government efforts such as this are said to be the first of their kind in Japan. Actually, Ms. Soman reported that when they attempted to obtain NPO corporate status "with an eye on the credibility and continuity as well as long-term care insurance" before the long-term care insurance system was begun, people around them were adamant in saying that "You cannot acquire a prefectural designation because you have children and also persons with disabilities" (Executive Committee for National Research Exchange Forum '99 for Old Folks' Homes and Group Homes for Elderly Persons with Dementia 1999: 68–69).

Under the circumstances, "Konoyubi Tomare" applied for the "special zones for structural reform" system to the government jointly with Toyama Prefecture and Toyama City and was designated as the "Special Zone for the Promotion of Toyama-style Day Care Services" in 2003. The system of special zones for structural reform is a system that deregulates laws and regulations exceptionally in some areas as part of measures for decentralization and regional revitalization. Designated as a special zone for structural reform under the system, care providers in Toyama became able to accept children (persons) with intellectual disabilities as an exceptional measure under the Long-Term Care Insurance Act. The exceptional measure became practical throughout Japan. The number of facilities operated under ideas such as the "Toyama-style day care services" exceeded 1,000 across the country in 2013 (Kobayashi 2013). The number of facilities regarded as inclusive service providers was 1,990 throughout Japan as of 2017.[4]

At present, the practice of Toyama-style day care services is treated as a successful model case in government meetings to discuss "a community-based inclusive society." It has been proved that the practice creates "a virtuous cycle" for children, elderly people, and persons with disabilities. For children, it has "an effect on their sound growth through living and relating to elderly people and others on a daily basis." For elderly people, it has "an effect on their prevention of care through having a role in parenting support." For persons with disabilities, it has "an effect on their independence and self-realization through having a place to play an active role".[5]

In this way, secondary effects have been discovered and shared in inclusive care. To be more specific, various people, such as elderly people, children, and children (persons) with disabilities, by sharing a place in a "natural" and "family-like" atmosphere and interacting in their lives, can improve their symptoms and maintain their cognitive and physical functions. Ms. Soman says that "Not Mr. XX with diabetes, but Mr. XX has diabetes" and "The person comes first and then the disease".[6] Along with her expressions, their motto is said to have a perspective of responding to users, including their symptoms and other characteristics, not to address diseases and disorders. Such ideas and practices can be said to function as a place for people who have requirements slipping through the conventional welfare supply systems separated according to the subjects and people who seek care in short supply.

Effects of inclusive care have been confirmed through existing research. Interaction that is effective for both elderly people and children is everyday life interaction. By living part of their lives together, a good and sustainable relationship can be extended between elderly people and children. Positive emotions such as senses of

[4]The facilities were those grasped based on "the information about inclusive service providers that the Ministry of Health, Labour and Welfare has grasped on a national basis and the information collected by making a reference over the telephone and a request for the provision of list information to the responsible division in respective prefectures" in a national survey conducted by a committee in which Mr. Hirano Takayuki serves as the chairperson" (Mitsubishi UFJ Research and Consulting 2018: 2).

[5]Descriptions in brackets were added by the author. The 142nd Subcommittee on Long-term Care and Benefits of the Social Security Council (July 5, 2017). Reference material No. 1 "Inclusive Services (Reference material)" https://www.mhlw.go.jp/stf/shingi2/0000170293.html. Accessed 15 December 2018.

[6]Descriptions in brackets are based on an interview video with Ms. Soman. "We're All Different and All Wonderful: Ms. Soman Kayoko talking about Toyama-style day care services" (Video of ninchisho-forum.com) https://www.ninchisho-forum.com/movie/n_028_01.html. Access 3 September 2018.

exaltation and personal fulfillment can be brought to elderly people through spontaneous intergenerational exchange on a daily basis (Hayashitani and Honjo 2012; Murayama et al. 2017). However, to produce those effects, staff must have a high level of performance and management skills and understand and share ideas: many challenges been identified. For example, the following are needed: consideration for exchange programs to promote everyday interaction, understanding of characteristics of elderly people, children, and children (persons) with disabilities, the presence of staff members and supporters who serve as coordinators, and precautions about safety and health as well as environmental improvement. In addition, some concern exists about privacy for elderly people who reside there (Hayashitani and Honjo 2012; Murayama et al. 2017). Kitamura (2003, 2005) points out the following problems. In some cases, no spontaneous interaction occurs in reality, even at facilities combining childcare and elderly care. Staff members bear a heavy burden of promoting interaction. Training courses for nursery teachers and care professionals are separated. Consequently, it is difficult for many expert staff to have knowledge of childcare and elderly care.

2.3 Process of Inclusive Care Becoming a Policy

As described in the preceding paragraph and earlier in the discussion, central and local governments came to understand the importance of inclusive care through the old folks' home movement. At the same time, friction arose between facilities that tried to maintain the function to be a place for various users to stay in the same space and welfare policies that have been developed according to the subjects. When it was difficult for the central government to formulate inclusive care as a policy, prefectural administration advanced it as independent subsidiary programs.

According to Hirano (2015), the formation of a policy of inclusive care succeeded in Toyama Prefecture in 1997. The idea of inclusive care came to spread widely through the starting of national network organizations, cooperation between practicing organizations and the central and local governments, and designation as "special zones for structural reform." Inclusive care became a policy in Nagano Prefecture in 2002 and Kumamoto Prefecture in 2004. In 2005, Saga Prefecture started a "community-based inclusive station program" that inherited the idea of Toyama-style day care services as a prefectural independent subsidiary program. Entering 2009, the Cabinet Office introduced the "Flexible Support Program" based on the preceding examples as the government's attempt to formulate a policy. Although the program had limitations, such as an expiration date for the subsidies, Kochi Prefecture used the program and started policy support for inclusive care in 2012 as support for settlements in hilly and mountainous areas. Subsequently, through the establishment of the headquarters for realization of "a community-based inclusive society" in the Ministry of Health, Labour and Welfare in 2016, "inclusive services" came to be introduced by the government in April 2018 (Fig. 1).

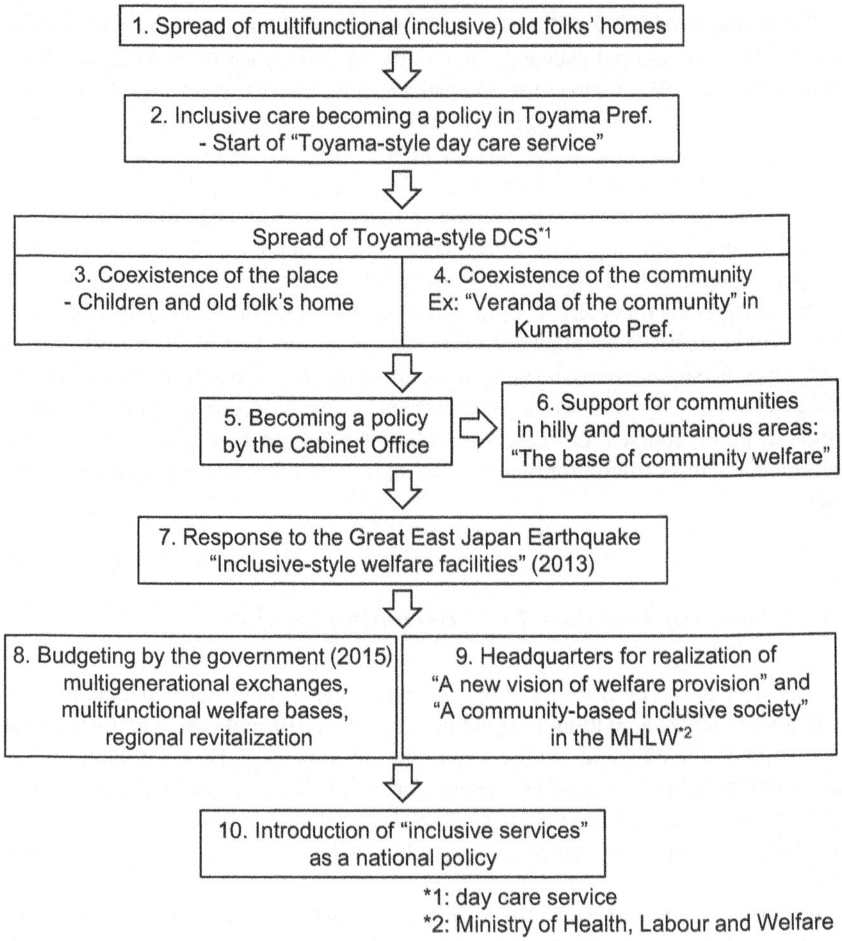

Fig. 1 Development of inclusive care becoming a policy (Referred and partially modified from Mitsubishi UFJ Research and Consulting 2018: 9)

"Inclusive services" were classified into the following two cases. New standards and fees were determined on that basis. Case 1: providers for long-term care insurance services accept persons with disabilities (Designated long-term care insurance service providers obtain the designation of welfare services for persons with disabilities). Case 2: providers for welfare services for persons with disabilities accept elderly people (Designated welfare service providers for persons with disabilities obtain the designation of long-term care insurance services). In the former case, exceptions have been made for home help services, visiting care for persons with severe disabilities, care for daily life, short-stay service, rehabilitation service, child development support, and day care services, such as after school care. To the extent that it is a provider of long-term care insurance, the provider can be designated as

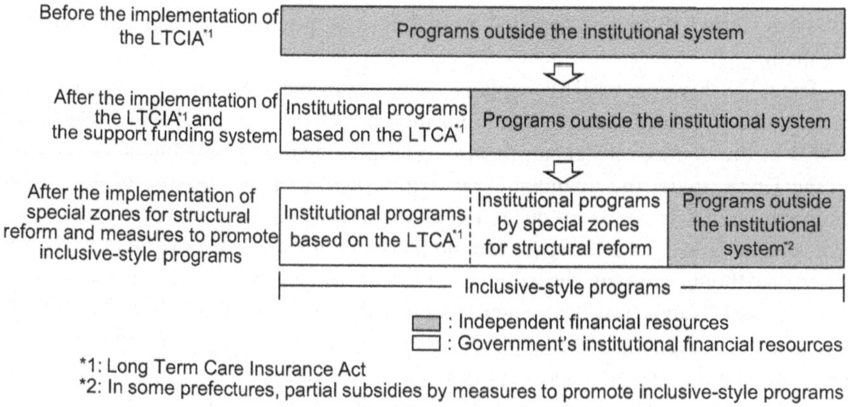

Fig. 2 Structure and financial resources of inclusive-style programs (Referred and partially modified from Nishiyama 2016: 197)

an inclusive provider of welfare services for persons with disabilities, in principle.[7] In the latter case, exceptions have been made for obtaining the designation of home care, day care services, and short-term stay for personal care under the long-term care insurance system. As long as it is a provider that has obtained the designation under the welfare system for persons with disabilities, the provider can be designated as an inclusive provider of long-term care insurance services, in principle.[8]

Nishiyama (2016) reports that the structure of inclusive-style programs and the financial resources for the programs have changed to date by particularly addressing the differences in institutional environments surrounding inclusive care (Fig. 2). Before the long-term care insurance system started, inclusive care had been implemented as programs outside the institutional framework using independent financial resources. After the long-term care insurance system and the government's inclusive

[7]In terms of the fees in this case, because they do not meet the original standards of welfare service providers for persons with disabilities, the units are set as separate from the original unit price of fees using existing services applicable to standards as a reference. For example, when a day care service provider under the long-term care insurance system provides care for daily life for persons with disabilities, 694 units are given, which is the same number of units as the existing services applicable to standards. On that basis, new additional units have been introduced for providers of inclusive care for daily life. For instance, the additional units are given for the placement of a service manager and the addition for strengthening the inclusive service system is given for the placement of a nursery teacher, a child counselor, or other personnel.

[8]Fees in this case are structured based on the idea that the level of fees for welfare services for persons with disabilities should be guaranteed and that it should be separated from the unit price of fees for long-term care insurance system providers. For example, when a provider for care for daily life for persons with disabilities provides day care services, the number of units should be multiplied by 93 percent of the fee for care. On that basis, when the provider places a life counselor and is engaged in activities contributing to the region, the provider is assessed with an additional 13 units each day.

program started, the programs came to be run using government financial resources for them.

Then, Nishiyama (2016) explains why inclusive-style programs had been promoted and subsidized as prefectural measures from the late 1990s through the 2000s in the context of decentralization and administrative and financial reforms. In the 1990s, under the financial crisis that ensued after the collapse of the bubble economy, social demands for decentralization and regional administrative and financial reforms grew. As a consequence, nonpartisan or reformist governors who were elected without receiving direction and recommendations from the existing major political parties appeared. These new leaders had to devise new measures. In addition, with the introduction of the long-term care insurance system in 2000, municipalities were obligated to develop a system providing care and welfare service programs to expand services that welfare users were able to choose. It is said that regional welfare measures, including inclusive-style programs, were devised when prefectures had to support the movement in their municipalities (Nishiyama 2016: 201).

Behind the recent legislation of inclusive care as "inclusive services" lies the problem of "the wall at 65 years old" in welfare for persons with disabilities and a shortage of care personnel in underpopulated areas. The problem of "the wall at 65 years old" among persons with disabilities is described as follows. When a person who has been able to use care services within a framework of the welfare system for persons with disabilities reaches the age of 65 year, the person becomes unable to receive the care services that the person has been able to receive until the age of 64 years because the services applicable to the person changes to those in the long-term care insurance system under the "principle of giving priority to long-term care insurance." By changing to long-term care insurance, persons with disabilities are treated as the same as "healthy persons" and some have experienced the termination, or a decrease in the number of times, of using previous services. Demand grew for some solution to these persistent difficulties.

3 Current Situation of Spread and Challenges of Inclusive Care: The "Community-Based Inclusive Station" Program in Saga Prefecture

As explained in the preceding section, inclusive care started from a grassroots practice by the private sector. It has since been supported and developed by independent programs of prefectures and other local governments. As of 2017, the percentage of local governments that implement an independent program to promote inclusive services is 13.8%.[9] Here we take up a "Community-based Inclusive Station Program"

[9]The figure is based on a questionnaire survey of prefectures, ordinance-designated cities, and core cities in Japan conducted by Mitsubishi UFJ Research and Consulting. The subjects were asked about the implementation of inclusive services. The survey was conducted during December 2017 – January 2018. In all, the distributed questionnaires were 115: 94 valid responses, yielding a valid

Fig. 3 Saga Prefecture

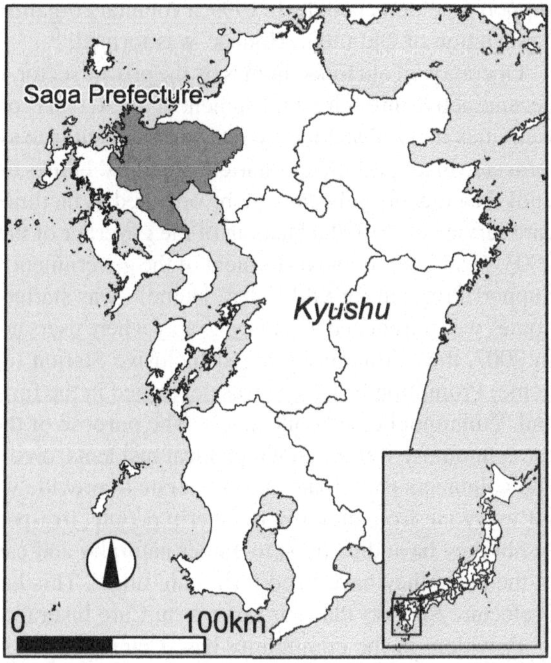

in Saga Prefecture as an example of independent support for inclusive care by a local government and examine the development and challenges in the spread of inclusive care (Fig. 3).

3.1 *Development of Inclusive Care in Saga Prefecture*

The "Community-based Inclusive Station Program" in Saga Prefecture originated in home care services and volunteer activities started in the 1990s. In the wake of the Great Hanshin-Awaji Earthquake, with social background such as attention to volunteer activities, establishment of the Act on Promotion of Specified Nonprofit Activities, and the birth of "old folks' home" by the private sector, old folks' homes in Saga started from mutual help activities in the local community particularly addressing

response rate of 81.7% (Mitsubishi UFJ Research and Consulting 2018). Local governments that do not implement an independent program to promote inclusive care gave the reasons. The most common reason is that "There is no department dealing with in the administrative organization," accounting for 45.6%. It exceeded the percentages for "There were no providers practicing the program in our jurisdiction" (26.6%) and "We were unable to secure financial resources" (2.6%).

support for elderly people. In 1999, a voluntary organization "Saga Prefecture Liaison Association of Old folks' Homes" was formed.[10]

Operation of old folks' homes by the private sector, which is one of those practices, became active through establishment of the Act on Promotion of Specified Non-profit Activities in 1998 and the long-term care insurance system in 2000. After the long-term care insurance system started, old folks' homes that had been run independently until then increasingly came to be operated as facilities with the designation of "day care services." With the backing of the governor of the prefecture who took office in 2003, in addition to the movement of the government, the "Saga Prefecture Opening Support Program for Old Folks' Homes" was started in 2003. In 2005, "Nukumoi home" was introduced as a place to stay where users are not limited to elderly people. In 2007, the "Community-based Inclusive Station (old folks' home and Nukumoi home) Promotion Program" was developed in the form of calling "old folks' home" and "Nukumoi home" collectively. The purpose of the program is to "create a base for community welfare in which local residents, civil society organizations (CSOs), and volunteers participate and cooperate to provide widely various welfare services for everyone from children to elderly people, irrespective of their age and whether or not they have disorders, to gather naturally and can live with a sense of security in the place they have lived for a long time." This has been positioned in the Saga Prefecture Support Plan for Long-term Care Insurance Services (Kato 2014).

Providers in the community-based inclusive station program comprise of "old folks' homes" and "Nukumoi homes." When a provider cares mainly for elderly people, the provider is classified into "old folks' home." When a provider provides inclusive care spanning multiple fields, such as elderly people, children, and persons with disabilities, not limiting people to be cared for, the provider is classified as a "Nukumoi home." The community-based inclusive station promotion program offers subsidies for the new establishment of a Nukumoi home up to 2,000,000 yen, for the new establishment of a Nukumoi home with an attached salon for exchange up to 2,500,000 yen, and for the conversion of the existing old folks' home to a Nukumoi home up to 1,500,000 yen.[11] As of March 2018, there were 185 "community-based inclusive stations" in Saga Prefecture, 80 of which were classified as "Nukumoi homes".[12] However, according to interviews with officers of the responsible department and the representative of the Saga Prefecture Liaison Association of Community-based Inclusive Stations, many facilities have become "private elderly care homes" because of prolonged overnight stays. In addition, even in "Nukumoi homes," only a few facilities actually have daily exchanges among elderly people, children, and children (persons) with disabilities sharing a space.

---

[10]This was reported during an interview with the representative of the Saga Prefecture Liaison Association of Community-based Inclusive Stations (in October 2018).

[11]This figure is based on the "Guidelines for Granting Subsidies of Saga Prefecture Community-based Inclusive Station Promotion Program".

[12]This figure is based on the "List of Community-based Inclusive Stations in Saga Prefecture (as of March 31, 2018)".

As the reasons that daily exchanges between diverse users are limited, the following three points were described. First, many providers secure the operational foundation through long-term care insurance services. Supplying services to people in addition to elderly people does not necessarily increase profits. Secondly, because strict standards have been established for facilities and personnel placement in long-term care insurance services that underpin the operation, providers are anxious about whether "inclusive" services which are a prefectural program can be understood by the entities that undertake the designation, guidance and supervision of care service providers and revocation of the designation under the Long-Term Care Insurance Act (e.g. the responsible department in the prefecture, municipalities, wide-area unions and partial cooperatives). Thirdly, staff members bear a heavy burden when caring for elderly people, persons with disabilities, and children while securing the safety of each. For these reasons, even if providers have set to provide services to children (persons) with disabilities and children, they often become reluctant to have exchanges "in the same space" and "on a daily basis."[13]

A provider who used to address inclusive care in Saga Prefecture says: "We feel a vague anxiety. Although we do not act contrary to the program subject to long-term care insurance, I wonder if it is really okay for us to continue the inclusive program. If we are subjected to an on-the-spot investigation, they might say something. Should they revoke the designation of a care services provider from us, what should we do? This scrutiny progresses continuously. As a result, a feeling of trying to undertake inclusive care is weakening little by little." The person also says that when providers ask to gain approval for services applicable to standards as Toyama-style day care services, they need energy and vitality to find out the departments in charge of inclusive care in municipalities and often go there to prepare documents and give explanations. It is difficult for providers who are very busy with routine work to bear such a burden.

3.2 *Actual Operation of Inclusive Care: Interviews of "Nukumoi Homes"*

Hereinafter, we present results of the interviews of two facilities that cooperated with us among facilities operating inclusive-style facilities as "Nukumoi homes" in Saga Prefecture.

3.2.1 Case of Facility A

Facility A was established in 2005 by the then representative after retiring from being a hospital nurse. The present representative is a staff member who used to be

---

[13]Based on an interview with a representative of the Saga Prefecture Liaison Association of Community-based Inclusive Station (October 2018).

a public health nurse. The former representative thought that it is difficult for large-scale facilities with a high capacity to keep an eye on individual users no matter how hard the staff work. Particularly, it is difficult to provide detailed care when users exhibit symptoms, such as a desire to return home. For this reason, she opened a facility capable of practicing care for a small number of people.

The former representative purchased a stand-alone private house built in 1979 that was on the market in the community. Several years later, the house went through a large-scale interior renovation, including firewall installation. However, she was conscious of leaving a "family-like atmosphere." For example, the original doorsteps were left on purpose. It is based on the following ideas: "This is not an elderly care home, but a facility to support users to live in a home" and "Since their own house is not completely barrier-free, we place importance on preserving users' body functions by moving their foot up and down. We want to avoid complete barrier-free surroundings." Based on the same intention, they prepare bedding according to the user's bedroom environment at home (bed or futon) at overnight stay when users are likely to fall into anxiety or confusion. Furthermore, in day care services, they adopt easy tasks in addition to recreation, such as vegetable preparation and sewing, which make use of users' experiences in addition to recreation because many of female users are from farming families. They intend to preserve finger functions and maintain a sense of self-dignity by giving a role to undertake their familiar tasks, including housework, and value that "users share a place and an experience among the members in the facility."

In addition to community-based day care services at long-term care insurance services, Facility A practices temporary overnight stay for elderly people, temporary childcare, and paid volunteer programs to address problems in the community as independent programs of the corporation.

At the time of the interview, the capacity of long-term care insurance services was 10 people and the number of users was 16 people, 15 of whom were certified in long-term care insurance; the remaining person was not. Although (living-out) day care services are fundamentally the main services, two people use overnight stay on a regular basis (twice a week) because of family reasons or to reduce care burdens. The overnight stay is also used irregularly. The users' required levels of care are low because the facility was created by remodeling a private house and therefore has not installed equipment in the water closet and the bathroom that people with severe physical disabilities can manage. The levels of care required by users certified in long-term care insurance are described as follows: 1 person is in need of assistance level 1; 4 people needing assistance level 2; 6 people needing care level 1; 1 needing care level 2; 2 needing care level 3; and 2 needing care level 4. The number of staff menbers was 12 at the time of the interview. The facility assigns 5.5 staff members a day.

In their independent programs outside long-term care insurance services, the day care program for children was introduced for the following reasons: When working as a public health nurse, the present representative heard this explanation of a difficulty from a parent in a medical checkup for infants saying, "I have no place to leave my child when the sudden need arises." Temporary childcare implemented in nursery schools had problems, such as parents had to book seats in advance and seats were booked up soon. She recognized a great need for temporary childcare that can care for children on the very day, irrespective of parents' employment status and reasons and introduced the program.

At the beginning, the facility cared for children in a circle of acquaintances. However, because they have been featured on parenting magazines in recent years, inquiries are increasingly coming. The facility accepts children only after their parents fully understand the following: a nursery teacher cannot be additionally placed for users of child day care; the facility is a remodeled private house and different from the standards of nursery schools; and children spend time with other users, including elderly people and persons with disabilities, in the same space. Because they do not place nursery teachers, the fee for use is 200 yen per hour, which is reasonable compared to the standard fee for temporary child care.

Childcare hours for day care are from 8:00 to 18:00, which is a period of time accessible for working parents. The childcare hours can be extended earlier or later after consultation. In 2017, five children aged 0–2 used a total of 136 times. In the past, elementary schoolchildren were cared for after school. The reasons for using temporary childcare vary. There are reasons in addition to those presented in Table 2. In one case, a parent said, "I am returning to my job and resuming full-time work in April. However, the nursery school where my child entered provides childcare only in the morning during the first half of April because of 'gradual entry.' So, I cannot pick up my child." In response to the request, a staff member of Facility A picked up the child from the nursery school and cared for the child until the evening when the parent finished work.

The paid volunteer program was started in 2015 to "help resolve small problems in the community." In preparation for the decrease of long-term care insurance reimbursement in the day care service program in the future, the facility started the program with a view to converting to a general program to support preventive care and daily life. In 2017, they handled 1,125 cases, including assistance for hospital

Table 2 Independent program in Facility A outside long-term care insurance

Child daycare	Paid volunteer
Care hours: 8:00–18:00	Subjects: All residents in the school district
Fee: 200 JPY per hour	Fee: 500 JPY per 1 time (approx. 40 min.)
Age groups of user: Aged 4 months to 2 years	Annual fee: 2,000 JPY
Main reasons for use: giving birth to a younger child, afterbirth, job interviews, mother's hospital visits	An additional 200 yen for wheelchair users
	Registered users in 2018 annual year: 47 persons
	Number of volunteers (March, 2018): 3 persons
	Main usage: assistance for hospital visits, support for shopping, and personal shoppers

Created based on interviews

visits and transportation for shopping in the area. However, the facility is short of volunteers in comparison with the number of requests and unable to increase volunteers from three people they gathered by searching for connections of acquaintances. It has been a challenge to increase the number of volunteers. Administration provides a subsidy for operating the program for three years from the start. Consequently, they will decide a policy of expansion according to future development.

With regard to inclusive care, the representative of Facility A says that she "sympathizes with the idea," but "it will be "easier" for the caregiver side to divide according to the subjects. Persons with disabilities have wide diversity in terms of characteristics of disorders and how they are addressed. Of course, with regard to physical disabilities, there are many points in common to the parts that elderly people develop because of aging. Still, the parts to which they should devote careful attention differ depending on whether the person has hearing disorders or visual disorders, for example. "We need to have a 'perspective' to devote careful attention in care. Caregivers who have cared for a specific group of people must study again to have a 'perspective' with which they are not familiar." Furthermore, foods that children can eat change according to the child's age in months, such as baby food, which requires consideration different from that paid to elderly people. Parents of children for whom they have cared many times and with whom they are familiar sometimes say, "A menu that is the same as that for elderly users will be fine." In that case, they prepare meals just as elderly people do, but parents are asked to bring their children's meals, in principle.

The representative talks about sharing a space between children and elderly people. "When babies come, even users who have a strong desire to return home comfort and play with them. It might get their mind off things. I was surprised because they even smiled sometimes." "For (elderly) users, babies are their great-grandchildren (in age), aren't they? I think it makes them imagine that living with great-grandchildren whom they seldom see on a regular basis might be like this." She mentions advantages and possibilities of inclusive care. At the same time, the representative says, "Sharing a space with children might be tiring for elderly users in a sense. Our facility is okay because we have no children every day, though."

The operational style of Facility A can be explained as follows. They set day care services for users with mild symptoms at the center of the business and develop temporary childcare and paid volunteer programs. However, they think that they will continue this over "the next 2-3 years." The reason is that because the long-term care insurance reimbursement has been revised since 2006, they feel anxious about the continuity of their business that has been operated, particularly addressing day care services for elderly people who require light levels of care. The representative explores the following directions to increase long-term care insurance reimbursement. One is a direction that stabilizes the operation as a small-scale multifunctional provider by improving equipment for elderly people who require higher levels of care and increasing the capacity. Another direction is that shifts to the general program, particularly addressing volunteer activities. However, she "thinks that this is for providers that have other business with stable operation within the corporation because the amount of subsidies is not very high." Therefore, this is not a realistic option for organization that runs only Facility A.

Facility A has been able to provide temporary childcare to date because they have created a condition in which children are supervised, involving users with mild symptoms. The representative reports uncertainty about whether they can continue the independent program. Specifically, the representative cites the difficulty of accepting risks of looking after children when staff members would be obligated to provide higher levels of care to users.

3.2.2 Case of Facility B

Facility B was established in 2006 by a representative who had worked at relief facilities for more than 30 years. She thought that "I want to create a place for elderly people where they can spend time with a sense of security." Based on the idea, the representative took early retirement in her late 50 s and launched the facility. She built a new stand-alone house on her own land of approximately 2,650 square meters. It cost about 80,000,000 yen to open, most of which was covered with her retirement allowance and a loan from a bank. Aimed at a facility where the representative herself and her family "wish to enter," she developed a system that can provide doctor's visits in cooperation with two hospitals to be able to attend a deathbed when they are in their later stages of life. On some occasions, parents of staff members enter the facility seeking terminal care. The representative takes pride in providing a facility "home away from home" and "a place to stay."

Although the day care service program in long-term care insurance was the main program at the beginning, the facility accepted children and children (persons) with disabilities as an independent program because there were needs. The representative knew that it would end up in the red because children with disabilities require care almost on a person-to-person basis. After the facility had implemented the service as an independent program for about six years, the city learned that they accepted children as well as persons and children with disabilities and invited it to apply for the city's independent subsidy system. The facility now uses public subsidies for part of the cost of temporary childcare and care for persons and children with disabilities. However, Facility B does not accept persons and children with disabilities solely under the framework of government's inclusive services. The reason is that system details are difficult to understand because it just started in 2018. They believe that it might be impossible to accept people flexibly as before. Government's inclusive services strongly recognized the implication of the system that users who reached old age are cared for as before in facilities for persons with disabilities. Consequently, the responsible person of Facility B does not consider that the system is deeply related to their facility that has run the business mainly for elderly people.[14]

---

[14] At a symposium held after introduction of government's "inclusive services" was decided in 2017, some pointed out the following with regard to inclusive-style services: 'Inclusive services' to be newly introduced are not positioned institutionally that a new form of services is introduced. (...)

Table 3 Independent program for children (persons) with disabilities and children in Facility B

Child daycare	Care for children (persons) with disabilities
Care hour: 8:30–17:00 (Extendable) Fee (1 day):	Care hour: 9:15–16:45 (Extendable)
• 600 JPY: babies under 1 year old • 500 JPY: infants aged 1–6 (preschooler) • 300 JPY: children in elementary school • An additional 200 JPY for non-residents • Meals and snacks: 300–500 JPY Fee (monthly contract): 30,000 JPY Extension fee (1 h): 100 JPY Age groups of users: children in elementary school and under Main reasons for use: unexpected work on Saturdays, Sundays, and holidays	Fee (1 day): (1) 2,500 JPY: Class 1 and 2 physical disabilities, Mental disabilities, Medical rehabilitation handbook (2) 1,500 yen except 1) Meal: 600 JPY, assisted bathing: 400 JPY, transportation service: 200 JPY (per one way, only for residents inside the same city) Extension fee (1 h): 500 JPY

Created based on interviews

Current users other than elderly people include a person with severe intellectual disabilities in her 40 s who uses the facility continually (one day a week). There are also requests for child day care and care for children with disabilities by parents who are unable to use usual authorized nursery schools on Saturdays, Sundays, and holidays. The facility accepted as many as 7–8 children and children (persons) with disabilities a day. Because All Facility B staff members are eligible to use child day care for 100 yen a day. In the past, some of them left their children for care at the facility.

Facility B has the following services for elderly people: day care service (9:15–16:45) as long-term care insurance services, day care service outside long-term care insurance, overnight stay, and an old folks' home (monthly contract). As an independent program for people other than elderly people, they provide care for children (persons) with disabilities (9:15–16:45) and child day care (8:30–17:00). Table 3 shows programs for people other than elderly people.

The capacity of Facility B in long-term care insurance services is 30 people a day. The number of staff menbers was 28 at the time of the interview. The facility assigns 10–11 staff members a day, 2 of which undertake cooking and 8–9 provide nursing and care. Four staff members have the qualification of nursery teachers and are asked to go on shifts when accepting children. The number of registered Facility B users is 45, 7 of whom are in need of assistance and 6 of whom are in need of care level 1, 7 needing care level 2, 10 needing care level 3, 4 needing care level 4, and 2 needing care level 5.

The representative of Facility B did not have a feeling of resistance toward inclusive care that accepts not only elderly people but also children and children (persons) with disabilities. Her motto is to accept children without refusal even if they have severe disabilities. However, she has an idea that it will be "safer" with the presence of

Therefore, the energy to promote them is not necessarily high at the prefectural or national level" (Mitsubishi UFJ Research and Consulting 2018: 54).

staff with qualification for work as nursery teachers. When having received requests or consultation for child day care or care for children (persons) with disabilities, particularly in the case of first-time users, the representative makes them understand through observation and a free trial that their children spend time with elderly people in the same space. She mentions that even facilities have been registered in "community-based inclusive stations" that the prefecture promotes as inclusive care, "only a few seem to practice 'coexistence' in the true meaning of spending time in the same space sharing a place of living." The representative gives the reasons as follows. "Spending time with children often has a positive effect on elderly people. However, in some cases, elderly people refuse to do so. Although it might be possible by separating their rooms and having exchanges at a set time, revenues will not increase accordingly. It can be forgiven that many providers might have second thoughts about positively introducing it, I guess."

As future prospects, the representative thinks that "If depending solely on long-term care insurance, we cannot continue our operation (in the future)." She is unable to decide which direction to take: either to develop a system to accept users who require high levels of care as a small-scale multifunctional facility or to increase exchanges with the community as a general program for preventive care and daily life support. However, investment in facilities is needed for expansion to increase exchanges with the community as a general program. In addition, the representative has concerns about securing volunteers. She also feels anxious about whether the facility can maintain its acceptance of children and children (persons) with disabilities and "a family-like atmosphere" when accepting users who require a higher level of care.

4 Conclusions

This chapter has investigated "inclusive services" that are about to be promoted as a government program. After understanding the idea of inclusive care and the development as a policy, we examined the current situation and challenges to operation in areas other than the preceding model areas.

The "Community-based Inclusive Station Program" in Saga Prefecture became a policy as a prefecture independent program with the backing of the governor of the time when the program started and addressed support for inclusive care as one of the earliest in Japan. Providers that practice inclusive care cross age groups, provide child day care that is useful for any reason and children (persons) with disabilities care that a supply shortage has been pointed out. They have become a recipient of the demands that it is difficult for the current systems to address. However, it was also suggested that a few providers practice daily exchanges in the form in which widely various users share a space.

Although the prefecture promotes inclusive care (Nukumoi homes), it has different entities, such as municipalities, which designate, guide, and supervise long-term care insurance services that form providers' economic base to maintain the operation.

Consequently, the determination of whether or not programs related to inclusive care contradict the standards of the long-term care insurance system depends on the understanding and stance of entities that guide and supervise matters related to long-term care insurance. In addition, because the standards (for facility and equipment and personnel placement) and the responsible departments differ even between the systems for children and children (persons) with disabilities, applying for the program as a separate program by partitioning a space in the facility will cause less friction in administrative procedures. Toyama-style day care services were designed to resolve such friction. However, providers must have a great deal of energy to make the departments in municipalities understand the standards. Providers who are very busy with routine work and paperwork for inspection and guidance of long-term care insurance services strive to avoid risks of incurring new paperwork as well as guidance and supervision by adopting an inclusive care program. Furthermore, they fear the possibility that the reality of inclusive care in which diverse people share a space might be judged as being in contravention of the designation requirements, including the area standards in the field of long-term care insurance and the standards for personnel placement. Such "anxiety" gives many providers second thoughts about it.

Of course, if they do not need public subsidies for the programs for children and for children (persons) with disabilities, then practical and mental friction will not be generated. However, for providers who stabilize their operations by providing long-term care insurance services, providing inclusive care services is not reflected in fees. Little economic incentive exists to maintain and start independent programs. Moreover, as earlier research has indicated, it is not easy to practice inclusive care in terms of "the ease" for persons engaged in care and "safety". Detailed care that large-scale facilities cannot provide is worth doing: Simultaneously, it imposes a burden.

With regard to programs provided for people other than the elderly in inclusive care, such as temporary childcare, the quantity of demand is unstable solely by holiday childcare and short-time or short-term childcare before returning to work. Consequently, it is a service that has no choice other than to provide after operational stability is secured by other programs. In inclusive-style programs that have developed based on the practice in old folks' homes, support for the demand became possible for the following reasons. Particularly after long-term care insurance started, providers stabilize the operation with the long-term care insurance reimbursement for elderly people day care services as the base. On that basis, they are able to anticipate the "cooperation" of users with mild symptoms who require light levels of care to a certain extent. However, for the providers examined for purposes of this paper, future prospects for the continuity of the independent programs remain unclear because of the decrease of long-term care insurance reimbursement for users who require light levels of care.

As described, at the field level challenges remain that hinder wider realization of the idea of inclusive care in reality that secures a place to stay while diverse users mutually interact in the same space. The challenges result from fixed "vertical sectioning" in administrative procedures and restrictions in training for persons engaged in care that has been provided according to the subjects. At the same time, the possibility exists that systems and policies to stabilize the operational foundation of programs that started as grassroots efforts make it rather difficult to realize the initial idea.

These points are consistent with the following findings and suggestions. In a survey conducted in 1999, when long-term care insurance has not started yet, the percentage of providing "two or more services" or "services for people in addition to the elderly" is likely to be higher in old folks' homes and group homes that do not receive public subsidies from local governments. "Receiving public subsidies might rather prevent providers from offering services tailored to users' needs" (Hirano 2000: 27). Ohara et al. (2018) report the following. Using the welfare systems divided according to users might result in a program across the subjects of elderly people, children, and persons with disabilities. However, a condition will be attached as a requirement for use that users must be "persons in need of care." Consequently, the function as "a place to stay where everyone can gather" will be restricted. Furthermore, extra energy to strengthen community support will diminish because attention tends to emphasize individual users' use of the system.

The problems described above might be resolved to a certain degree by the spread of government's "inclusive services" introduced in 2018. For example, the following difficulties are expected to be resolved by clarifying responsible departments and sharing the idea through the national program: responsible departments are unclear. Some differences exist in understanding of inclusive care between departments. Moreover, providers bear a great burden of troublesome administrative procedures. However, providers who address inclusive care show a wait-and-see attitude toward the shift to government's "inclusive services" at this time. Results of this study suggest that the system should be well informed and suggest that guidance and supervision should be examined to avoid degradation of practices of inclusive care to date. In terms of difficulties at the field level in caring for widely diverse people in the same space, we insist that more intangible support should be given, including training and management of persons capable of dealing with inclusive care.

References

Executive Committee for National Research Exchange Forum '99 for Old Folks' Homes and Group Homes for Elderly Persons with Dementia (eds) (1999) Report and proposal. Executive Committee for National Research Exchange Forum '99 for Old Folks' Homes and Group Homes for Elderly Persons with Dementia, Sendai (in Japanese)

Hayashitani H, Honjo M (2012) Pregnant women's ability to request support: Factor analysis of focus group interviews. Acad Pap Sonoda Women's Univ 46(1):69–87 (in Japanese)

Health and Welfare Department of Toyama Prefecture (2013) Community inclusion in Toyama: a 20-year history and future of Toyama-style day care services. Toyama Prefecture, Toyama (in Japanese)

Hirano T (2005) Activities and support for inclusive care: From a survey of Toyama-style "Konoyubi Tomare." CLC (Community Life Support Center), Tokyo (in Japanese)

Hirano T (2015) Process of inclusive care bases becoming a policy and future political challenges. Jpn Intercult Acad Municipal 86(1):17–22 (in Japanese)
Hirano T (eds) (2000) Current situation and support for old folks' homes and group homes: Exploring a way small-scale care should be from a national survey. CLC (Community Life Support Center), Tokyo (in Japanese)
Hirano T, Okuda Y (2005) Developmental stages of old folks' homes and group homes. In: Takurosho and group-home network Study Group on Small-scale Multifunctional Homes (eds) Takurosho and group-home white paper 2005. CLC (Community Life Support Center), Tokyo (in Japanese)
Hiroi Y (2003) Care studies: toward care crossing the borders. Igaku Shoin, Tokyo (in Japanese)
Ikeda M (2010) Reasons old folks' homes do not choose the entry for private elderly care home. TOSHI MONDAI (Municipal Problems) 101:64–71 (in Japanese)
Kato N (2014) Development and problems of living together-type care in Saga prefecture: From interviews of "Saga prefecture's community living together station." Res J Care Welfare 21(1):16–26 (in Japanese with English abstract)
Kitamura A (2003) Effort of intergenerational exchanges in facilities combined childcare and elderly care: Possibility of inclusive care for children and elderly people in welfare society. Life Design Report, Dai-ichi Life Research Institute Inc 2003 August. 4–15 (in Japanese)
Kitamura A (2005) Effort of intergenerational exchanges in facilities combined childcare and elderly care (2): emphasizing cases combining daycare service facilities and nursery schools. Life Design Report, Dai-ichi Life Research Institute inc 2005 January 4–15 (in Japanese)
Kobayashi H (2013) Effort in the special zone for promotion of coexistence welfare society in Toyama. Soc Secur Rev 2526:14–23 (in Japanese)
Kukimoto M (2019) Current situation and challenges in diffusion of "inclusive care": A case of the "Community-based Inclusive Station Program" in Saga Prefecture. Bull Grad Sch Soc Serv Admin Oita Univ 11:15–32 (in Japanese)
Mitsubishi Research and Consulting (2018) 2017 Subsidies to Promote Healthcare programs for the Elderly, Health Promotion Programs for the Elderly: Programs to raise awareness and extend inclusive services further. Mitsubishi Research and Consulting, Tokyo (in Japanese)
Murayama Y, Takeushi R, Yamaguchi A, Yamagami T, Kaneda T, Tago M, Fujiwara Y (2017) Possibilities and problems of the intergenerational exchange in complex facility for children and elderlies. Jpnese J Gerontol 38:427–436 (in Japanese with English abstract)
Nishiyama Y (2016) The implication of the symbiotic welfare service in the community welfare policy by prefectural governments. Ann Public Policy Stud (Hokkaido University Public Policy School) 10(1):187–212 (in Japanese with English abstract)
Ohara M, Sugioka N, Hatakeyama A (2018) Financial analysis of an adult day care home (Takurosho) business. Hokusei Rev Sch Econ 74(1):47–63 (in Japanese with English abstract)
Soman K (2009) A large family in the caring community: Looking back on fourteen years of our day care service in Toyama. Jpnese J Gerontol 23:379–383 (in Japanese)
Tanaka N, Asakawa S, Adachi K (2003) Frontier of caregiving NPOs. Minerva Shobo, Kyoto (in Japanese)

Formation of Comprehensive Community Welfare Bases in Urban Areas

Ryo Koizumi, Teruo Hatakeyama, and Hitoshi Miyazawa

Abstract In this chapter, we examine the three local governments of Nabari City (Mie Prefecture), Toyonaka City (Osaka Prefecture), and Yokohama City (Kanagawa Prefecture) as examples of the formation of comprehensive community welfare bases in urban areas. No program provided by the local governments limited eligibility to elderly people, but applied the benefits to widely diverse people, including persons with disabilities and parenting households. Yokohama City started addressing community welfare as early as the 1970s. This effort is regarded as laying the groundwork over which the current system has spread into the region. It seems possible to uncover hidden needs by linking the area of responsibility to the sphere of daily life, making use of connections among residents in the local community. However, it seems impossible that the existing bases single-handedly deal with rapid population aging resulting from the imbalance in age structure of the regional population. Such a situation is observed in areas with large apartment complexes in the suburbs of Yokohama City. Cooperation will be needed with related NPO corporations and medical institutions located in surrounding areas. Under the circumstances, comprehensive community welfare bases taking on challenges in the community in a one-stop manner will become increasingly important in the future as the first consultation service for residents.

Keywords Community-based inclusive society · Comprehensive community welfare base · Large urban area · Long-term care insurance system · Parenting support

R. Koizumi (✉)
Kanagawa University, Yokohama, Japan
e-mail: rkoizumi@kanagawa-u.ac.jp

T. Hatakeyama
Naruto University of Education, Naruto, Japan

H. Miyazawa
Ochanomizu University, Tokyo, Japan

H. Miyazawa and T. Hatakeyama (eds.), *Community-Based Integrated Care and the Inclusive Society*, International Perspectives in Geography 12,
https://doi.org/10.1007/978-981-33-4473-0_15

1 Introduction

The community-based integrated care system brought up before the long-term care insurance was introduced as a medical-welfare cooperative model to support elderly people's life in the community after discharged from the hospital, using the hospital as the starting point (Wake 2018). It was redefined in the "Act on Promotion of Reforms of Establishment of Sustainable Social Security Systems" established in 2013 as a comprehensive system that includes an array of support such as housing, social interaction, preventive care, and livelihood support. Furthermore, in 2017, the "Act on Partial Revision of Long-term Care Insurance Act, etc. to Strengthen Community-based Integrated Care Systems" was promulgated, which was aimed at strengthening community-based integrated care (see Chapter "Welfare Regime in Japan and Recent Social Security Reform").

Wake (2018) mentions the increase in the following cases as the background behind which such a community-based integrated care system is required: cases in which challenges to households have become complex; cases in which adequate support is not provided by any legal system or institution in the space between systems; and cases in which necessary support and information are not delivered because of being isolated or excluded from the community. Traditional consultation support systems have been established by registration upon application and professionalism, dividing the subjects into specialized categories of old age, disability, child, and so on. However, the limitations of the systems have been pointed out in recent years because of an aging population with a declining birthrate, the increase of nuclear families, and the weakening functions of mutual aid provided by communities and families.

In light of the circumstances, the idea of building a comprehensive support system in the community to help realize a "community-based inclusive society" began to be envisioned. The system is designed to provide support not only for elderly people, but for all people living in the community while mustering various resources of the community, not limited to legal services, such as the long-term care insurance system, and securing and developing diverse leaders (Wake 2018).

A key point in such a comprehensive support system is to hold consultations at a single counter to provide support for problem-solving. As described earlier, conventional social welfare systems in Japan have a historical background: they have divided the subjects into specialized categories such as old age, disability, and children. Although this practice heightens specialization in each area, a difficulty is apparent by which the response to households with complex challenges is divided. Moreover, not all households in need can receive support as a result of sectionalism or a lack of cooperation (Wake 2018). In view of the above, the formation of "comprehensive community welfare bases" to realize a community-based inclusive society is characterized by the fact that many professions and many institutions work together to organize a support system. One area in which challenges related to cooperation are most readily apparent is urban areas. For this reason, comprehensive community welfare bases have been formed in urban areas.

Fig. 1 Locations of study cities

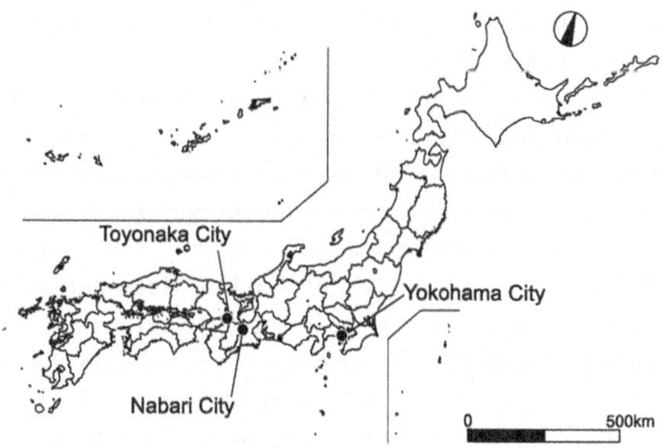

Taking up "comprehensive community welfare bases" in several advanced urban local governments, this chapter clarifies the respective characteristics. The local governments studied are Nabari City in Mie Prefecture; Toyonaka City in Osaka Prefecture; and Yokohama City in Kanagawa Prefecture (Fig. 1). All cities have abundant achievements in terms of the operation of comprehensive community welfare bases. The next section outlines the comprehensive community welfare bases in the three local governments and reveals the political and regional backgrounds of the service commencement and the characteristics of the services in respective cities. On that basis, Sect. 3 and the following sections specifically examine Yokohama City, which operationalized comprehensive community welfare bases going ahead among the three examples. They clarify the actual state of various activities in community welfare bases and assess the effectiveness and potential of the creation of comprehensive community welfare bases.

2 Examples of Local Governments Addressing Formation of Comprehensive Community Welfare Bases

2.1 *Case of Nabari City, Mie Prefecture*

Nabari City in Mie Prefecture has established a number of "Machi no Hokenshitsu (public health room in the community)" as a comprehensive community welfare base. Nabari City is a local government located about 60 km east of Osaka City with a population of 78,896 as of 1 January 2019 (Basic Resident Register in Japan). The city has developed as the suburbs of the Keihanshin metropolitan area. The aging population increased and the young population decreased considerably after the late 1990s. The population started declining after 2000 influenced by people returning to the central cities of the Keihanshin metropolitan area. Therefore, key challenges in Nabari City are to secure the productive age group population and to promote community welfare.

As its challenges, Nabari City has addressed the promotion of intra-regional decentralization and the formation of comprehensive community welfare bases. In terms of the former, the city has delegated part of the authority and funds of administration to community development organizations that are formed, covering roughly an elementary school district since 2003. The funds at their disposal, called "Yume-dukuri Chiho Kofukin (Dream-Making Regional Grant)," are granted for community development organizations to support their activities. At present, community development organizations expand town development activities led by residents in 15 districts in the city. "Machi no Hokenshitsu" have been established in each of the 15 districts as a comprehensive community welfare base. "Machi no Hokenshitsu" was established based on the Nabari City's "community welfare plan" using existing facilities such as community centers and citizen centers for three years since 2005. "Machi no Hokenshitsu" became a branch of comprehensive community support centers in 2006. Since 2014, "Machi no Hokenshitsu" has been a satellite of comprehensive support centers for childrearing generation.

Each "Machi no Hokenshitsu" has a public health nurse or a nurse and 2–3 professions, such as a certified social worker, a certified care worker, or a long-term care support specialist, as the city's contract employees. The staff members provide consultation support from pregnancy to childbirth and childcare on a continuous basis by receiving specialized training. They are also welfare counselors for mothers and children called "children's partners." The primary role of "Machi no Hokenshitsu" is to serve as a familiar general consultation counter to provide consultation on nursing care and parenting as well as advocacy. They cooperate with commissioned welfare and child welfare volunteers, child support centers, community development organizations, and community activity groups, in addition to comprehensive community support centers and comprehensive support centers for childrearing generation, to provide support to connect clients with those professions and services. The secondary role is to investigate eligibility for long-term care insurance and to apply for welfare services on behalf of users as well as to give information about check-ups and to hold health classes as a base for community welfare. Community centers and citizen centers with attached "Machi no Hokenshitsu" hold various classes and courses. They are bases for the activities of community development organizations. At the centers, community activity groups hold various salon programs to promote health and preventive care and to support parenting. Salon programs in Nabari City are conducted using existing facilities such as vacant shops and houses. Support for the implementation of those salon programs is also an important task for "Machi no Hokenshitsu." As described, "Machi no Hokenshitsu" in Nabari City is a comprehensive community welfare base to promote the health and welfare of local residents of all generations, from children to elderly people, together with community development.

2.2 *Case of Toyonaka City, Osaka Prefecture*

Toyonaka City, Osaka Prefecture has established "Fukushi Nandemo Sodan Madoguchi (consultation counter for everything in welfare)" in each elementary school district as a comprehensive community welfare base. Toyonaka City is a local government located northward of Osaka City with a population of 406,593 as of 1 January 2019 (Basic Resident Register in Japan). The city has Japan's first large-scale new town, Senri New Town. It has been about 50 years since the towns were formed in the area. For this reason, growing concerns arise about aging of the population residing there. The Great Hanshin-Awaji Earthquake in 1995 caused widespread damage to the southern part of Toyonaka City and forced numerous disaster victims to move to temporary housing in the northern part of the city. Although it was originally an area in which residents had a strong sense of mutual aid, the earthquake disaster made it much stronger. The residents' mutual cooperation forms the foundation of community welfare network activities. The networks provide support not only for elderly people, but also for widely diverse people and difficulties, including persons with disabilities, parenting, hoarding houses, and social withdrawal. The support system has been developed in the way that resident volunteers take responsibilities, community social workers of social welfare councils serve as coordinators, and concerned departments in the city government back them up. The base for the system is "Fukushi Nandemo Sodan Madoguchi."

In all, 36 "Fukushi Nandemo Sodan Madoguchi" have been established in 38 elementary school districts in Toyonaka City. Commissioned welfare volunteers and welfare committee members in the school districts provide advice through various consultations from residents (about introduction and cooperation of welfare systems, services and specialized institutions). The program is one effort based on the "community welfare plan" in Toyonaka City and the "community welfare activity plan" in Toyonaka City Council of Social Welfare. Operation of the counters was started in January 2005 when the social welfare council was commissioned by the city. The counters have been run to serve as a base for community activities where local residents can gather, contact, and exchange in a friendly manner.

Consultations from residents are informed to responsible departments in the city government or comprehensive community support centers in terms of the part that can be dealt with by the system. The part which cannot be dealt with by the system is undertaken by the social welfare council (Tanaka 2015). In the social welfare council, community social workers consider necessary support and concrete methods to solve problems in cooperation with resident volunteers and local service providers.

Toyonaka City has "Toyonaka City Life Safety Net" as a blueprint to collect and examine difficulties in daily life and to make efforts for the solution. There had been the "general coordination committee for life safety nets" in which representatives of groups concerned with children, elderly people, and persons with disabilities gather to address challenges in the region. The committee has served as a base for community welfare plans.

Toyonaka City has established a community welfare network committee in each of the seven spheres of daily life. The committee has a subcommittee on disabilities. It has also established subcommittees on old age and on children in each sphere of daily life to seek cooperation between the fields. People from many different fields participate; they join community activity groups, welfare facilities, administration, commissioned welfare volunteers, and social welfare councils to discuss regional challenges. The subcommittee on old age described above is hosted by comprehensive community support centers as equivalent to the community care conference. The conference organized in each sphere of daily life takes up and discusses even those challenges that cannot be solved in "Fukushi Nandemo Sodan Madoguchi."

As described, Toyonaka City has established "Fukushi Nandemo Sodan Madoguchi" in elementary school districts as the base closest to residents. It summarizes the collected consultations in the community welfare network committee in each sphere of daily life and further in the life safety net meeting and Toyonaka City liaison meeting on community welfare that covers the entire city. Based on the system, a circular system has been built by which policies are reflected between multilayered scales and by which feedback is given through support.

2.3 *Case of Yokohama City, Kanagawa Prefecture*

Yokohama City, Kanagawa Prefecture has established the "community care plaza" as a comprehensive community welfare base. Yokohama City is a local government with a population of 3,745,796 (1 January 2019, Basic Resident Register in Japan), which is the largest population among the cities in Japan. It also has a vast area of 437.4 km^2. Therefore, the city divides its municipal area into 18 wards as an ordinance-designated city. It provides administrative services according to regional features. Yokohama City has developed as a major city and the suburbs of the Tokyo metropolitan area. Although the total population is growing continuously, the natural increase and decrease of the population started decreasing in 2016. The city includes suburbs of the metropolitan area. Therefore, the proportion of the first baby boomers in the population is high. The baby boomers will be 75 years or older in 2025. Consequently, key challenges in Yokohama City are community development and the promotion of community welfare to respond to rapid population aging.

Yokohama City originally addressed community welfare from an early time. Under a reformist city administration, the city established "Yokohama City Fukushi No Fudo-dukuri Suishin Iinkai (Promotion Committee for Welfare Climate Creation)" in 1974 and has developed community welfare across the city using regional features in collaboration and cooperation with the social welfare council in each ward. During the accumulation, the basic guidelines for community-based care systems were formulated in 1991. The community-based care system in Yokohama City is defined as "a system to provide community care services combining health care, medical care, and welfare for everyone, including elderly people, children and persons with disabilities, persons suffering from intractable diseases, and persons with mental disabilities who need support at home to continue to live independent life with a sense of security in the community in which they have lived for a long time" (Yokohama City Health and Welfare Bureau 2017). The goals are the following three: "comprehensive understanding of needs and integrated provision of services," "citizens' participation and network formation," and "improvement of services."

The "community care plaza," Yokohama City's original facility, has been established as a comprehensive community welfare base in the community-based care system. The first "community care plaza" was opened in 1991 as a "home care support service center." In 1994, the name was changed to "community care plaza." "Community care plazas" have been established in 138 locations as of May 2018 under the development goal, one in each junior high school district. The operation is commissioned to 57 social welfare corporations under the designated manager system.[1] Each community care plaza has five professional full-time staff members, each of which is engaged in specialized work. Their roles are, for example, regional exchange activity coordinator, public health nurse, certified social worker, chief care manager, and director.

Article 1 of the city ordinance related to the establishment of "community care plaza" prescribes that "For every citizen in the City to be able to live healthy life with a sense of security in their communities, the City establishes community care plazas to provide welfare services and health care services comprehensively in their familiar places, as well as to promote welfare activities and health care activities in their communities." The community care plaza has the following four major roles based on the purpose: (1) to support local residents' welfare activities and health care activities and provide facilities for the interaction of the activities; (2) to hold workshops and courses related to welfare and health care; (3) to provide consultation and information about welfare, health care, etc.; and (4) to coordinate the provision of welfare services and health care services.

The community care plaza has more than one function. The first function is as the comprehensive community support center. The comprehensive community support center is a facility based on the long-term care insurance system. Similarly, it provides services related to adult day care, care management, and preventive care support in long-term care insurance (Social Security Institute ed. 2010). As described, if it is based on the roles shown in the establishment ordinance, then people eligible to use community care plazas are not limited to elderly people. However, Yokohama City specifically responds to rapidly progressing population aging. Consequently, the community care plaza places importance on the roles mainly for elderly people. It is a base for nursing care and preventive care for elderly people for the time being. However, the community care plaza also has a function as a place to conduct community activities and to have exchanges. It has been a base for independent programs in the specialty fields of corporations commissioned with the operations of community care plazas and as a base for community welfare activities by residents.

[1] The commissioned corporations are far fewer than the number of facilities because, in many cases, one corporation is commissioned with the operation of more than one community care plazas.

The contents and eligible people range widely. On this point, it is a comprehensive community welfare base that is not limited to elderly people.

Hereinafter, we specifically examine details of the characteristics of community care plazas in Yokohama City. Based on the example, the effectiveness and potential of the creation of comprehensive community welfare bases are considered.

3 "Community Care Plaza" as Comprehensive Community Welfare Base in Yokohama City

3.1 Outline of Community Care Plazas

As described above, the number of community care plazas in Yokohama City was 138 as of 2018. We examine details at the years of opening, the commissioned corporations, whether or not they have attached facilities, and the features of the areas of responsibility as characteristics of the community care plazas.

Many community care plazas opened during the late 1990s and the early 2000s (Fig. 2). We divided the years after 1991 into five-year periods and added up the number of facilities. Results showed that 20 facilities opened during 1991–1995 (Period I), 49 facilities in 1996–2000 (Period II), 29 facilities in 2001–2005 (Period III), 21 facilities in 2006–2010 (Period IV), and 15 facilities in 2011 and after (Period V): more than half of the present facilities opened by 2000. More than 20 years have passed in many communities since their community care plazas opened, which suggests that the community care plazas have become established in the communities. To examine a regional balance in the opening of facilities, we mapped the time of opening for each 5-year period described above (Fig. 3). The map clarifies that community care plazas were established across the city in an unbiased manner.

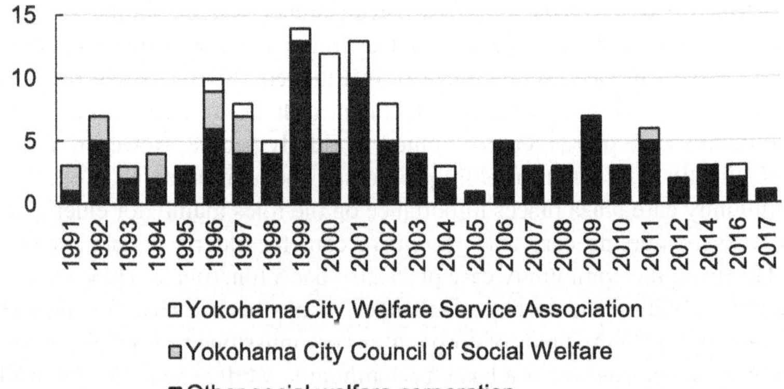

Fig. 2 Years of opening of community care plazas (Created based on the homepage of Yokohama City)

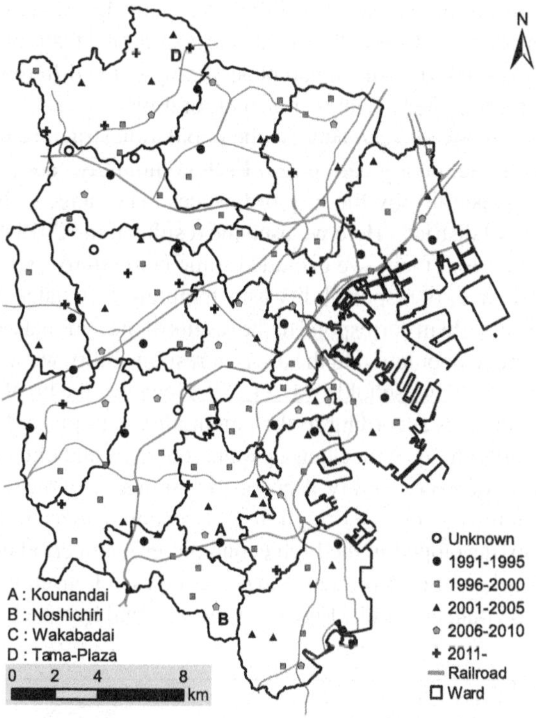

Fig. 3 Time of opening of community care plazas (Created based on the homepage of Yokohama City)

The operation of facilities is commissioned to 57 social welfare corporations under the designated manager system. Among them, two social welfare corporations, Yokohama-City Welfare Service Association and Yokohama City Council of Social Welfare, run a total of 34 community care plazas. The former conducted mainly home-help programs from the beginning. Now it is a social welfare corporation that has established more than 55 elderly care centers in Yokohama City and which provides wide-ranging services. Each of the two social welfare corporations has a history of more than 40 years from its foundation and has developed activities rooted in the community. In other commissioned corporations, as many as 40 corporations specialize in the field of elderly care. However, some corporations specialize in the fields of welfare for persons with disabilities, child welfare, and medical care. To be more specific, 21 corporations deal with welfare for persons with disabilities, 14 corporations handle child welfare (mainly nursery centers), and 6 corporations have a medical care division. Twenty-three corporations provide services in more than one field.

A majority of the community care plazas, 72 facilities, had attached facilities, 21 of which were city district centers or community centers. This fact implies that the community care plaza functions as a base for the district. The number of facilities with an attached intensive nursing home for elderly people and the number of facilities with an attached vocational facility or residential facility for persons with disabilities are 9 facilities, respectively. As described above, this fact further emphasizes the possibility that the operational features of respective community care plazas will become apparent according to the difference in the specialization of the commissioned social welfare corporations.

Finally, one can address the population structure in the area of responsibility of each community care plaza. Each community care plaza is specified in its area of responsibility by the city ordinances. The range is based, overall, on junior high school districts. Here we used the results of the 2015 National Census as population data. From them, we excluded some community care plazas from analysis because their areas of responsibility were minutely designated. For that reason, it was difficult to make them correspond to the results of the national census. Results revealed that the average population in the areas of responsibility was 25,533 and the average number of general households was 11,229. However, the breakdown by age showed regional differences according to the community care plaza. Figure 4 shows the percentage of households with elderly people (65 years and older) for general households in the areas of responsibility. A cursory examination of the results makes it readily apparent that the percentage is high in the residential areas in the southern part of Yokohama City developed in the high economic growth period and is low in the northern part. Similarly, Fig. 5 presents the percentages of households with children aged six years and younger. Unlike Fig. 4, it shows a trend by which the percentage is higher in the

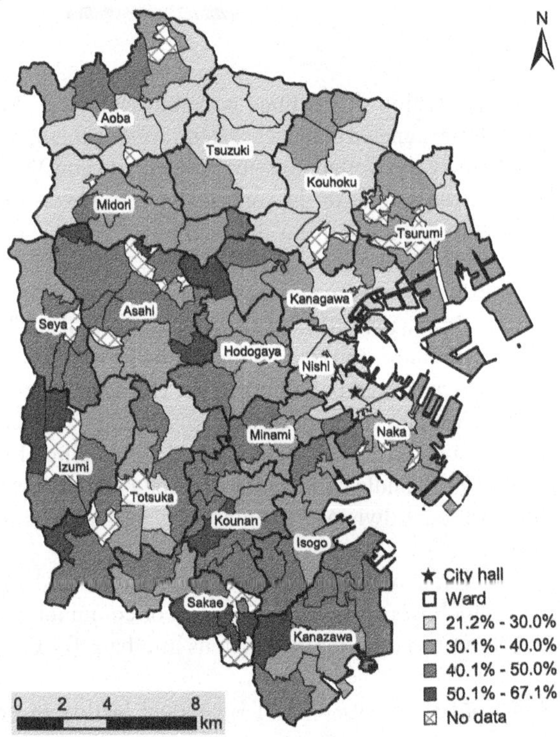

Fig. 4 Percentages of households with elderly people (65 years and older) in the jurisdictional areas of community care plazas (Created based on 2015 Population Census of Japan)

Fig. 5 Percentages of households with children aged six years and younger in the jurisdictional areas of community care plazas (Created based on 2015 Population Census of Japan)

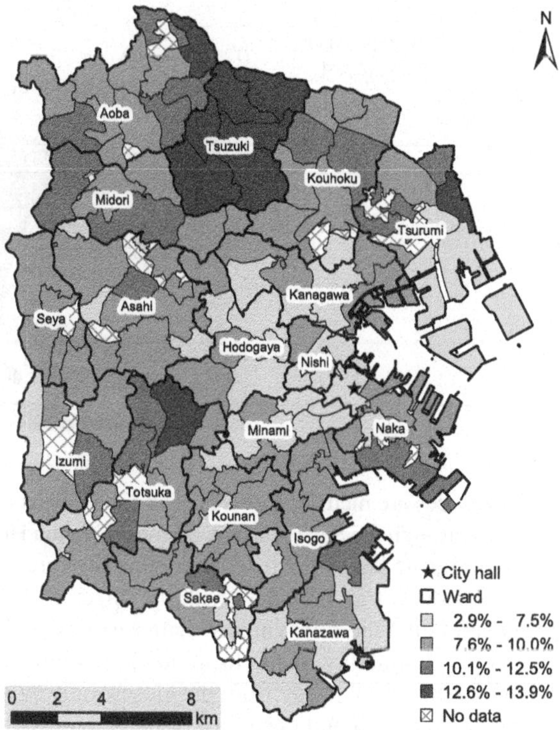

new town areas in the northern part of the city developed after the 1980s. However, there are areas with a value of 10% or more in addition to areas in the northern part, which indicates that parenting households have moved into large-scale apartment buildings constructed on the former sites of factories. In this way, the population structure differs greatly depending on the area even within Yokohama City. This difference suggests that the roles required for community care plazas as a base for welfare in the community also differ.

3.2 *Example of Community Care Plazas: Konandai Community Care Plaza*

To explain community care plazas in Yokohama City overviewed above with the use of a specific example, we present the Konandai community care plaza (Fig. 6). The community care plaza, located in the southern part of Yokohama City, opened on November 1, 1994 (A in Fig. 3). Operation of the facility is commissioned to the Kanagawa branch of Social Welfare Organization Saiseikai Imperial Gift Foundation. The corporation, which is the largest social welfare corporation in Japan, has developed health, medical, and welfare services in 40 prefectures across the country. Its

Fig. 6 Konandai community care plaza (Photograph taken by Ryo Koizumi in August 2018)

main services are medical services. The Kanagawa branch mainly provides medical services through the Saiseikai Yokohamashi Nanbu Hospital. The corporation serves as the designated manager for four community care plazas, including the Konandai.

We conducted an interview with the director of the Konandai community care plaza and found the following as characteristics of the community care plaza. The first is the desirability of the location. The community care plaza is located only a few minutes' walk away from the closest railway station. A nearby bus terminal supports the arrival and departure of many buses along their routes. Furthermore, the street from the railway station to the community care plaza has wide sidewalks and no slopes. This construction implies that it might not be so difficult for elderly people and persons with disabilities to visit the community care plaza.

The Konandai community care plaza is located next to the Saiseikai Yokohamashi Nanbu Hospital, which brings various benefits to operation of the community care plaza. It is apparent in the fact that doctors of the hospital and persons engaged in medical care are involved in independent programs held at the community care plaza. An example of the independent programs held there are lectures given by an ophthalmologist on the appropriate time of operation for cataracts that occur mostly in elderly people and by a physiotherapist on efforts to increase the health life-span. Although each program has a high seating capacity of 50, all seats are occupied in some programs. Reflecting on the good accessibility, participants in the programs are not only from Konan Ward, but also from neighboring wards.

The number of groups registered for rent of the Konandai community care plaza reaches 103, which indicates that the community care plaza has been used actively. The groups' activities are divisible into those for elderly people, for persons with disabilities, for patients of the nearby hospital, for children and their parents, for members of self-government associations, for welfare-related professionals, and for other members of the general public. The registered groups' activities for elderly people are mainly for the prevention of social withdrawal and dementia through the encouragement of exchanges and for the maintenance of health. Activities for persons with disabilities consist mainly of programs providing a place for them such as music therapy, dining together, and playing games. Mental care programs are provided for parents of persons with disabilities. For patients of the nearby hospital, exchanges are made between patients suffering from the same diseases, such as kidney diseases and Parkinson's disease, and their families. As activities for children and their parents, Boy Scout activities and postpartum physical and mental care are provided in addition to parenting support, such as picture book reading and play parks. Activities for members of self-government associations include holding salons and meal gatherings in apartment complex self-government associations and preparing boxed meals for elderly people living alone. Welfare-related professionals, such as care managers, and those who address consultations and activities for adult guardianship independently organize social gatherings and study meetings. Furthermore, widely diverse activities are offered to general residents, including yoga, chorus, and flower arrangement. The frequencies of the groups' activities differ, ranging from about 90 min once a month to about five hours once a week, depending on the type of the activity and the group characteristics.

The director described the features of residents living in the vicinity as a factor for why various groups actively use the Konandai community care plaza. He said that self-government association activities were originally active in the Konandai district and that numerous independent efforts had been observed. The director thinks that the community care plaza is being used well because some activities use it as a place for their activities. In addition, the background that it is used actively is considered that the designated provider displays its specialization in the holding and planning of independent programs in community activities and exchanges.

4 Activities in Community Care Plazas for Realization of Community-Based Inclusive Society

Compared to comprehensive community welfare bases in other cities, distinctive features of community care plazas in Yokohama City are explained below. As described above, while mainly functioning as welfare for elderly people, including comprehensive community support centers, the community care plazas also have established exchange divisions. This division enables them to form a comprehensive community welfare base through various independent programs provided there and community welfare activities by residents. That base is not limited to elderly people. Activities such as health promotion and preventive care classes and parenting circles are conducted using a multipurpose hall in the facilities. In some cases, local residents started up a group and undertook service activities that are necessary for the community. Hereinafter, we introduce some examples of efforts in the community activity exchange division in community care plazas. On that basis, we examine the roles that community care plazas play in forming a community-based inclusive society.

4.1 Support and Promotion of Health for the Elderly

In the population composition in the area of responsibility of each community care plaza presented in Fig. 4, the percentage of households with elderly people was the highest in the Noshichiri community care plaza, accounting for 67.1% (B in Fig. 3). The Noshichiri community care plaza is located in Sakae Ward, the southernmost part of Yokohama City. It is in charge of the suburb residential area created in the first half of the 1970s. Operation of the community care plaza is commissioned to Social Welfare Corporation Shinko Fukushikai. The corporation has a long history of social welfare activities. For example, it established an unauthorized child care center in 1976 and the first private elderly care home in 1986. The social welfare corporation also provides services such as group homes and employment support for persons with disabilities.

The community care plaza offers numerous programs, among which are health promotion and volunteer courses. For elderly people, a 90 min health promotion exercise program has been conducted once a week to improve posture. In addition, a 60 min exercise program to maintain leg strength is offered once a week. To cope with population aging in the community, it is necessary to train residents who support elderly people. From such a perspective, the community care plaza runs courses for the general public. The courses address how elderly people or dementia patients should be treated, how personal information should be handled, and the basics of care technology. The community care plaza also runs an "end of life planning" course that deals with how care facilities should be chosen, along with contents related to wills and inheritance. In addition to the courses, it holds a handicraft class that is intended to prevent dementia by moving of the fingers and a health mahjong class that has drawn attention as a means of communication.

The content of consultations that comprehensive community support centers received was analyzed in areas where population aging has progressed; there were many consultations on elderly people's meals. As a result, the community care plaza holds meal gatherings and implements a meal delivery service that delivers boxed meals to elderly people living alone in cooperation with the community activity exchange division. The meal delivery service checks on the safety of elderly people simultaneously. In this way, establishing the community activity exchange division attached to comprehensive community support centers makes it easier to connect consultations with activities.

However, because the increase of elderly people in need of care and support is becoming severe, a limit exists to the role that community care plazas can play independently. For example, in the Wakabadai district in Asahi Ward, a community welfare base was established by an NPO corporation in addition to a community care plaza (C in Fig. 3). There is Wakabadai Danchi in the district, which is a large-scale apartment complex into which people started moving in 1978. The population of the apartment complex was more than 20,000 people at its peak in 1993. However, because the age structure of residents was heavily imbalanced, the rate of population aging has been rising rapidly in recent years. The percentage of households with elderly people is 64.4% in the area of responsibility of the Wakabadai community care plaza. With population aging such as this, it is increasingly difficult for community care plazas to single-handedly accommodate the needs of people in need of care and support in their areas of responsibility.[2] For that reason, the Wakabadai NPO opened the "Wakabadai Community Salon Himawari (sunflower)" in the district in the spring of 2016. The NPO, which was established in 2009, has opened and run a space for residents to interact, along with support facilities for parenting, and community workshops for persons with disabilities in the Wakabadai apartment complex. The background against which such an NPO opened new facilities is that an external agency pointed out the insufficiency of the in-home medical care system in the Wakabadai apartment complex. In response to the indication, the NPO and the management association of the apartment complex held discussions. As a result of the discussions, they were to provide consultation related to residents' medical and nursing care to connect it to appropriate institutions, as well as to make care plans involving neighboring medical institutions (Kanagawa Prefectural Housing Supply Corporation 2016). This event indicates a limit to the role that community care plazas can play independently. Community care plazas are required to pursue cooperation with other bases for community welfare, as in this example.

4.2 *Support for Parenting*

In the population composition of the areas of responsibility of community care plazas presented in Fig. 5, one district with a high percentage of parenting households is the Tama-Plaza district, Aoba Ward, Yokohama City (D in Fig. 3). A suburban residential area developed along the Tokyu Den-en-toshi Line and many parenting households exist in the district. The percentage of households with children aged six years and younger is 12.5%.

The "Tama-Plaza community care plaza" is located in the building directly linked to Tama-Plaza Station. Many of regional exchange activities in the community care plaza are related to parenting support, reflecting the features of the regional population described above (Table 1). The Tama-Plaza community care plaza opened on March 1, 2013, with its operation is commissioned to Social Welfare Corporation Ryokuseikai. The corporation is an organization associated with Ryokuseikai Medical Association, which runs the Yokohama General Hospital.

As parenting support in the Tama-Plaza community care plaza, for example, there is a consultation salon for mothers raising two or more children and a program for parents and children to enjoy rythmique[3] together. Study meetings are also held for parents of children with developmental differentiation and their supporters. The programs are held about once a month. The seating capacity is about 15. Some of

---

[2]Asahi Shimbun morning edition (Yokohama version) of June 18, 2014.

[3]The method of music education that Swiss music educator d'Arc Rose invented. A method to let child grasp the rhythm by physical exercise.

Table 1 Regional exchange activities related to parenting support at the Tama-Plaza community care plaza (October 2018)

Content	Frequency	Time	Subject
Family play experiencing English	Once a month	10:30–11:20	Toddlers—pre-kindergarten children and their guardians
Aoba toy square	Once a month	10:00–15:00	Children with developmental concerns and their caregivers
Chat salon for moms with two kids	Once a month	10:00–11:00	Moms raising two or more children (including those expecting a second child)
Study group on support for children with developmental differentiation	Once a month	10:00–11:30	Parents of children with developmental differentiation and their supporters
Family rythmique	Once a month	10:30–11:15	Toddlers—preschool children and their caregivers
Story telling	Once a month	14:30–15:30	Preschool children and their caregivers
Parenting backup course for new dads	Once a month	10:00–11:30	Babies able to hold the head up—children younger than two years and new dads
Family playground	Tues., Thurs.	9:30–16:30	Children younger than preschool age and their caregivers

Created based on materials provided by the Tama-Plaza community care plaza

the programs are hosted by commissioned welfare volunteers, not by the community care plaza. One program, the "Utsukushigaoka Kosodate Hiroba (Utsukushigaoka parenting square)" program, is open from 10:00 to 13:00 on the third Wednesday every month. There is no need to apply for the program. Participants are allowed to enter and leave the room freely and to have meals in the venue. It is provided as a place where parents and children can feel free to drop by.

Aoba Ward, Yokohama City, where the community care plaza is located, provides a program called "Donichi-Hiroba (Weekend Play Room)" that Aoba Ward and private enterprises work together under a public-private partnership. This is a program that Aoba Ward invites service providers to work with and which implements events that parents and children can enjoy together on Saturday or Sunday. The program is aimed at providing services that neither administration nor service providers can provide by themselves and uses the public-private partnership system in Yokohama City. The contents of programs are those making use of the characteristics of each cooperative enterprise (Table 2). Although the entry fee to the programs is free, actual expenses might need to be paid in some cases. The Tama-Plaza community care plaza functions as a place to implement the programs. A flyer for recruiting participants says that the venue is convenient because it is linked directly to Tama-Plaza Station. It is readily apparent that its good accessibility is one of its strengths.

Table 2 Programs of "Donichi-Hiroba" (2018)

Name of program	Open	Cooperative provider
Solid foods course/Baby food course	April, July, October, January	Asahi Group Foods
Games with songs and languages in the world	Every other month	LEX Institute
Raise a child to be a book lover	June, July, October, December, February, March	FAMILIE
Ball stair class	Every other month	R Sports Network
Family friendly dietary education	September	Asahi Group Foods
Family exercise class	Every month	MEGALOS Ichigao
Family hula dance	Even month	MAHANA hula studio

Created based on materials provided by the Tama-Plaza community care plaza

4.3 Measures Against Child Poverty

In the "Konandai community care plaza" introduced earlier, NPO corporation "Santa No Ie (Santa's house)" runs "Santa's kitchen (Santa Kodomo Shokudo)." The children's cafeteria has been open on the first Friday evening of each month since May 2016.

The NPO corporation "Santa No Ie" is a nonprofit organization established by Mr. Noboru Yoshida (a resident of Konan Ward, Yokohama City) who has been engaged in volunteer activities not only in Japan but also abroad. Mr. Yoshida, a Christian, has been involved with many volunteer activities, including YMCA. The facility characteristics are well reflected in the momentum that he started the children's cafeteria in the community care plaza and in the acquisition of members who were to be engaged in the operation (as a volunteer activity). Mr. Yoshida was shocked by the severity of "child poverty" in newspaper reports. He decided to run a "children's cafeteria" by himself and sought a venue. The cost of renting a venue needed to be low to continue operations of the children's cafeteria. However, most places proposed for a venue charged a fee for rent; finding a venue was not going smoothly.

On one occasion, meals were provided as a volunteer activity to support persons with disabilities held in the Konandai care plaza in which Mr. Yoshida had been involved. At that time, he noticed that a kitchen was attached to the community care plaza. Mr. Yoshida devised the use of the community care plaza as the venue for a children's cafeteria and consulted with staff members of the community care plaza about his idea of opening a children's cafeteria. After receiving approval of the idea and advice on operations from the staff, he started running a children's cafeteria. Although the cost for renting the venue had been a concern, it came to be provided free of charge because the children's cafeteria program was recognized and positioned as part of welfare services.

The fees for meals provided in the children's cafeteria are free for children younger than junior high school students and 300 yen for junior high school students and

Fig. 7 Meals provided at "Santa's kitchen" (Photograph taken by Ryo Koizumi in August 2018)

older. Although the meal fee for children younger than junior high school students was charged during its first year, the meals were provided without charge from the second year of starting. Objections were raised against dispensing with the meal fee for children from members of the steering committee and the community care plaza, but it was executed by Mr. Yoshida's determination. Actually, he was worried about dispensing with the meal fee for children from an operational perspective. However, by dispensing with the fee, they succeeded in acquiring a subsidy from Yokohama City. The reason for acquisition of the subsidy was said to be that the city appreciated not only that the meal fee for children was free, but also that the children's cafeteria has been operated continuously. The number of meals provided used to be 30 meals or so, which has increased to the current number of about 50 meals (Fig. 7). On the day on which we conducted an interview, parents and children came in one after another before the opening hour of 17 o'clock. The prepared meals were all served before the expected time.

Students of prefectural high schools in the Konandai district participated in the operation on the days of the event as volunteers. The trigger for their participation was posters for recruiting volunteers posted in their schools. As just described, community care plazas function as a place for voluntary activities in the community and make it possible to gather people of different generations.

4.4 Support for Persons with Disabilities

Efforts for persons with disabilities and their families living in the community are also programs provided by community care plazas. The programs are divided broadly into those for persons with disabilities and those for their families. An example of the former is "Youth Class 'Night Cruise'" at the Konandai community care plaza. This program was started based on the idea that persons with mental disabilities working in vocational facilities did not have sufficient opportunities for leisure and interaction. The program is held on Friday night once a month, in which participants have dinner together and hold events. Students of the neighboring nursing school participate in the events as volunteers for cooking and other tasks. Although some variation exists in the number of participants, around 20 people are said to participate. The staff in charge pointed out that the closeness of the facility to the railway station and a bus terminal facilitated their participation in the program after work.

As an example of the latter, there is a course for preparing after parents' death held for parents of children with disabilities at the Aobadai community care plaza, Aoba Ward, Yokohama City. The course deals with, among other points, an explanation from a manager about living life in group homes for persons with disabilities, an explanation from a specialist about the outline and reality of the adult guardianship system and guardianship activities, inheritance and will, and family trust. The seating capacity of explanatory meetings is about 40. However, they say that because of having received too many applications for specific content, they increased the number of sessions for the content. Although held at the Aobadai community care plaza, the course is hosted by a facility attached to the community care plaza called "Aoba Chiiki-katsudo Home 'STEP'." The facility is run by the social welfare corporation "LE PLI," which is the designated manager for the Aobadai community care plaza. The corporation was born in 2017 by merging three social welfare corporations specialized in welfare for persons with disabilities. The three pre-merger corporations are now positioned as operational headquarters directly under the corporation. The designated manager for the Aobadai community care plaza is one corporation. The operational headquarters have run residential facilities and short stay for persons with mental disabilities and community care plazas mainly in the northern part of Yokohama City. At present, they serve as the designated manager for the Aobadai and other community care plazas and run nursery centers.

Nevertheless, residents staged a protest campaign against the construction of a facility run by the operational headquarters, which was a single social welfare corporation at that time. The facility was a combination of three facilities: residential facilities for persons with mental disabilities, short-stay facilities for persons with disabilities living at home, and a community care plaza. Residents opposing the construction obstructed the construction work of the facility. They said that "Yokohama City, which is granting a building permit, has not provided sufficient explanation to local residents." and that "The facility ignores the human rights of persons with disabilities." The lawsuit of the opposing residents vs. Yokohama City and the social welfare corporation over the construction work was not settled for more than three years. The protest campaign continued after the facility was completed. There is not a huge protest going on now. Even so, some newspaper articles reported the circumstances after the opening of the facility as follows: "The community care plaza is well used. Some residents stated that 'It is difficult to visit a facility where a protest took place'."[4] The problem could be the manifestation of NIMBY, by which welfare facilities are regarded as "nuisance facilities" by local residents (Suzuki 2015). Challenges of this kind exist in forming a community-based inclusive society.

5 Conclusions

This chapter took up three local governments as examples of the formation of comprehensive community welfare bases in urban areas. No program provided by a local government limited eligibility to elderly people, but applied to widely diverse people, including persons with disabilities and parenting households. Among the local governments, Yokohama City started addressing community welfare as early as the 1970s, which served as the groundwork for present community care plazas to spread into the region. Some designated managers for community care plazas were corporations that have been active over such a long period of time and which are rooted in the community. They ran community care plazas in cooperation with the community. Furthermore, they are attached to comprehensive community support centers and care facilities facilitated cooperation of various kinds. Specialized services and information were provided, making use of the features and specialization of the designated managers. Cooperation among many professions, such as public health nurses, certified social workers, and long-term care support specialist staff, made it possible to provide comprehensive service, support, and consultation for various challenges faced by people in the community in a one-stop manner. The jurisdictional range of one base was based on the elementary or junior high school districts in each local government. This establishment of appropriate service areas is important to provide services in accordance with different regional features in the city. The school district is often the community sphere, such as neighborhood associations in the district. For this reason, it might be easy for local residents to imagine with which base they should consult. It also might make it possible to uncover hidden needs while making use of connections among residents.

All three local governments addressed in this chapter are those in urban areas. The absolute number of elderly people is large in metropolitan areas (particularly in the suburbs) in Japan into which the first baby boomer generation moved during the post-war high economic growth period. Consequently, a large volume of diverse needs in welfare and life is expected to surface: as demonstrated in the Wakabadai district case described in Sect. 4.1, the absolute volume of needs cannot be dealt with solely by community care plazas. Cooperation will be needed with related NPO corporations and medical institutions located in the surrounding areas. Such a situation will become apparent around 2025 when the first baby boomer generation will be 75 years or older. Under the circumstances, comprehensive community welfare bases taking on challenges in the community in a one-stop manner will become increasingly important in the future as the first consultation service for residents.

[4] Asahi Shimbun morning edition (Kanagawa version) of March 14, 1999.

Those bases must continue programs in a stable manner and over a long period of time. Continuation of the consultation service beyond the boundaries of fields is anticipated in Japan, a super-aging society.

References

Kanagawa Prefectural Housing Supply Corporation (2016) Town development by cooperation between residents and promotion of community-based integrated system model. https://kouseikyoku.mhlw.go.jp/kantoshinetsu/houkatsu/ documents/dainikaikanagawasiryou.pdf. Accessed 1 March 2019 (in Japanese)

Social Security Institute (ed) (2010) Ultimate goal and future prospects of comprehensive community support centers: community care plazas in Yokohama City. Long-term Care Insur Inf 11(1):20–24 (in Japanese)

Suzuki K (2015) "Nuisance facilities" from perspective of NIMBY. Toshi Mondai (Municipal Problems) 106(7):4–11 (in Japanese)

Tanaka S (ed) (2015) A guide to successful community-based integrated systems. Medica shuppan, Osaka (in Japanese)

Tsutsui T (2018) Challenges in collaboration between health care, long-term care and social care in community based integrated care system: From implementation to further development of integrated care. Jpnese J Gerontol 39:415–425 (in Japanese)

Wake J (2018) Transformation from the community integrated care system to the community integrated society. Jpnese J Gerontol 39:452–459 (in Japanese)

Yokohama City Health and Welfare Bureau (2017) Guidelines for cooperation between community care plazas (in Japanese)

Development of Community Welfare Activities with Resident Participation and Their Importance in Hilly and Mountainous Areas

Akihito Nakajo

Abstract This chapter explains the regional background and significance by which community welfare activities with resident participation are developed, particularly addressing hilly and mountainous areas where the major characteristics are those of areas in which disadvantaged people reside. In hilly and mountainous areas, elderly people's livelihood problems have become severe. Adequate response to them is a pressing need. Hilly and mountainous areas were originally local communities oriented to the elderly people. Therefore, social relations are thought to have been maintained based on traditional social norms. Moreover, expectations for "a community-based inclusive society" are likely to be raised. Nevertheless, the society is rapidly aging in hilly and mountainous areas, making it increasingly difficult to secure persons who can take responsibility to practice mutual aid among local residents that a community-based inclusive society aims at. Consequently, life challenges cannot be addressed as intended. The formation of communities to secure persons who take responsibility for community welfare is needed. In addition to local residents, it is necessary to involve children living separately and those who come and go between settlements where they live and settlements where they are from on a regular basis.

Keywords Community welfare activity · Elderly people · Giving and receiving of support · Hilly and mountainous area · Resident participation

1 Introduction

Community welfare activities have set a goal of solving welfare challenges without disturbing the rhythm of people's daily life and making them leave the area where they have lived for a long time. The primary response to sudden illness and injury depends greatly on local residents. In the local community, there exists a function to prevent circumstances from getting worse by discovering welfare challenges when they are

A. Nakajo (✉)
Shizuoka University, Shizuoka, Japan
e-mail: nakajo.akihito@shizuoka.ac.jp

H. Miyazawa and T. Hatakeyama (eds.), *Community-Based Integrated Care and the Inclusive Society*, International Perspectives in Geography 12,
https://doi.org/10.1007/978-981-33-4473-0_16

small and addressing them at an early stage. It is effective to address the challenges with resident participation, such as the giving and receiving of support between residents. The idea of building such residents' connections in local communities has been promoted by government policies as "a community-based inclusive society."

In hilly and mountainous areas, families live in a spatially dispersed arrangement compared to urban areas. An increasing number of households have difficulty in the giving and receiving of support within a household or between aged parents and children. Furthermore, the increase in elderly households seems to make it difficult to give and receive support within a local community (settlement). Private providers cannot be expected to enter the areas because the threshold related to service supply is small. For this reason, local residents have been required to cooperate beyond the spatial scale of households or settlements.

In recent years, such community welfare activities have been incorporated as a key function of "region management organization (RMO)" that the government has prompted in measures for regional revitalization. With fiscal austerity and diverse residents' needs, it is difficult for the public sector such as the administration to supply effective and efficient services independently. Consequently, various entities, including administration and local residents' organizations, have been required to work together to provide the services. In hilly and mountainous areas, social relations based on ties to the land or ties of blood and settlement organizations based on the relations have acted on livelihood challenges with which individual households cannot be satisfied. Some people point out that local residents' organizations formed beyond the spatial framework of the existing settlement organizations and settlements share roles according to the function (Odagiri 2013). We must consider how the element of community welfare is acting on the formation of a new community.

This chapter reveals the regional background and significance by which community welfare activities are developed by local residents in hilly and mountainous areas. At the same time, it is intended that challenges of "a community-based inclusive society" be considered.

2 Support Networks Surrounding Elderly People in Hilly and Mountainous Areas

2.1 Sources of Support Surrounding Elderly People and the Geographical Characteristics

Hereinafter, we consider the regional significance of welfare activities by local residents, particularly addressing welfare activities surrounding elderly people in hilly and mountainous areas. First, we conceptually organize "sources of support" surrounding elderly people and describe the geographical features. "Sources of support" generally refers to people and organizations that provide an array of support to help elderly people's lives.

Based on Table 1, sources of support are divisible into "the informal sector" and "the formal sector." Informal sources of support include family (relatives), neighborhood, and residents' organizations undertaking welfare activities. The formal sector can be divided further into the public sector and the private sector. Local governments and social welfare corporations fall into the former category, with enterprises and profit-making corporations in the latter. The sources respectively provide support based on blood ties, ties to the land, public assistance, and market principles.

Informal sources provide support to elderly people through the building of social relations with them. Fragility in specialization and continuity have been pointed out because informal sources are not included in the framework of the long-term care insurance system. However, it is possible to develop support based on urgency and awareness because they are close to elderly people's daily life. It has also been pointed out that when residents provide welfare support by forming an organization for welfare activities, the emotional stability and social solidarity of residents' activities can be achieved through the activities.

However, most formal sources of support exist in the framework of the long-term care insurance system. Such sources are highly specialized and therefore provide necessary support when elderly people's physical ability has declined. Although the public sector has the benefit of high fairness by which costs to be borne by individuals are small and even low-income earners cannot be excluded, criticism of its inefficiency is ever-present. By contrast, the private sector can be evaluated as efficient, but it has a difficulty of continuity of support because the sector will withdraw when demand is low and the business becomes unprofitable because of its high threshold.

When examining the geographical features of support, we can find regional differences in formal support according to the locations of providers: although the high demand density facilitates more profit-making corporations to enter cities, the low threshold hinders profit-making corporations to enter hilly and mountainous areas and although social welfare councils are the main entities for service supply. Formal support has an aspect by which the amount of users' needs prescribes the supply mainly in the private sector, which is reflected in the locations of providers.

Informal support is created by social relations that elderly people and sources of support build. There are "instrumental support" and "emotional support" in informal support. The former is physical support for elderly people such as help and errands in daily life and temporary nursing when in bed with a cold. The latter is mental support, such as to be a companion or advisor. Because the former requires face-to-face contacts with elderly people, the spatial distance between elderly people and sources of support has influence on the giving and receiving of support. The latter does not necessarily need face-to-face contact because it can be addressed through a means of communication such as phone calls, emails, and letters. Comparison of the two shows that "instrumental support" is regarded as directly related to the execution of daily life. Geographical conditions must be considered in the giving and receiving of support.

Networks of informal sources of support are important for elderly people to maintain independent life. However, the provision of support is readily influenced by

Table 1 Categories and features of sources of support for elderly people

	Source of support		Connection principle	Long-term care insurance system	Contents of support	Administrative benefits	Administrative shortcomings
Informal support	Family (Relatives)		Blood ties	Out of coverage	Instrumental/Emotional support	• Flexible support can be given and received • Actions can be taken without consideration of profits • It has human warmth, which is likely to build the solidarity of community	• Continuity and specialization might be absent • Economic base is weak
	Neighborhood		Ties to land				
	Welfare activity organizations						
Formal support	Public sector	Local governments Social welfare corporations	Public assistance	Within coverage	Home-based/Institutional care services	• Fair and universal • Low-income earners cannot be excluded because personal burdens of expenses are small	• Inefficient and rigid: Personnel costs are high
	Private sector	Enterprises Profit-making corporations	Market principles			• Efficient: Expenses can be reduced • Consumer-oriented: Support will be provided if there is demand	• Possibility of withdrawal if it becomes unprofitable; continuity might be unreliable • Low-income earners are highly likely to be excluded

Partially modified from Nakajo (2007)

geographical features of sources of support, which are expected to be notable in hilly and mountainous areas. In that sense, hilly and mountainous areas are areas that readily exhibit the importance of community welfare activities.

2.2 Shrinking Households and Hollowing Out of Support Networks

Growing concerns exist for the shrinking of households in hilly and mountainous areas, which is a problem that affects elderly people's life directly because the family has great meaning for elderly people as the main provider of physical and mental support. The shrinking of households implies a situation in which the family cannot directly address all the various challenges including welfare needs. Especially when it becomes difficult for elderly people to cope with challenges by themselves, they must rely on sources of support for the resolution such as children living separately, neighborhoods, and local governments.

However, with an increasing number of elderly people whose children have left them, support by children living separately has limitations because the giving and receiving of support between aged parents and children is influenced by the spatial relationship between them. It also seems difficult to give and receive support within the neighborhood amid an increasing number of elderly households because it is difficult to believe that a family which cannot handle challenges by itself can afford to support elderly people of other families.

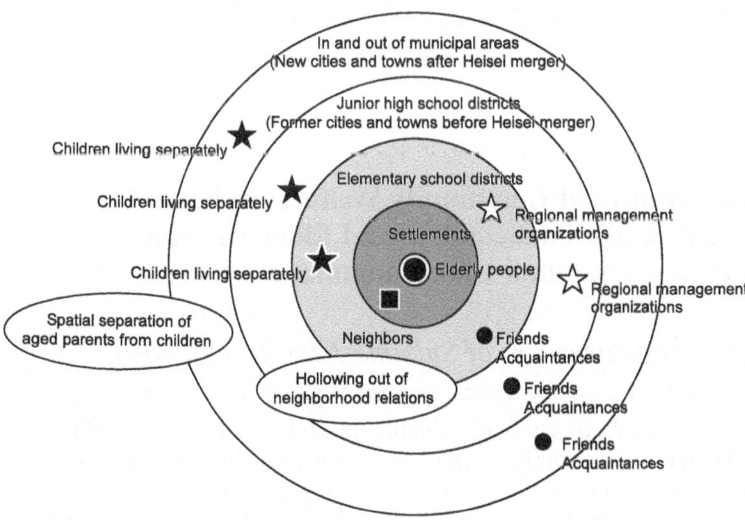

Fig. 1 Spatial location of support resources near elderly people in depopulated areas

Figure 1 is a schematic depiction showing the networks of informal sources of support in hilly and mountainous areas according to the spatial location and the change in the giving and receiving of support with aging for each source of support. The greater the spatial distance of the source of support from elderly people becomes, the weaker the development of social relations becomes with aging and the more difficult the giving and receiving of support becomes. Although many elderly people in hilly and mountainous areas live separately from their children, support is given and received because the relationship between them persists, irrespective of aging. However, the frequency and quality of the giving and receiving of support are influenced depending on the spatial distance of the relationship between aged parents and children. Neighbors living in the sphere of neighborhood where elderly people live are likely to be connected with strong bonds, irrespective of aging. Among neighbors, the next-door neighbors located within the visible range from elderly peoples' houses assume the function of watching over elderly people's daily life and are positioned at the pivot of support networks as the most stable source of support.

Having said that, the shrinking households have caused the hollowing out of support networks. The increasing number of elderly households and the accompanying weakening of social relations serve to stagnate the giving and receiving of support in the neighborhood. When it becomes difficult for elderly people to handle challenges independently, they must rely on informal sources of support such as children living separately and in the neighborhood, or formal sources of support for solutions. Nevertheless, support by children living separately has limitations because of spatial factors. In addition, formal sources of support cannot supply sufficient support because of the low demand density and the reduction of social security expenditures. Therefore, relying on them is difficult. Welfare activities with resident participation draw attention as a new source of support to complement the hollowing out of such sources of support. They are an informal source of support that is independent of marketability, profitability, and long-term care insurance systems. They are regarded as able to function irrespective of the area in which the family lives or the site where the welfare provider is located.

3 Development of Community Welfare Activities with Resident Participation and Elderly Peoples' Life: Practice in Sakuma-Cho, Hamamatsu City

3.1 Regional Overview of Sakuma-Cho, Hamamatsu City

We clarify the development of community welfare activities with resident participation and functions for elderly peoples' life, while providing some examples. Specific emphasis was given to Hamamatsu City, which incorporated wide hilly and mountainous areas, including depopulated mountainous villages, and which became an ordinance-designated city as a result of "the Heisei Municipal Merger." The area to be studied is Sakuma-cho, Tenryu Ward (former Sakuma Town, Iwata district), which is an incorporated mountainous village where local residents develop efforts to address population aging and where the rate of population aging is particularly high: elderly people accounted for 55.2% of the population as of 2015.

Sakuma-cho (Fig. 2) is located about 50 km northwest of the city center, such as Naka Ward, Hamamatsu City, as the crow flies and about 30 km distant from Futamata, which is the main settlement in Tenryu Ward. It is a mountain village located in the western part of the Hokuen mountainous district, occupying the basin of the main course of the Tenryu River and its tributaries. The town is distributed through mountainous areas stretching to the Akaishi Mountains. The highest altitude of the mountainous region is Mount Yatake-yama, of 926 m, and Mount Atago-yama,

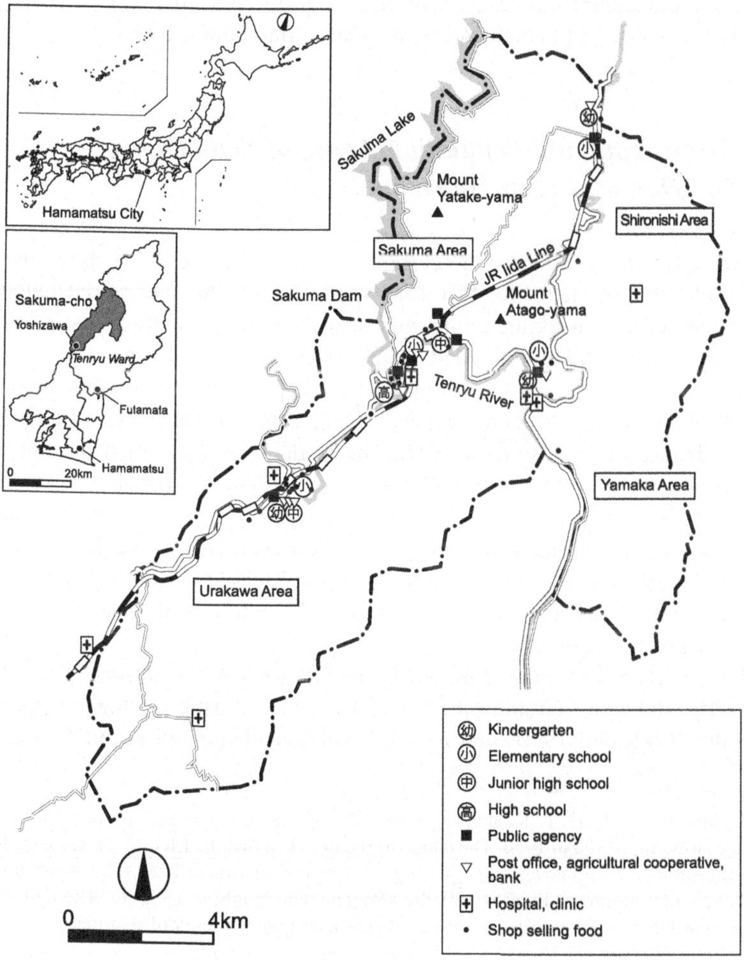

Fig. 2 Overview of Sakuma-cho, Hamamatsu City

of 800 m. The mountainous regions extend 500–600 m above the coastline. The forest area accounts for 90% of the town areas, with inclines of over 30 degrees. Flatlands and easy slopes are formed by erosion and sedimentation of fault zones such as the Median Tectonic Line, and by rivers in the vast mountainous region, where settlements are located.

Former Sakuma Town was absorbed into Hamamatsu City by "the Heisei Municipal Merger" in 2007 along with four villages and towns of the former Iwata district. It comprises Tenryu Ward, Hamamatsu City with the former Tenryu City. The former Sakuma Town was originally formed in 1956, immediately before "the Sakuma Dam"[1] was completed, by the merger of four villages and towns: the former Urakawa Town, former Sakuma Village, former Yamaka Village, and former Shironishi Village (Fig. 2). The Showa Municipal Merger is said to have been influenced strongly by huge revenues from fixed property taxes that the former Sakuma Town was to receive by the construction of the dam and power plant (Machimura 2006).

3.2 Downsizing and Population Aging of Settlements and Downsizing of Households

Next, we specifically examine the current state of settlements, which are the basic units of elderly people's life. In recent years, society has been confronted with concerns about the shrinking, aging, and disappearance of settlements in hilly and mountainous areas. All are regarded as important social problems.

According to a survey of settlement status in depopulated areas conducted by the Ministry of Land, Infrastructure, Transport and Tourism (Ministry of Land, Infrastructure, Transport and Tourism 2016), the settlements for which the number of households constituting the settlement were fewer than 20 households was 20,071 out of 75,662 settlements nationwide (26.5% of total). The settlements in which the percentage of population aging was 50% or more were 15,568 settlements (20.6% of total). The survey uses the threshold of 20 households as an indicator because it has been pointed out that when the number falls below that, settlement functions deteriorate precipitously.

Sakuma-cho, a depopulated mountainous village that was absorbed into Hamamatsu City under the circumstances experienced a drastic decline in population during the 1960s and 1970s and is faced with shrinkage and population aging in

[1]The Sakuma Dam, which is located on the border of Sakuma-cho and Toyone Village, Aichi Prefecture, was completed in 1956. The dam was the first dam that the Electric Power Development Company established with government funding started construction in 1953. The maximum output of the Sakuma Hydropower Plant is 350,000 kW. The dam height is 155.5 m. The dam length is 293.5 m. It was constructed using the best civil engineering technology of its time.

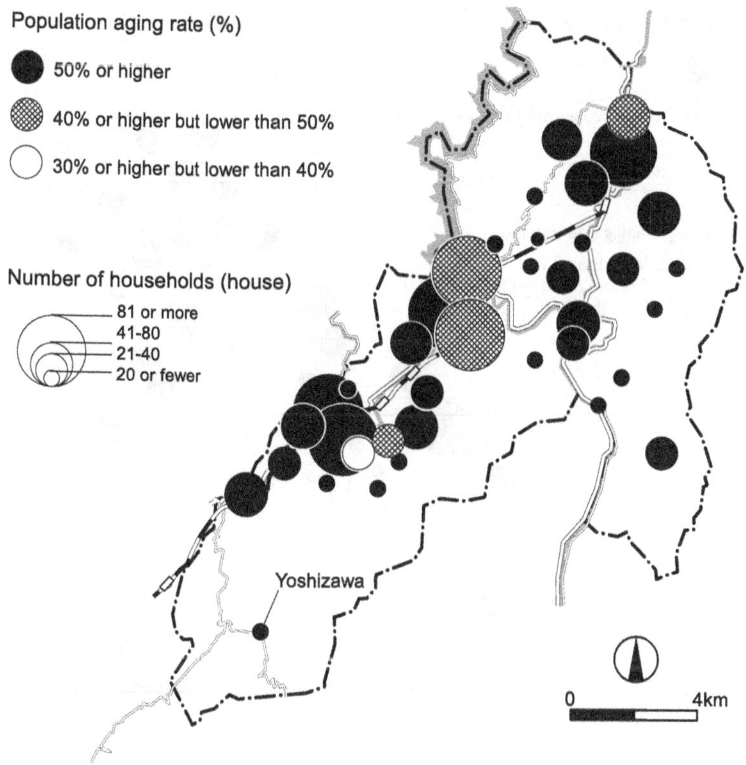

Fig. 3 Population aging rate and the number of member households in settlements in Sakuma-cho, Hamamatsu City (Created based on 2015 Population Census of Japan)

settlements. At present, there are 37 settlements in Sakuma-cho, 15 of which are settlements for which the number of households constituting the settlement is fewer than 20 households (Fig. 3). In fact, seven settlements have fewer than 10 households; settlements with few households are noticeable. They are characterized by grouping of the central settlements located along the railway (JR Iida Line) with numerous households and marginal settlements on a smaller scale. The settlements for which the percentage of population aging exceeds 50% are 32 settlements. Six settlements have population aged 75 years and older accounting for 50% or more. Population aging is progressing in all parts of the former Sakuma Town, except in a part of the central settlements. We further assess the relationship between the percentage of aged population and the average number of household members per household (Fig. 4). The average number of household members per household is fewer than 3.0 people in all settlements. The percentage of population aging exceeds 50% in most settlements. These imply that the settlements consist either of elderly married-couple households or elderly single-person households. Seven settlements have households fewer than 2.0 people, on average. The percentage of population

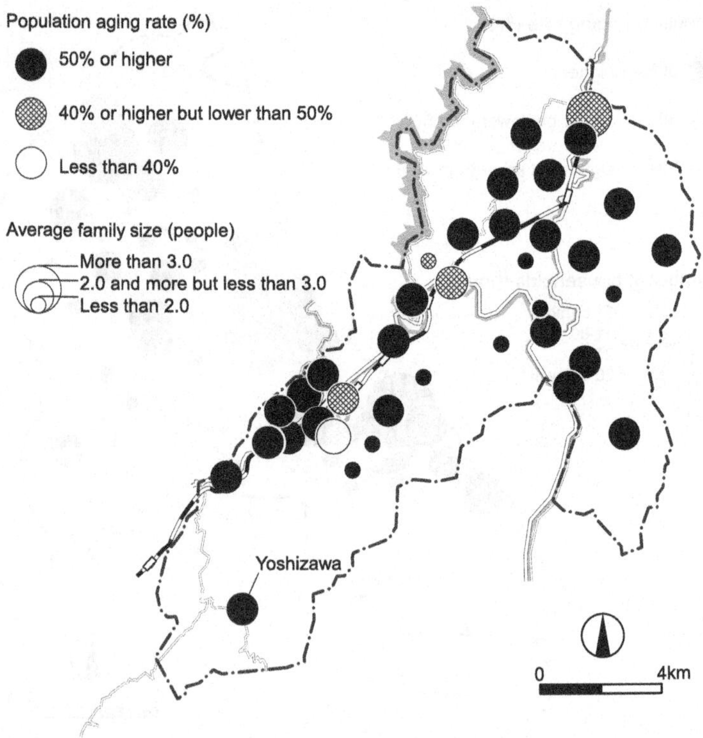

Fig. 4 Average family size and population aging rate in settlements in Sakuma-cho, Hamamatsu City (Created based on 2015 Population Census of Japan)

aging exceeds 60% in marginal settlements. It is readily apparent that the population shrinking and aging of households are markedly progressing throughout the area, although there are differences among settlements. A decrease in the number of household members signals a decrease in the sources of support within households, which are fundamentally important for elderly people's life. In other words, it has become difficult to give and receive support within households.

3.3 Efforts of Small Community Welfare Activities in Sakuma-Cho

In Sakuma cho, small community welfare activities have been practiced to maintain and create regional social relations between elderly people as a result of approaches by the Sakuma Area Council of Social Welfare. They are commonly known as "salon activities" aimed at complementing and newly creating social relations by elderly people and supporters gathering together in one place.

As of 2018, 20 salons have been distributed in Sakuma-cho. By spatial unit, there are 14 salons in "settlements" and 5 salons in "districts" (former elementary school districts) that are larger than settlements. As presented in Table 2, the numbers of members vary from dozens to more than 600. The activities range in frequency from several times a year to several times a month. Because many users use day care services in addition to salon activities, an activity frequency of around once a month is not too low: salon activities play a role in complementing the services.

Table 2 Small community welfare activities ("salons") in Sakuma-cho, Hamamatsu City (2018)

Covered area						Operating group	
Area	Name of salon	Spatial range	Number of households	Rate of population aging (%)	Frequency of holding activities	Name of group	Number of members
Urakawa	A	Area	747	54.7	Four times a year	Volunteer Group	22
	B	Settlement	12	79.2	Twice a month	↓	
	C	Settlement	44	64.1	Once a month	↓	
	D	Settlement	27	74.0	Once a month	↓	
	E	Settlement	61	50.4	Once a month	↓	
	F	Settlement	34	45.3	Once a month	↓	
	G	Settlement	64	53.7	Once a month	↓	
Sakuma	A	Area	250	57.0	Once a month	Mizumaki Group	18
	B	Area	543	51.0	Once a month	Himawari Group	15
	C	Settlement	18	57.1	Once a month	↓	
	D	Settlement	16	57.6	Once a month	↓	
	E	Settlement	9	61.9	Once a month	↓	
	F	Settlement	124	48.7	Once a month	↓	
Yamaka	A	Area	411	54.2	Once a month	Yamabuki Group	31
	B	Area	411	54.2	Once a month	Kaori Group	ND
	C	Settlement	21	50.0	Once a month	↓	
	D	Settlement	26	71.7	Once a month	↓	
	E	Settlement	16	65.8	Once a month	↓	
Shironishi	A	Settlement	697	55.5	Once a month	Volunteer Group	ND
	B	Settlement	47	40.6	Twice a year	↓	
	C	Settlement	89	61.9	Twice a year	↓	
	D	Settlement	70	58.0	Four times a year	↓	
	E	Settlement	62	55.9	Once a month	↓	

Arrows (↓) in the table signify that members of the volunteer group in the area are active in the resident settlement. Groups with shading are salons that were active as of 2011, but which had suspended activities by 2018. ND denotes no data.

The activities facilitate mutual communication among elderly people. For example, users have mutual chats or conversations with volunteers and receive training offered by the social welfare council or the local Sakuma Hospital. Particularly, salons established in "districts" based on the former elementary school districts are commonly known as "Oshokuji salon (Meal salon)." It is held once a month, with volunteers preparing lunch and participants enjoying it all together.

Those who take responsibility for salon activities are female volunteer groups organized in each district. Each group comprises about 15–30 volunteer women who are in their 60 s and 70 s and who live in the district. They prepare meals and run the salon. Most salon users are in their 80 s or older. Volunteers are residents who are slightly younger than the users; there is not much difference in their ages. These people participate in regular annual events implemented by the social welfare council as the management side. The current situation is that they are fixed as actors in Sakuma-cho. It can be rephrased as districts in which a salon is formed are districts that have secured actors.

3.4 *Development of Community Welfare Activities Based on the Settlement Unit*

3.4.1 Households Constituting Settlements, and Village Communities

Next, we take up "Yoshizawa Syakunage Kai (Yoshizawa Rhododendron Association)," which is active in the Yoshizawa settlement, Sakuma-cho to assess the actual state of social relations at the settlement level and their functions.

First, we look at the actual state of the existing social relations in the Yoshizawa settlement where the association exists. The settlement is located in the Urakawa Area, Sakuma-cho and is a marginal settlement about 8 km distant from Urakawa Station on the JR Iida Line, which is the center of the area (Fig. 3). The Yoshizawa settlement consists of five groups of Aikawa, Kouchi, Nishi, Higashi, and Shinzo, forming a moderately dispersed settlement along the Aikawa River, which flows through the settlement. The groups are small settlements composing the Yoshizawa settlement, distributed at 1–2 km intervals.

The author conducted field surveys in 2011 and 2019. As of 2011, when complete enumeration was conducted, the percentage of population aging was 75.0%; the average number of household members was 1.7, which demonstrated the progress of population aging and shrinking of households. Herein, we confirm the outline of households constituting the settlement in 2011.

Table 3 shows that all neighbor groups had fewer than five households and that they consisted of single-person households or married-couple households, mainly elderly people in their 70 s and 80 s. In Aikawa, Kouchi, Nishi, and Higashi groups, some households had children living with parents who could be classified into the category of the second generation. Children living separately who most frequently visited

Table 3 Compositions of households constituting the Yoshizawa settlement

Group	No. households	No. household members	1st generation		2nd generation	
			Male	Female	Male	Female
Aikawa	1	3	80 (farmer)	78 (none)	46 (farmer)	
	2	1		80 (none)		
Kouchi	3	1		84 (none)		
	4	1		80 (farmer)		
	5	2	88 (none)	86 (none)		
	6	4	81 (self-employed)	76 (self-employed)	50 (self-employed)	47 (self-employed)
	7	2	65 (farmer)	61 (home-maker)		
Nishi	8	1		78 (none)		
	9	4		88 (none)		
	10	4		8 9 (none)		
			81 (none)	76 (home-maker)	40 (none)	
	11	3	80 (farmer)	75 (home-maker)		
	12	2	84 (none)	80 (home-maker)		
Higashi	13	3	90 (none)	84 (farmer)	50 (farmer)	
	14	1	80 (farmer)			
	15	1		81 (none)		
	16	1		81 (none)		
Shinzo	17	1	62 (potter)			
	18	1	65 (none)			

(continued)

Table 3 (continued)

Group	No. Households	Residence of children living separately with whom elderly parents most frequently exchange	Coming or going to visit	Car owner
Aikawa	1	Living together		Yes
	2	Oldest son: 90 min by car (once a week or so)	Coming to visit	No
Kouchi	3	Oldest son: 90 min by car (once a week or so)	Coming to visit	No
	4	Oldest son: 120 min by car (3 times a year or so)	Coming to visit	No
	5	Oldest daughter: 30 min by car (more than once a week)	Coming to visit	No
	6	Living together		Yes
	7	Oldest son: 90 min by car (once a week or so)	Going to visit	Yes
Nishi	8	Oldest son: 60 min by car (once a week or so)	Going to visit	Yes
	9	Oldest daughter: 60 min by car (once a week or so)	Coming to visit	No
	10	Living together		
				Yes
	11	Oldest son: 90 min by car (everyday)	Coming to visit	Yes
	12	Oldest daughter: 60 min by car (more than once a week)	Going to visit	Yes
Higashi	13	Living together		Yes
	14	Oldest son: 60 min by car (several times a year)	Going to visit	Yes
	15	Oldest son: 60 min by car (once a week or so)	Coming to visit	No
	16	Oldest daughter: 60 min by car (more than once a week)	Going to visit	No
Shinzo	17			Yes

(continued)

Table 3 (continued)

Group	No. households	Residence of children living separately with whom elderly parents most frequently exchange	Coming or going to visit	Car owner
	18			Yes

Created based on data from interviews conducted in August 2011 their parents most commonly lived 30–120 min away by car. The most common frequency of "visits made by" children living separately from aged parents' homes was about once a week. In some cases, aged parents "made a visit" to children living separately. Seven households did not possess a car; 11 households did. Most households without a car were single-female elderly households, which demonstrates that there are constraints on traveling by gender.

Next, we confirm the activities of the village community. "Kumicho-kai (group head meeting)" as a meeting of the residents' association is held once every two months. Raising funds for facility management costs has become a challenge for them. The residents' association membership fees paid by Hamamatsu City have been reduced from 328,000 yen a year to 80,000 yen a year as a result of the merger. Labor exchange in farm work is based on the group unit. Residents who can bear farm work go to help rice planting and reaping. There is the "Yasaka Jinja" as the shrine of the local tutelary deity, in which Shinto rituals are held at the festival on August 25 and the first shrine visit of the year on New Year's Day. When repairing the buildings of the shrine, they use the 2,500,000-yen profit from the sale of lumber of the settlement's common forest for expenses to continue the management.

An important matter in elderly settlements is "soshiki-gumi (funeral team)," which are organized in each neighbor group. On the occasion of a funeral, it is often the case that children living separately return home and conduct a funeral on behalf of their aged parents because of the aging of those involved. It is noteworthy that children living separately have little knowledge and experience with funeral services. Therefore, a manual related to funeral services has been prepared in advance. When the need arises, based on the guidance from the manual, communication to the family temple is made and roles are divided among neighbor groups in accordance with the tradition of the settlement. Of course, it is sometimes difficult for neighbor groups to conduct a funeral by themselves because of aging or because children living separately cannot return home. In such a case, a funeral is to be conducted with the involvement of a funeral director and elderly people who are acquaintances of the aged parents. Although Koshin-ko and Akiba-ko exist in the religious group as a faith group, the limitations of a rotation system of toya (house on duty) became apparent. It was suspended in approximately 2005.

3.4.2 Activities of the "Yoshizawa Syakunage Kai"

"Yoshizawa Syakunage Kai" is an organization of small community welfare activities whose members are elderly people living in the Yoshizawa settlement. Because it is a settlement in which mainly elderly people live, the association consisting of 17 members as of 2011 is practically the main body to take responsibility for settlement activities.

On the second and fourth Thursdays of each month, the activities are held in the classrooms of the former Yoshizawa Elementary School buildings located in the center of the settlement. The members spend time from 9:00 to 17:00, including lunch time, discussing challenges in the settlement, listening to lectures given by lecturers who are dispatched from the social welfare council, doing recreation, and having small talk. Member Household No. 7 (male, 65 years old), who migrated from Toyohashi City, Aichi Prefecture, plays a key role in the operation of the association. He picks up and drops off members who live far away from the venue and goes to a caterer's shop in Toei Town, Aichi Prefecture about 8 km away to receive box lunches. It is said that the male president (Household No. 11) serves miso soup at lunch time and that participants look forward to it.[2] Furthermore, social gatherings were held as frequently as three times a year (June, December, and January).

Other activities include cleaning-up of streets and mountain paths and implementation of repair work called "michi-bushin (road repair)" mainly in the summer; and acceptance of a traveling clinic from Sakuma Hospital once a month.[3] Particularly, male members take care of michi-bushin to their possible extent. For example, they repair a bridge over a mountain stream damaged from heavy rains and do repair work on water pipes to bring water from a mountain spring. Arrangements have been made for support in an emergency. Contact addresses of children living separately to be informed and settlement residents to handle the situation have been determined in advance.[4]

To date, we have explained the current situation in which settlements are shrinking and aging and the activities of settlement organizations by which elderly local residents formed by newly adding welfare functions to them. As presented in "Syakunage Kai," the activities are led mainly by elderly people. This maintained or created connections in a village community and kept an ability to address challenges by the strength of the settlement while refraining from depending on the administration to the greatest extent possible. However, it is difficult to infer that the present situation will continue in the medium term to long term because elderly people face mental and physical changes as a result of aging. It can be inferred from the fact that they receive support from a source of support outside the settlement, children living separately, with regard to the management of funerals.

3.5 *Formation and Functions of "Ganbaramaika Sakuma"*

An organization at the settlement level has limitations to cope with the aging of actors. Then, Specified Nonprofit Corporation "Ganbaramaika Sakuma" was established in

---

[2]Although the expenses necessary for the operation of the association (public utility charges, money for tea, etc.) are paid from the annual membership fee of 12,000 yen, the money for a boxed meal, 500 yen, is collected separately. If an elderly person attends the association twice a month, then the money for boxed meals will be 1,000 yen. The expenses will be 2,000 yen a month together with the membership fee. However, reportedly, no one particularly raises a complaint.

[3]Resident doctors working in Sakuma Hospital participate in a traveling clinic from the hospital. They are said to have undergone numerous training sessions related to medical care in depopulated areas.

[4]Settlement residents who possess a car are in charge of handling the situation.

2005 under the slogan of "Whole Resident Participatory NPO" using the former Sakuma Town Area as the unit.

The movement toward the establishment of the NPO became severe after December 2004, when the signing ceremony for a merger agreement between former Sakuma Town and former Hamamatsu City was completed. Political and administrative leadership groups in the former Sakuma Town who were concerned about the marginalization after the wide-area merger pushed the organization forward. The New Hamamatsu City merely subsidized events that had been hosted by former Sakuma Town. Therefore, the NPO intended to assume some of the roles. The formation of the organization is positioned as the "region management organization," which has been declared as one recent government policy for regional revitalization. The former Sakuma Town can be said to have responded ahead of it.

As of the end of 2016, the members are 1,170 households, accounting for 67% of the total number of households in the former Sakuma Town. Each household pays the annual membership fee of 1,200 yen. As Fig. 5 shows, Ganbaramaika Sakuma comprises the board of directors, the administrative office, seven activities committees, and supporting members consisting of all residents. The seven activities committees are the following: the general affairs committee; the health and welfare activities committee; the regional development activities committee; the culture, sports, and social education activities committee; the environment creation activities committee; the women's activities committee; and the intergenerational exchange activities committee. In all, 625 members, called activities members, belong to respective committees (as of the end of 2016).

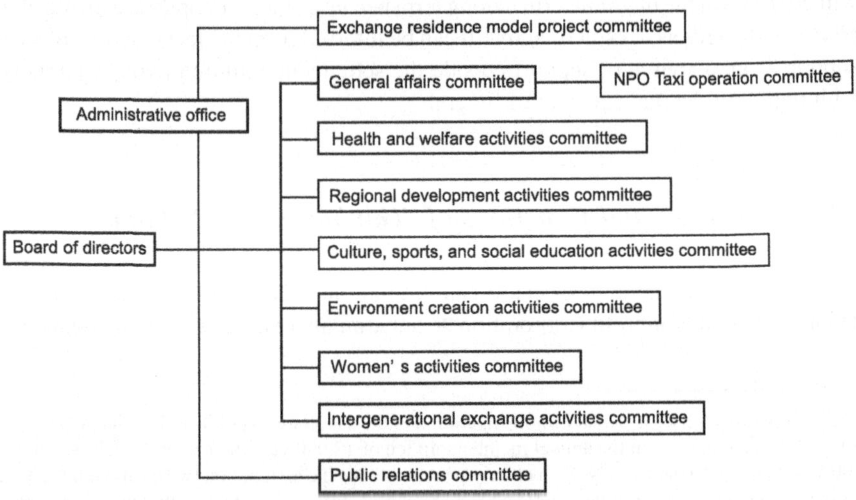

Fig. 5 Organizational chart of "Ganbaramaika Sakuma" (Created based on materials of NPO corporation "Ganbaramaika Sakuma")

Each committee develops activities centering on events passed down from the time of former Sakuma Town.[5] The operations of on-demand taxi "NPO Taxi" and restaurant "NPO no Omise Idobata (NPO Restaurant Well Side)" can be described as activities that particularly contribute to residents' lives.

The former activities were started with two taxies in 2007 by activities members of the NPO Taxi operation committee as operators. It is possible to receive the service if a reservation is made the day before. The starting fare is 300 yen for the first 1.5 km, with 500 yen added for every additional 2 km. The service was used in 6,728 cases in 2010, which is an average of 561 cases per month. The latter opened a restaurant near the former town office in 2007. The women activities committee offers products using homegrown buckwheat. Three members of the committee work each day for 6 h at 500 yen per hour. In addition, the restaurant sells agricultural processed products and invites single elderly people living in Sakuma-cho for meals. We can find that the operations of the taxi and the restaurant create jobs for activities members who are local residents.

The participation rate in "Ganbaramaika Sakuma" (Fig. 6) shows that the rate is high mainly in marginal settlements. By contrast, a low participation rate prevails in settlements with numerous houses facing major roads. Those settlements have easy access to various facilities. Welfare activities in "Ganbaramaika Sakuma" are limited to taxi services at the moment. Many households are presumed to participate in the activities to secure a means of transportation. It can be inferred that people in marginal settlements join the activities to gain "a sense of security," assuming cases in which they can no longer deal with challenges in their own settlements. In settlements where population shrinking and aging are progressing, people might expect spatially larger regional organizations to take new measures because, even if it is currently possible to address challenges by individuals and local communities, it will become increasingly difficult in the future.

4 Conclusions

This chapter presented consideration of the regional background and significance by which community welfare activities are developed with resident participation, particularly addressing hilly and mountainous areas where the major characteristics associated with the area are those of disadvantaged areas.

In hilly and mountainous areas, elderly people's livelihood problems have become severe and the response to them is a pressing need. Hilly and mountainous areas are originally elderly oriented local communities. Therefore, social relations are thought

[5]"Ganbaramaika Sakuma" has established "the exchange residence model project committee" corresponding to the movement of return to rural areas, in addition to the implementation of events on behalf of the administration and services related to livelihood support. Here, they are efforts to attract families who can move into the former Sakuma Town through exchanges with people living in urban areas, such as a program to introduce vacant houses and a partnership program for making soba (buckwheat noodles).

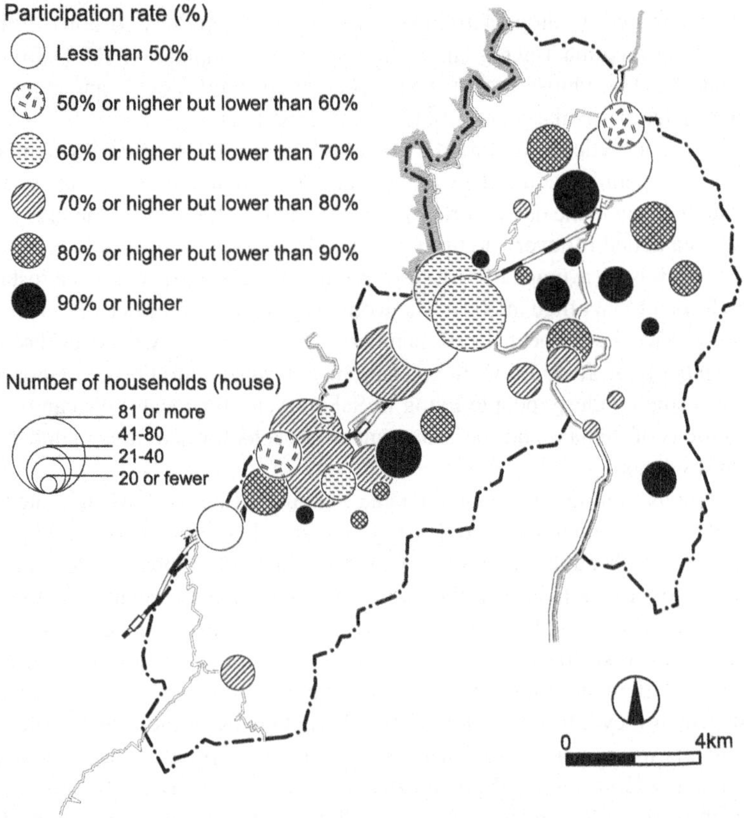

Fig. 6 "Ganbaramaika Sakuma" participation rate by settlement (Created based on materials of NPO corporation "Ganbaramaika Sakuma")

to have been maintained based on traditional social norms. The relations are denser and stronger than those of cities. Expectations for "a community-based inclusive society" are likely to be raised. Particularly, settlements that have been the units of the local community in hilly and mountainous areas from earlier times can be a model for that. However, in modern settlements, combined with the shrinking number of households, aging of the population, and the shrinkage of households as a result of spatial dispersion of family members, concerns persist about the weakening of traditionally maintained social relations. In actuality, social relations can serve as the basis of the giving and receiving of support for elderly people's life. Even if not giving support to elderly people, modern hilly and mountainous areas create social relations that can serve as a source of support. It is particularly an important challenge in merged mountain villages where the functions are being deprived.

This chapter specifically described a merged mountain village of Sakuma-cho (former Sakuma Town), Hamamatsu City to elucidate the importance of community welfare activities with resident participation. In Sakuma-cho, the shrinking and aging of settlements of the population have progressed to a considerable degree. Social relations in settlements are assumed to be weakening. In response to the changes, Sakuma Area Councils of Social Welfare have developed small community welfare activities called "salon activities" to maintain or create social relations in settlements. The Yoshizawa settlement, which we cited as an example, has rearranged salon activities at the center of settlement activities and has strived to strengthen social relations.

However, the medium-term to long-term maintenance of livelihood has become a challenge in hilly and mountainous areas where population aging is progressing because elderly people face mental and physical changes as a result of aging. In Sakuma-cho, an NPO organization was formed after the merger with Hamamatsu City to assume regional governance on the former town scale. The main activities are taxi services to secure a means of transportation, restaurant management to provide a place for residents, and response to migrants. All can be said to complement the livelihood needs accompanying the aging of society. The participation rate in the NPO is high in marginal settlements where population shrinking and aging are progressing. It can be considered that people in marginal settlements join the activities to gain "a sense of security," assuming a case in which they can no longer cope with challenges in their own settlement.

It is difficult to say that growing welfare needs of local residents with the progress of the aging socicty can be addressed within the settlements independently. As the case study areas demonstrate, new local residents' organizations have been formed by extending the spatial scale for securing actors from the settlement scale to the former town scale. This shows that some sectors cannot cope with challenges in settlements, which was readily apparent in the welfare sector.

In recent years, "region management organizations" have been developed as political efforts in hilly and mountainous areas where the settlement functions are weakening. It is to resolve challenges that are difficult to solve on a settlement basis or challenges that wide-area implementation will be efficient by subsuming other settlements to secure actors or the threshold. An effort of community welfare, "an inclusive local community," can be positioned as one model of the local community that regional management organizations realize by functioning. However, as described earlier, the society is rapidly aging in hilly and mountainous areas. Therefore it is increasingly difficult to secure persons who take responsibility to practice mutual aid among local residents that the community-based inclusive society aims at. Consequently, life challenges cannot be confronted as intended. Specifically, because life challenges are mainly those confronting elderly people, the formation of a community to secure persons who take responsibility for community welfare is needed. In hilly and mountainous areas, new communities will be formed by the development of welfare activities as the organizational principle.

Those who are assumed to be actors in the welfare activities are of course local residents participating in the activities. However, they can involve children living separately who come and go between settlements where they reside and settlements they are originally from on a regular basis. It is unrealistic to expect migrants who have drawn attention with the movement of returning to rural areas because of the quantitative scale of those who come from urban areas. It will be realistic to give the role of actors to the related population, mainly children living separately who repeat coming and going while forming a connection with their local community. In fact, in Sakuma-cho, Hamamatsu City, which is taken up as the case study area, it became apparent that welfare activities have neither been developed nor been used on a daily basis, except at some settlements, which indicates that support has been given and received through children living separately. Their presence must be actively used and encouraged.

References

Machimura T (ed) (2006) Constructing postwar development at multiple levels: The Sakuma dam project and its regional consequences. University of Tokyo Press, Tokyo (in Japanese)

Ministry of Land, Infrastructure, Transport and Tourism (2016) Outline of survey of the current status of settlements in depopulated and other disadvantaged areas (in Japanese)

Nakajo A (2007) Support networks of elderly and welfare activity of residents in hilly and mountainous area. Geogr Sci 62:79–92 (in Japanese)

Odagiri T (ed) (2013) Challenging reconstruction of rural villages: Theory to practice. Iwanami Shoten Publishers, Tokyo (in Japanese)

The manufacturer's authorised representative in the EU is Springer Nature Customer Service Centre GmbH, Europaplatz 3, 69115 Heidelberg, Germany. If you have any concerns regarding our products, please contact ProductSafety@springernature.com

Printed and bound by CPI Group (UK) Ltd, Croydon, CR0 4YY
30/09/2026
02225330-0013